Ancient Oaks in the English Landscape

Ancient Oaks in the English Landscape

Aljos Farjon

With contributions on biodiversity by
Martyn Ainsworth, Keith Alexander and Pat Wolseley

Kew Publishing
Royal Botanic Gardens, Kew

First published in 2017 by the Royal Botanic Gardens,
Kew, Richmond, Surrey, TW9 3AB, UK
www.kew.org

ISBN 978 1 84246 640 7

Distributed on behalf of the Royal Botanic Gardens, Kew
in North America by the University of Chicago Press,
1427 East 60th St, Chicago, IL 60637, USA

British Library Cataloguing in Publication Data
A catalogue record for this book is available from the British Library.

Copyeditor: Sharon Whitehead
Proofreader: Alison Rix
Design and page layout: Ocky Murray
Production Management: Andrew Illes

For information or to purchase all Kew titles please visit shop.kew.org/kewbooksonline
or email publishing@kew.org

Kew's mission is to be the global resource in plant and fungal knowledge, and the world's
leading botanic garden.

Kew receives about half of its running costs from Government through the Department
for Environment, Food and Rural Affairs (Defra). All other funding needed to support
Kew's vital work comes from members, foundations, donors and commercial activities,
including book sales.

Printed and bound in Malta by Melita Press

Contents

Foreword

Twenty-five years ago, I bought a piece of ancient woodland across the valley from Hergest Croft. It had never been part of our estate but was a remnant of the ancient royal hunting forest of Kingswood that had covered a much more extensive area. How far its history stretched back, I have no idea but one place name, Pembers Oak, suggests that it may even date back to the Kings of Mercia, possibly referring to Penda, King of Mercia in the seventh century. It was certainly in existence at Domesday in 1086. None of the original trees survive, but their descendants from either acorns or stools still grow there. The woodland had been felled in the First World War to provide timber for the war economy and my father remembered this; I was keen that he should see it replanted during his lifetime, and we achieved this before he died in 1997. Below the woodland, on the banks of the River Arrow, grow alders whose coppiced forebears provided the charcoal for the gunpowder for Nelson's ships and later the clogs for millworkers in Lancashire, maybe these stools are as old as the oaks above them. Herefordshire contains the largest number of ancient oaks in England, possibly because it was no longer at the forefront of English history after the end of the Glyndwr uprising in 1405, because it was too far away from the sea to provide timber for the Royal Navy, or because it survived the depredations of the Civil War.

History and mythology show the central place in the English psyche of oaks and oak woodland. It rings all sorts of emotional bells from English sailors, 'Hearts of Oak' to Shakespearean black and white buildings including the all too flammable theatres; oak was the mainstay of 'Ye Olde' construction and 'Mock Tudor' suburban buildings are the 1930 descendants even if not much oak is involved.

There is a positive industry in trying to date oaks back a thousand years, almost always owing more to romantic dreams than to any connection with reality. We need to remember that many oaks large enough to be noted in ancient writings are likely to be long dead today and any that remain are almost certainly hollow, so ring dating is impossible. Thomas Pakenham got it right when he wrote in a recent book, "How old is an old tree? Perhaps people would rather not know." Ironically, the timber of ancient buildings can be dated from dendrochronology and one of Hergest Court's beams dates back to 1268, which fits the historical record exactly. The timber may even have come from our own oak woodland.

Aljos Farjon has applied scientific rigour to discussing these issues and this work is a major contribution to understanding them. He modestly says, "Some issues are disputed and I am far from pretending to have all the right answers...Nevertheless, I hope that this book will, at the very least, have made it abundantly clear how significant and special the ancient oaks in the English landscape really are." It certainly does this and it puts them firmly in the context of the English landscape and its long history. This book will, I am sure, encourage debate and controversy which will enable us to clothe the mythology surrounding our oaks with more scientific and historical facts, though the myths have their place as well. I can only end by saying that the author's recent lecture at the Linnean Society of London is believed to be the best attended ever, which shows what an exciting topic this is.

Lawrence Banks
Kington, Herefordshire
April 2016

Introduction

This book is about the 'English oak'. This native tree (*Quercus robur*), also referred to as the 'British oak', is considered to be the national tree of England. The other oak species native to England, *Q. petraea*, is very similar, although less common in most parts. In this book, the two species (and their putative hybrids) are treated as one entity: the native oaks of England. I am not interested here in their taxonomy, or in how they are related to any of the other approximately 530 species of *Quercus* known to botany. Both *Q. robur* and *Q. petraea* are widely distributed across Europe and beyond, and in this respect, there is nothing particularly English or British about them. The 'Englishness' of oaks has to do with their symbolism in English tradition, in which they are thought to represent strength and endurance. Numerous English pubs are called 'Royal Oak' and together with 'Oak Apple Day' commemorate the escape of Charles II at Boscobel House in Shropshire, where the king hid in an oak tree to evade the Parliamentary troops searching for him after the Battle of Worcester in 1651. There are numerous other histories and tales involving English oaks, often resulting in names for individual trees. This book is not about these, either. There is no shortage of publications which deal with these stories, and the better ones do far more justice to this cultural aspect of the English oak than I would be capable of.

Oaks have played an important part in the history of England and the UK as their timber has been the primary construction wood for cathedrals, churches, castles, manor houses, palaces and ships. It was much the same in other European countries where this tree is abundant. This utilitarian aspect is also not the subject of this book, although sometimes there may be the briefest reference to it. Oak timber, in this country at least, was mostly grown as standard trees in coppiced woodlands and also in hedgerows. In rare instances, oaks grown in woods were allowed to develop into large tall trees, which provided the great beams for cathedrals, palaces or sometimes huge barns. Most oaks were cut when still much smaller, to be used to build timber-framed houses. Woodlands managed for this oak timber and for hazel (*Corylus avellana*) and other coppice woods are also not the subject of this book, because although the woodland may be ancient, the trees in it are not and never were. What can be ancient is the coppice stool, also excluded from discussion here.

Despite all these no-go subjects, there is still a lot left to write about England's ancient oak trees. How old are they anyway? And if very old, how did they survive? Why were they not cut down for the Navy or to build more cathedrals? Were they of no use at all, and if not, what was the rationale behind preserving them in an age long before the concept of nature conservation was on anyone's mind? Why were they not cut for firewood, which was often in short supply? England still has a large number of ancient oaks, far more than any other country in Europe where *Quercus robur* and *Q. petraea* are also common. If nature conservation was not taking place, why is this so? Was England's climate or soil more suitable for the survival of oaks than those in other countries? Or were the ancient oaks destroyed in greater numbers elsewhere in Europe, and if so, why? Has the English veneration of oaks played a role in their preservation, or is this just a romantic idea? Why are ancient oaks significant and why should we continue to preserve them? How can we ensure that successor oaks will live long enough to replace ancient trees when they die? These are the questions that I am interested in and which this book attempts to answer.

What is an ancient oak? This question is not as easy to answer as it may seem. 'Ancient' is a rather subjective concept to apply to a tree. Perhaps it pertains to the life stage of an oak, but at what stage does a tree become ancient? What is the relationship between a tree's relative age in terms of its life stage and its age measured in years? To help us answer these questions, Chapter 1 discusses the life of an oak tree, from seedling through maturity, old age, senescence to ultimate death. What are the physical characteristics of an oak tree during each of these successive stages, assuming that it lives through all of them? In the field, tree recorders make observational notes that describe an oak's physical characteristics and they measure its girth. The oak database that I used as a basis for the work described in this book now comprises the *Ancient Tree Inventory* (ATI), which is maintained by a partnership between the Woodland Trust, the Ancient Tree Forum and the Tree Register of the British Isles. For this database, the decision on whether a tree met the criteria necessary to define it as 'ancient' is not made by the recorder, but by a data verifier, an assessor trained by the Woodland Trust. Often, the verifier had to revisit the tree if there was any doubt. A photograph of the tree or a visit to it helped me to agree or disagree with this judgement in many cases. 'Ancient' remains a subjective concept that was not applied consistently in the database, despite the use of measurements and criteria, to all oaks across the whole of England.

For the purposes of this book, I am actually more interested in oaks that are ancient in absolute terms, as measured in years, although their relative age as assessed by life stage is relevant, for instance, in relation to biodiversity. In Chapter 2, the assessment of the actual age in years of oaks is treated in some detail. When a relatively simple count of annual growth rings in the trunk is impossible, how can we approach this without losing ourselves in conjecture and indeed wishful story telling? We have to look at 'proxy data' substituting for the annual rings which are missing. Many people have occupied themselves with this subject and if done on a scientific basis useful insights have been gained. Nevertheless, I believe that it is not possible to determine the exact age of most ancient oak trees, or even to obtain a best estimate of a particular tree's age. Instead, I focus on estimating the average ages, in years, of oaks of a minimum girth and of greater size, up to the largest oaks still alive. If we can obtain such estimates, we can link the ancient and veteran oaks to the history of the English landscape and its various ownerships and uses in which they occurred in the past and from which time many survived to the present. I had to decide on a date in English history after which new oak trees are less likely to be ancient. I have taken this date to be the year 1603 (1600 in a rounded number) when, with the death of Elizabeth I, the Tudor reign ended. The best-fitting minimum girth for this age was 6.00 m; although all oaks of this size will not

have survived since 1600, on average, a 6.00 m oak could be that old and larger oaks are likely to be older.

I am interested in what happened in the Middle Ages and the Tudor Age, when the truly ancient oaks now alive originated. From Domesday Book (1086) to 1603, we have a lot of historical information about the English landscape which I have been able to analyse in relation to the oaks because we know the exact location of all recorded ancient oaks in existence today. We are looking for clues that explain why these trees have survived, and a first approach is to find patterns in their distribution. In Chapter 3, we produce maps of ≥6.00 m girth oaks. These maps show where there are concentrations of ancient oaks and where these oaks are spread very thinly, and allow us to begin to speculate as to why they are not distributed randomly. We already know that there is a relation between ancient oaks and certain forms of land use and types of landscape, but with my annotated database, which has information on land use types for almost every oak tree, I could now test how strong these relationships are. Having done this on a national scale, I could now move on to assess these historical relationships in more detail. Deer parks existing before 1600, which I separate into medieval and Tudor parks, are the landscape and land use context of the ancient oaks investigated in Chapter 4. These parks are of greatest significance to the presence of ancient oaks today, even though 75% of all medieval parks and their ancient oaks have now gone. All other types of land use relevant to ancient oaks are investigated in Chapter 5, among these are Royal Forests and wooded commons. Together these categories of land use explain >85% of the present ancient oaks in the English landscape; leaving only about 15% 'unexplained', which really only means that I have not been able to determine what the land ownership and use was before 1600.

We now turn to the question of why more ancient oaks survived in England than elsewhere. For this, we first take a brief look in Chapter 6 at the ancient oaks of Europe outside England. Many countries are recording their ancient and veteran trees, just as in England, even though this recording is incomplete and has been more thorough in some countries than in others. So we only compare the largest oaks, country by country, with the same in England, of which a list is given in Chapter 5. These biggest oaks are less likely to have been missed or ignored by the recorders. The results of this inventory are revealing. In Chapter 7, I then attempt to explain why England has most of the ancient oaks in Europe. We are getting into history again and the discussion ranges from deer hunting in parks and forests to the development of modern forestry, and from wars and revolutions to private land ownership.

What is the significance of the preponderance of ancient oaks in England? For anyone English (or British), it may be pleasing to know that their country is the champion for ancient oaks in Europe. But beyond national pride, is there perhaps a greater, more universal significance? Having established, with abundant evidence, the landscape and land use histories of the locations where ancient oaks remain, what, if anything, are the commonalities that have allowed these oaks to grow and survive? To understand this better, we go back much further in time in Chapter 8. After all, the ancient oaks of today, having passed through only a few tree generations, go all the way back to prehistory. They were not planted but grew spontaneously from acorns dropped by earlier oak trees, and so back to the time when oaks and other trees recolonised the land after the last Ice Age. In what type of woodland did these earliest oaks grow, and is there perhaps a continuity to the present? Or did these woodlands change with human occupation and use, especially in the context of pre-modern agriculture? We can only surmise what the distant past was really like. Nevertheless, we have in England a good number of sites with ancient oaks where the past is well preserved, allowing glimpses into the landscapes where oaks grew in medieval times and perhaps even beyond. This is a difficult and sometimes controversial topic, and my conclusions will not be agreed with by all, but that is to be expected in a scientific discourse.

In Chapter 9, I have evaluated nearly all the sites in England that have significant numbers of ancient and veteran oaks. Out of these, 23 were considered 'most important', not only for their oaks but also for providing a glimpse into the past. I have visited all and descriptions of these sites comprise much of this chapter. This landscape approach enhances the significance of the individual oaks, historically but above all ecologically. In Chapter 10, three guest authors, specialists in their fields of research, write about the biodiversity associated with ancient oaks in England. It turns out that ancient oaks in an ancient landscape host more invertebrates, fungi and lichens than those standing in a modernised landscape. The link with the distant past is obvious here. No other trees have the same significance for biodiversity as ancient oaks. This is an additional incentive for their preservation besides history and, indeed, admiration of their natural beauty. In Chapter 11, conservation issues are addressed. Many threats have been thrown at the ancient oaks in the past, causing the untimely demise of many trees. Which of these threats are still current and are there perhaps new calamities looming? Are we gradually losing the ancient oaks, as some fear? What can we do to protect them better? In this final chapter, I am again trying to find some answers, but here too, some issues are disputed and I am far from pretending to have all the right answers. Nevertheless, I hope that this book will, at the very least, have made it abundantly clear how significant and special the ancient oaks in the English landscape really are.

Aljos Farjon
Kew, April 2016

1

The life of an oak

The two native species of oak, pedunculate oak (*Quercus robur*) and sessile oak (*Q. petraea*), are among the longest living trees in England. Of native trees, only the yew (*Taxus baccata*) may surpass it in longevity, with evidence-based estimates for the largest (*c.* 10.00 m girth) living churchyard yews in England in the range of 1,100–1,300 years (Bevan-Jones, 2002 and references therein). There are estimates for the ages of coppice stools of native broadleaved trees, such as small-leaved lime (*Tilia cordata*), that may exceed 1,000 years (Rackham, 2003b, 2006) and there are clones with similar or greater age claims (but see Farjon, 2015); both categories are excluded here. In Chapter 2 I have outlined the problems inherent in estimating the ages of oaks, yet I concluded that a range of 800–1,000 years for the oldest oaks is within reason and backed by at least a measure of scientific evidence. We will take this as the maximum age for the largest ancient oaks in England. In this chapter, I describe how such a tree may have grown and developed from seedling to senescence and death, and how this growth is affected by many and often-changing external and internal factors. We cannot reconstruct the life of an individual oak over nearly a millennium, but it is possible to infer from the available evidence how such a tree may have grown and how it reacted to its environment, human interference, damage, pests and diseases to survive for so long. Like any tree, an ancient oak can also be viewed as a complex of interactions of many organisms, including fungi, invertebrates and others, existing in one place, with the tree as the main structural component. The environment influences all trees, yet oaks survive longer than most, so there must be attributes that aid survival in oaks that other trees do not possess to the same extent. These characteristics may have a genetic basis, and if so, are inherent in oaks, but there may also be external factors that apply to oaks and not to other trees.

Acorn production and germination

The germination of an acorn in its (semi-)natural environment is an unlikely event. In so-called mast years, of which 2006 and 2013 were examples in England, many but not all mature oaks produce thousands of acorns. Mast years occur at regular intervals of 2–5 years and there are occasional years without a crop on nearly all oaks. The native oaks bear their first good crops of acorns when between 40 and 80 years old and are in full production when between 80 and 120 years (Penistan in Morris & Perring, eds., 1974). In mast years, younger oaks may also produce acorns. At between 80 and 120 years of age, oaks growing in open conditions and left undisturbed

Figure 1-1. Oak seedling (*Quercus robur*) in Farnham Park, Surrey

reach their greatest crown expansion and their mature life stage (White, 1998); such trees may therefore produce the maximum possible numbers of acorns. Figures from 50,000 to 90,000 acorns have been estimated for mature trees (Tyler, 2008). Mean weight and viability tend to be higher for acorns produced in mast years (Brookes & Wigston, 1979). In these years, the ground under many oaks is densely littered with fallen acorns from November into the winter months.

Many acorns are taken by wood pigeons (*Columba palumbus*), jays (*Garrulus glandarius*) and rodents, or are invaded by boring insects. Many others simply rot in the leaf litter. In the past, commoners had the right of pannage and the acorns were fed to pigs, either after being collected or by driving the animals into the forests. For the purposes of germination, only predation by jays is useful because, in addition to eating some acorns, these birds plant others. This activity, which is peculiar to jays, may be significant in the success of seedlings as it enhances the chance of being established in more open ground away from the canopy of the parent oak trees.

Acorns that germinate under the parent tree's canopy usually fail to grow beyond the early sapling stage. Successful sapling growth under the canopy of trees that block even more light than oaks is practically absent, and the same is true for growth under those shrubs and ground flora that prevent sunlight from reaching the seedling. Oaks therefore tend to germinate and grow in un-grazed or very lightly grazed grassland adjacent to woods and, depending on the composition of vegetation, in woodland clearings (Figure 1-1). Thorns, but also birch (*Betula*) and Scots pine (*Pinus sylvestris*) — the latter no longer native in England but widely planted and naturalized — allow sufficient light penetration for acorns to germinate and develop, albeit with a lower success rate than that in open un-grazed grassland or heath. Downy birch (*Betula pubescens*) is usually the indigenous tree species that invades open ground in forest clearings on acid and/or sandy soils. This tree is followed by rowan (*Sorbus aucuparia*), alder-buckthorn (*Frangula alnus*) and other woodland shrubs, then by oak and eventually by more shade-tolerant trees if present in the area. Even before these arrive, holly (*Ilex aquifolium*) may spread into storm gaps and suppress further development of oak, as I observed recently on Ebernoe Common, West Sussex. Only where seed oaks are close by and competition from other trees is low can oaks be among the first trees to invade open grassland or heath. Oak thus seems to be at a disadvantage to other tree species in terms of seedling establishment rates, as was stressed by Vera (2000; see Chapter 8), but due to its longevity, only a few offspring per mature tree have to succeed to reproductive age over a period of several centuries to maintain the presence of the species. Even the most ancient oaks are seen to produce healthy acorns, extending the reproductive age of such trees to 800 years or more.

Seedling establishment

Once a seedling is established it has to survive and grow beyond a certain size to have a chance to become an oak tree. Where there are grazers and browsers in abundance that chance is very small, but when grazing is light or protection from animals is provided, for instance by surrounding thorn shrubs, the seedling can grow and within a few years become large enough to be minimally damaged or able to recover from browsing activity. With the unfortunate introduction of grey squirrels (*Sciurus carolinensis*) into Britain, an additional threat is present in areas where this animal is abundant. Squirrels can climb, whereas deer, sheep and cows cannot, and so this menace remains present for much longer in the life of an oak than does the threat from other predators. The grey squirrel strips off bark from relatively young branches and stems; if this stripping occurs below the crown and around the stem of a young oak, the sapling can be killed. In extreme cases, I have seen young oaks killed by repeated bark stripping. Otherwise, this action deforms the tree (especially from a forestry viewpoint), which will survive and develop new branches from dormant buds. Other possible setbacks include defoliation by the caterpillars of a number of moths that are specialized to eat oaks. This does not kill the trees but slows their growth, at least for the year in which they are defoliated, as the oak has to form a new crop of leaves later in the season. At some stage in its life, usually after a few decades, an oak tree becomes more or less immune to these ravages. The foliage is out of reach of grazers and browsers, much of the bark has become too thick to be stripped by squirrels, and the root system and crown are both large enough to allow a quick recovery from defoliation, which is now mostly partial. Dry summers, which might have caused trouble for saplings or pole trees on vulnerable sites, are also much less likely to kill an oak of this age and size. Oaks take longer than most native trees to arrive at this stage, but once there they are more resilient. The real formative stage (see also Chapter 2) has now begun.

The formative stage

According to calculations by White (1998) this formative stage lasts about 100 years on average. This calculation applies to maiden oaks that are free standing, or at least not impaired by surrounding trees that have similar height,

Figure 1-2. Mature oak tree in Farnham Park, Surrey

crowns and root systems. During this stage all parts of an oak expand: the root system, the trunk and major branches, the crown (foliage) and usually the height of the tree. These systems work together to make that expansion possible. A spreading root system will be able to take up more water and nutrients. An increasing bole and branches will put on more wood vessels for transport of this water, and are thus able to serve an expanding crown and a growing number of leaves. This greater leaf surface, in turn, allows for greater photosynthetic productivity and hence the production of more wood and more roots.

There are some genetically influenced differences in the crown shapes of oaks, both between the two species and within them. Pedunculate oak tends to have more or less ascending branches, sessile oak more spreading branches. German foresters distinguish several forms in these oaks; the most common, at least in pedunculate oak, is 'Zwieselwuchs' or 'forked growth', which develops main branches at various levels that spread out and are more or less contorted (Krahl-Urban, 1959). Alongside genetics, damage and reiteration undoubtedly play a role in determining this shape. Natural damage can occur from windbreak, lightning, shading or

other causes, but during this stage of its life, the tree has a maximum capacity for repairs.

Most of the repairs in the formative stage would involve restoration of crown capacity (roots are typically less likely to be affected). Branching increasingly becomes a matter of reiteration, that is, the activation of secondary buds to form new shoots replacing those that were lost or failed to grow (Hallé, 2002). If a large branch breaks off, restoration may be in the form of compensation, that is, increased branching involving dormant buds in more intact parts of the crown. Not only do oaks produce many dormant buds, but these buds are thought to be long-lived, in the region of 100 years (Read, 1999). Restoration can occur as lammas growth of twigs later in the growing season, or sometimes as the

Figure 1-3. Stag-horn oak in Hatchlands Park, Surrey

development of secondary branches in the damaged area long after the event.

Wounds in the trunk or major branches resulting from breakage can be healed by a callus-like growth of wood and bark that closes the gap, which at this stage is usually not too big for this. However, this takes time, and breakage of larger branches exposes the heartwood to the weather and to entry of water. Trees do not have a fixed life span and the life stages of an individual oak can expand or contract (Read, 1999). If fungi have already consumed much of the heartwood at a relatively young age, the mature stage which comes next may be very short before senescence takes hold and the tree becomes a 'veteran' or 'ancient' oak under the criteria adopted by the Woodland Trust's Ancient Tree Inventory (ATI). Given that trees are observed at a point in time and most recorders would not be confident to infer the life history of a tree from a combination of size and evidence of dead wood and hollowing of the trunk, the ATI assessment procedure may be 'best practice', but it does mean that trees can be and are recorded as 'ancient' when they are not very old in years. Truly ancient for a tree is obviously relative to the maximum possible life span of individuals of the same species; an ancient oak, for instance, would be older than an ancient beech (*Fagus sylvatica*). In other words, damaged oaks are not always ancient, but they may be veterans (see Chapter 2).

The mature stage

The mature stage is reached when root and especially crown expansion have reached a maximum barring damage (Figure 1-2). Every growing season, a new ring of wood is still added to roots, trunk and branches, so the tree still grows bigger, but not in height or width and depth of the crown. In oaks, this stage can extend over a longer period than the formative stage and can last a few centuries. The annual increment on the trunk and major limbs becomes narrower than that in the formative stage, because while the crown no longer expands to give more leaf area or photosynthetic productivity, the surface area to be covered with new wood increases each year. Eventually, there is a net increase of dysfunctional tissue, mainly dead wood in the form of heartwood and dead branches which are no longer fully replaced.

A stag-horn oak shows substantial crown retrenchment, often with the longest branches or branch ends dying while adventitious branching lower in the crown partly restores total leaf surface. Stag-horn oaks (Figure 1-3) are a common phenomenon in parks with oak trees of 200–300 years old, and are normally not dying unless adversely affected by disease or root damage. What we are seeing is a tree which is in the process of attaining a new balance between the volume of new wood to be laid down and the leaf surface in the crown. For reasons of physiology (related to water availability, water evaporation and length

Figure 1-4. Decaying trunk of oak with callus, Bradgate Park, Leicestershire

Figure 1-5. The bracket fungus *Fistulina hepatica* on oak pollard, Ashtead Common, Surrey

the environment. The living sapwood is too wet for wood-decaying fungi to begin wood rot. Nevertheless, chemical compounds that are deposited within the sapwood will select for certain fungi once this wood becomes dead heartwood. Some of these fungi begin to decay the heartwood core while others remain latent until conditions change. This is how veteran and ancient oaks become hollow in the trunk, in the largest branches and in the root bases while remaining alive and healthy.

The veteran stage

The veteran stage is considered to have been reached when the signs of retrenchment and decay have become prevalent. At this stage, there is much dead wood in the crown. Sections of bark on the trunk are no longer functional because the reduced crown is no longer capable of maintaining a complete cover of new wood around the trunk and branches. In those areas, there is no annual wood increment and the cambium is dead, as is the bark and wood it used to produce, giving opportunities for access and exit to wood-boring insects and fungi. The trunk becomes filled with brown and/or white rot wood in which ever larger cavities develop, eventually leading to a hollow trunk (Figure 1-4). Brown rot is common in oaks; it degrades the cellulose but leaves the lignin intact, often causing the wood to break into cubical pieces. The fungi beefsteak fungus (*Fistulina hepatica*) (Figure 1-5), oak mazegill (*Daedalea quercina*) and chicken-of-the-woods (*Laetiporus sulphureus*) are mostly responsible for brown rot in oaks. White rot is less common in oak heartwood, its fungi also decompose some of the cellulose; a frequent species is *Phellinus robustus*.

of growing season), the crown cannot expand as it should with an increasing amount of wood to be produced. Instead wood production slows down or ceases altogether in some large branches and the crown reduces in size through dieback, so that less wood needs to be produced.

Fungi colonise living tissue as well as dead wood; they may be endophytes or may enter the tree through the old dead taproot or through broken branches. Woodpeckers and wood-boring beetles may deepen the cavities through which the fungi entered, and moisture from rain collecting in these cavities stimulates the fungal growth. As the fungi continue to grow, fruit bodies of bracket fungi begin to appear. If these are at the base of the tree, the fungal hyphae have spread through the heartwood either down to the root base or starting from a dead taproot. The cambium is impervious to most wood-inhabiting bacteria and fungi and produces a protective cellular wall, so that it isolates new wood from

The reduced crown of a tree that is infected with rot is still able to function, producing not only annual foliage but also adventitious branching, and thereby maintaining photosynthesis and growth of wood. Oaks have a remarkable capacity for damage repair (also known as

Figure 1-7. King Offa's Oak along Forest Road, Windsor Forest, Berkshire

wound occlusion) which is often prominent at this stage. When a large portion of the tree, such as a huge limb, has broken off, cambial growth is directed from the edges of the damage towards the middle, enclosing the gap. If extensive, this redirected growth also reinforces the structural strength of the trunk. I have seen veteran oaks almost split in half (perhaps from a lightning strike) being restored and held upright in this way. Other examples are large ancient oaks, usually pollards, of which the bolling has come apart into several sections. These include the Gospel Oak in Herefordshire (Figure 1-14), the Marton Oak in Cheshire (Figure 2-5) and an ancient oak in Danny Park near Hurstpierpoint, Sussex (Figure 2-6). Damage-isolating growth sometimes occurs to such an extent that

the bole sections superficially look like entire trunks, giving some people doubts about the singularity of the oak tree. Ancient oaks in this condition can still produce crops of acorns and are in no way yet senescent; the living parts are in fact as vital as those of a young oak. When a severe storm breaks off major limbs or the trunk, severely reducing the crown, healthy veteran or even ancient oaks may grow a new but smaller crown with reiterated branching. Such natural damage is in fact very similar to pollarding and will have similar effects on growth and longevity. Examples of natural pollard oaks are present at Helmingham Hall and Ickworth Park in Suffolk, Langley Park in Buckinghamshire, Parham Park in Sussex and elsewhere. A particularly striking example is an ancient natural pollard oak in High Park (Blenheim Park), Oxfordshire, where almost all ancient oaks are maidens (Figure 1-6).

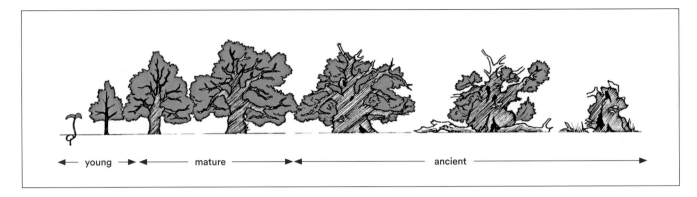

Figure 1-8. The life stages of an oak from seedling to death

The senescent stage

Finally, if not destroyed by accident, the oak will reach what is called by some the senescent stage. It is now in terminal decline. Brown (or white) rot wood inside the bole and largest limbs has largely decomposed and is being fragmented through burrowing and digestion by numerous invertebrates. This activity recycles nutrients into the soil which may in part be taken up again to support the remaining growth of the tree. Saproxylic invertebrate activity is at its peak, with wood borers active in the heartwood as well as in any sapwood that is dead. Remarkably, an outer shell of mainly dead wood in the trunk and largest limbs, of 10–20 cm thickness on average, is not affected by rot and stays firm and strong, although large gaps often occur presumably from breakage. This shell contributes in no small way to the longevity of ancient oaks.

The crown will be substantially reduced in comparison with the volume (including hollowing) of the bole and limbs and will consist mostly of reiterated branches. Dormant buds under the bark on the bole and large limbs are often activated, producing epicormic shoots. If the cambium surrounding these buds is hyper-activated due to traumatic stress from damage elsewhere, it will form burrs as the bark grows with the wood, burying most of the initiated shoots. Burrs may play a role in closing gaps, but this needs more investigation.

A distinction must be made between necrosis, which is dieback purely from damage, and senescence, which is dieback through genetically programmed cell death. In oaks of 20–25 years in age, some of the sapwood begins to die, forming dead heartwood. This means that, by volume, most of an oak tree is dead material, regardless of its apparent state of health or vigour. Damage can shorten the life of an oak insofar as it may hasten the decay process by giving more access to water and saproxylic invertebrates. It will

be less relevant the older the tree becomes, unless it leads to collapse. Oaks can apparently survive in a senescent state for a very long time, with minimal growth overall and large parts of the tree standing dead (Figure 1-7). The greater strength and hardness of oak wood compared to that of most other native trees allows oaks to stand, keeping the foliage aloft, for a comparably long time. Many living ancient oaks are little more than shells sprouting foliage branches. Such trees are, however, vulnerable to relatively small stochastic events, because if living parts are affected, there may be no other parts to keep the tree going.

We consider a tree to be dead when no parts of it are alive, that is, producing twigs and foliage. In some cases, the end of an oak tree may occur 800 or perhaps even 1,000 years after it germinated from an acorn (Chapter 2). Death may happen suddenly, as has been observed for several very ancient oaks that were photographed alive only one or two years previously. Alternatively, it may be a slower process: I have seen essentially dead old oaks which sprouted a few epicormic shoots as a final but feeble effort. A current (2014) example of an almost dead oak is the Sidney's Oak at Penshurst Place in Kent; this tree has only one propped up limb sporting a few branches with foliage (Figure 11-2). Without the props, the limb would break off and the oak would be dead.

Pollarded trees

Pollarding, it is assumed, prolongs the life of an oak. Several mechanisms for this improved longevity have been proposed; for example, the delay of crown development that occurs as the result of pollarding might prevent the total leaf surface area from outstripping the root system's ability to provide sufficient water and minerals; compartmentalisation of a pollard tree, with its multiple primary branch systems, might limit the spread of pathogens throughout the tree (Lonsdale in Read, ed., 1996; Read, 1999); or repeated cutting that limits the crown size and height of a pollarded tree might reduce the likelihood that it will be thrown by a storm, for instance, the Great Storm of 1987 in most cases only broke off branches from pollards, but toppled vast numbers of maiden trees. Nevertheless, there is no experimental evidence to support these hypotheses, and it could be that they are based on experience with willows, which have a shorter life span than oaks.

It is likely that a pollard oak will grow more slowly than a maiden (see Chapter 2), so that of two oaks with the same girth the pollard is probably older than the maiden tree, but this does not mean that the pollard will live longer. It is not true that all ancient oaks are pollards, some of the greatest in England are not, and in Eastern Europe, most of the largest ancient oaks are maidens. We cannot accurately calculate the age of the Fredville Oak, a maiden by all accounts, any more than that of the Bowthorpe Oak, a tree of similar girth that is undoubtedly a pollard.

Pollarding involves cutting off main branches, roughly at one level, causing wounds into which water and damageing organisms can enter. This amounts to necrosis, which potentially may lead to dieback; on the other hand, oaks have a remarkable capacity to close these wounds with inward growth of cambium, producing covering layers of wood and bark. In order to do this quickly, part of the crown must be retained. Typically, pollarding results in dieback only when it results in structural failure, usually a hollowing of the bolling, leading to collapse.

The pollarding of oak trees ceased up to 200 years ago, or even earlier in some cases, and has only been

Figure 1-9. Pollard oaks on Ashtead Common, Surrey

resumed piecemeal in recent years, so we still know little about its long-term effects on oaks. We may assume that in the Middle Ages pollarding was begun at a relatively young age, well before the oak was fully mature in the sense decribed here, that is, before the tree had attained a maximal crown. Nevertheless, the tree's capacity to close the damage by occlusion must usually have been sufficient to overcome the trauma and a tree's survival was rarely threatened, even in later stages when it was cut once again. We do not know exactly how long the pollard cycle for oaks was, but there is little doubt that trees were pollarded repeatedly. It is likely that pollarding had a longer and more irregular cycle than coppicing because of the greater effort, and indeed risk to life and limb, involved. Many of the veteran pollard oaks seen on Ashtead Common (Surrey) have widening bollings near where now the few remaining huge limbs begin, caused by excessive wood formation around repeated cuts of branches (Figure 1-9). We may assume that, in oaks, not all re-growth after the last pollarding was cut at once. Experiments with pollarding beeches and oaks (see contributions in Read, ed. 1996) seem to indicate that this minimises risk of death, while

willows (*Salix*), hornbeam (*Carpinus betulus*) and ash (*Fraxinus excelsior*) seem to recover regardless.

The callus-like growth that develops around cuts (branch collars) contains adventitious buds from which new shoots develop. As this growth is much more superficial than that from dormant buds, many new shoots last only for a short while, but some will develop into new branches with full connection to the root system of the tree. This may not happen if drought follows cutting or if oak mildew affects the new foliage, and failures in recent experiments with oak pollarding were often associated with these events. Ancient pollard oaks are often severely burred, and this hidden epicormic growth may well be a reaction to the pollarding of the past. Again, without the experiment of renewed pollarding, we will not know for certain what the causes are. New oak pollards are now being created in several places.

Figure 1-10 (below). Ancient pollard oak with prop in Windsor Great Park, Berkshire
Figure 1-11 (opposite). Large pollard oak in Bentley Priory NR, Greater London

Unanswered questions remain regarding senescence in ancient oaks. Does senescence actually cause the death of these trees? Are the parts of the tree that are still alive undergoing senescence? Field observations seem to suggest that the living parts, with newly reiterated branches, foliage, flowers and seeds, the most visible parts, are not senescent but rather still vigorous and with 100% functional activity. They contrast with the many dead parts of the tree, which are also very visible. The presence of lots of young tissue on an old dead framework could be interpreted as a way of 'hanging on to life' that relies on the 'colonial' nature of tree growth (Hallé, 2002). These ancient trees have 'indeterminate morphogenesis', meaning that they do not actually replace leaves, twigs and branches, but rather create new ones in different parts of the tree. An ancient oak is dead and alive at the same time, something that is impossible in an animal, which replaces cell for cell in order to maintain overall function. What the living parts of ancient trees need is maintenance of the living connection between foliage and roots. Although fallen trees can live on as long as this connection exists (or is created anew), an upright position provides a better chance of maintaining adequate exposure of the foliage to sunlight, which is important for shade-intolerant oaks. Therefore, the relative resistance of the outer dead wood of oaks to decay, perhaps in part due to its tannin content and to its ability to maintain a position off the ground where it becomes air dried, helps to prolong the life of the still-functioning parts of the tree. If the short, squat bolling of an oak pollard somehow maintains the connection between foliage and roots while being less prone to wind throw, its chances of long-term survival may be greater than that of a taller maiden trunk (Figure 1-10). Therefore, I do not think that there is much evidence for intrinsic physiological causes for the longevity of pollard oaks. By pollarding oaks in the past, we may have merely saved many of them from fatal accidents

The cessation of pollarding around 200 years ago has caused many ancient oak pollards to grow massive limbs. Commonly, a number of branches that originally sprouted from the bolling have died back (atrophied) over the years, leaving a limited number to grow on, sometimes only two or three, although 14 large limbs were counted on a 7.28 m girth ancient pollard oak in Bentley Priory Nature Reserve just north of London (Figure 1-11). There have been several consequences of this lack of 'management' of oak pollards. One is that the crown of such pollards may have expanded to a size equal to or larger than what could have developed

Figure 1-12. The Queen Elizabeth I Oak in Cowdray Park, Sussex

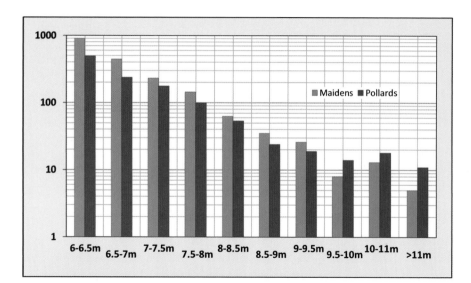

uncut growth to the hollow bolling. Breakage will result when this weight exceeds the holding capacity of the stem cylinder, which is often no longer entire or has cracks and gaps. This is the only reason why props have been installed under many ancient oaks, perhaps most notably in support of the Major Oak in Sherwood Forest (Figure 2-12), which is too famous a tourist attraction to be allowed to fall apart. First the creation of oak pollards and then the cessation of pollarding have resulted in many ancient and veteran oaks that may not have much further life expectancy if left uncut. With few exceptions, storms tend to be a destructive force causing more irreparable damage to the heavy branches of such neglected oaks than that which results from secondary 'self-pollarding'.

There are more very large pollard oaks in England than maiden oaks in the same size class. The oaks in the dataset (n=3,017) contained a total of 1,870 maidens and 1,147 pollards. 'Natural' pollards and other tree forms were excluded. If these are divided into 10 size classes and compared with the numbers in each class on a logarithmic scale (Figure 1-13), we find greater numbers of pollards in the three largest size classes. This despite the fact that the dataset contains substantially more maidens than pollards. If there is a positive correlation between girth and age, then the largest pollard oaks of England are, on average, older than the largest maiden oaks. This does not mean that pollard oaks in the same size class are always older than their maiden counterparts, but because there are more of them still alive in the size classes 8–10, we may conclude that pollards have had better chances of survival and thus

if the tree had been left uncut. Instead of one bole with branches, there are now two or more. Lapsed pollards with large crowns will produce more wood and may grow faster than they used to do when still regularly cut. Much of that wood, however, goes into the limbs rather than into the bolling, especially when the latter is fragmented or its bark no longer surrounds the entire stem. It is conceivable that those pollards which have not developed these large limbs, or which have lost them, may live longer. Some oaks are known, from old drawings and photographs, to have been reduced by such secondary 'self-pollarding'; examples include the Queen Elizabeth I Oak (12.67 m circumference) in Cowdray Park, Sussex (Figure 1-12). Another consequence of the cessation of pollarding is the increased burden of

that the life expectancy of a pollard oak is greater than that of a maiden oak.

Multi-stemmed trees

Finally, there is a question that the recorders of ancient and veteran trees are asked to address: what is the 'tree form' of the tree being recorded. The two options we need to discuss for oaks are: multi-stemmed or single-stemmed (including pollard forms). The two native species of oak are essentially monopodial trees. When an acorn germinates, the radix develops downwards to form a single taproot and the hypocotyl expands upwards to form a single aerial stem (Tyler, 2008, figure on p. 36). The basic architecture of an oak is a monopodial stem above an underground root system with alternating lateral branches. Therefore, if an oak appears to be multi-stemmed, something has happened to it that has led to abnormal growth. Two possibilities can be considered: a) there were initially two or more germinating acorns close together and the stems grew together and became fused, and b) damage of some kind broke up the single stem into parts that continued growing.

Fusion of stems occasionally seems to occur but is seldom complete in oaks. Most examples are from trees in which partial fusion occurred much later in the development of the trees. In natural circumstances, acorns

Figure 1-16. Coppiced oaks in Hatfield Forest, Essex

are unlikely to germinate in close proximity; acorns are too big to fit more than one in a jay's crop, so these birds almost always plant just a single acorn at a time. Cases of later fusion usually involve an older (larger) trunk and a smaller younger one. A remarkable example can be seen in the Old Park at Chatsworth, Derbyshire. Here two large oaks have grown side-by-side over boulders, developing bulging root bases that eventually fused the tree bases above the rocks (Figure 1-15).

Much more common are multi-stemmed trees resulting from damage. Hollowing trunks, especially the shorter bollings of pollards, eventually tend to develop lateral gaps and may eventually split into several sections. The capacity to isolate this damage by the growth of new wood and

bark around the margins of the sections can result in the appearance of separate stems from the ground up. Because this is a late development in the life of a single-stemmed tree, it is correct to record these trees as (remnants of) monopodial trees and to estimate the circumference of a bole as if it were intact. During this process of multi-stem formation, an individual tree will obviously have passed through many stages, from a single gap in the bole to extreme separation, with or without broken sections lying about, as can be seen in King Offa's Oak in Windsor Forest, Berkshire (Figure 1-7).

The other important form of damage that leads to multi-stem formation involves coppicing. This is man-made damage and amounts to cutting off a tree near ground level, as opposed to pollarding which is done well above ground level. Normally, trees are first cut at a young age and then often and regularly. As a result, a coppiced oak never develops a monopodial stem but rather several stems from a stool (or 'moot' in oaks, Rackham, 2003b). Measuring the circumference of a stool may be an indication of a tree's size but, unlike that of a monopodial tree trunk, the relation of stool circumference to age in years is extremely contentious and goes no further than the statement: a large coppice stool is likely to be older than a small one. Coppiced oaks (Figure 1-16) are correctly described as multi-stemmed and the correct measurement of circumference is therefore of the individual stems on the coppice stool, just like one would measure the stems of a suckering elm (*Ulmus*). The common practice of recording an oak as 'multi-stemmed' yet giving its circumference as if it was a monopodial tree is therefore contradictory. We have to separate superficial appearances from an understanding of how the tree grew and came to appear 'multi-stemmed'. Given the rarity of two or more merged stems, unless a coppiced oak, all apparent multi-stemmed ancient and veteran oaks can be assumed to have grown up in an earlier stage as monopodial trees.

2

The age of ancient oaks

The monarch oak, the patriarch of trees, shoots rising up and spreads by slow degrees;
Three centuries he grows and three he stays, supreme in state, and in three more decays.
John Dryden (1631–1700)

A clear and scientifically based understanding of the approximate age of living ancient native oaks in England is important for a number of reasons. First, the age of these trees links them with the history of the countryside. Second, their duration as (micro-) habitats, often isolated from other ancient trees, influences the biodiversity that is dependent on them. And third, their maximum age can tell us something about the physiological capacities to overcome adversity inherent in oaks as opposed to those in tree species of shorter longevity in the same locality.

In Chapters 4 and 5, the current distribution of ancient oaks in England is explored in relation to forms of land ownership and use dating from the Middle Ages, but which began to change in Early Modern time and to a large extent disappeared in the eighteenth century. In Chapter 8, I discuss the occurrence of ancient oaks in pasture woodland, as this appears to have been the habitat in which most attained their great age. During the Middle Ages, the trees in forests, woods, commons

and deer parks grew naturally; they were not planted. Among the many rules and regulations issued under Forest Law during the Middle Ages until as late as the reign of Henry VII (1485–1509), "there appears to have been none enjoining planting" (Loudon, 1838, Vol. III p. 1750). The earliest reference to the planting of oaks dates from 1589 (Mitchell, 1966) and it seems unlikely that any were planted in England before 1550. If the truly ancient oaks now alive are older than 500 years, or even 400 years, they have not been planted, which has important implications for our concept of the medieval wooded landscape (Rackham, 1976) and the earlier landscape from which this evolved (Chapter 8). It is increasingly well documented that ancient trees, in particular oaks, are among the most biodiverse (micro-) habitats in this country (Green, 2001, 2010b; Harding & Wall, eds., 2000; Tyler, 2008; Chapter 10). This biodiversity increases with the age of the (standing) oak tree.

The physiological attributes of oaks differ from those of other native trees and some of these may contribute to longevity (Longman & Coutts in Morris & Perring, eds., 1974; Buck-Sorlin *et al.*, 2000; Chapter 1). Detailed studies of the physiology of ancient oaks, including the growth rates of various parts, the replacement of damaged cambium and the reiteration of foliage could contribute not only to general understanding of how oaks function,

Figure 2-1. Oak trunk cross section *c.* 130 years old, Stoneleigh Park, Warwickshire

but could also assist with the very question asked in this chapter: how old are these ancient 'veterans'?

When it comes to the age of the ancient oaks, there is no shortage of conjecture, anecdotes, and folklore, some of it downright nonsense. Even relatively serious publications about trees often cheerily endorse traditions that link some of these trees to King John, William the Conqueror or even King Offa. Such claims have no scientific or historical basis. "Planted by William the Conqueror" is an anachronism even if the tree were to turn out to be old enough for such an act by the first Norman king of England (1066–1087). The idea projects notions from later centuries about planting forest trees onto an age in which these notions were not held and were unwarranted: much of the 'wild' countryside was still full of trees. Asserting that King Offa's Oak in Windsor Forest was a notable tree in the time of his reign (757–796) adds several hundred years to the time that has

elapsed since his death, resulting in an absurdity. The figure 1,000 seems to have a special fascination, perhaps linked to ancient millenarian beliefs endorsed by Christianity. Many ancient oaks are thus claimed to be "a thousand years old" without the slightest evidence. Such mythical ages of ancient oaks can be dismissed as mere folklore; mentions of them as probable ages of ancient oaks in publications, in print or online, are misleading. As Mitchell (1966) correctly pointed out, the ancient oaks of England are of unknown age and "a precise age can never be found for them, since they are all hollow and lack most of the annual rings that could have been counted." What we can do is to attempt to estimate their age by evaluating circumstantial evidence,

including proxy data for annual increment such as diameter of trunk, growth rates during various life stages of a tree, and archaeological and historical evidence. With all of these there are caveats, irregularities, pitfalls and sources of error, which tend to increase with the actual age of the ancient tree. In other words: the older the tree, the more fragmentary and uncertain the evidence becomes. These issues need to be elucidated first, before we can even begin to make estimates of ages for some of the ancient oaks mentioned in this book. One important consideration is

terminology; we need to be clear what is meant by the terms 'ancient oak', 'veteran oak', 'ancient' and 'circumference' or 'girth' and how informative they are in relation to estimates of the ages of ancient oaks.

Ancient versus veteran oaks

The Woodland Trust has given guidelines to the volunteer recorders of ancient and veteran trees as part of the Ancient Tree Inventory (ATI) project; among these is a definition of ancient (or veteran) trees (www.ancient-tree-hunt.org.uk/

Figure 2-2. Hollow oak trunk with brown rot in Richmond Park, Surrey

potential life span of a particular species. Among other native trees, only yew surpasses the two species of oak (pedunculate oak and sessile oak) in potential age. We can be reasonably certain of this even though determining the age of an ancient yew is even more hazardous than determining that of an ancient oak. The Woodland Trust gives further guidelines; for oaks these include the following indicators (here slightly amended or combined):

- Trees with a girth of more than 6.00 m are likely to be truly ancient
- Major trunk cavities and progressive hollowing; decay holes (e.g. of former branches)
- Physical damage to trunk; sheltered crevices in bark or under branches
- Large quantities of dead wood in the canopy
- Bark loss, with bark on the trunk no longer present all around
- Fungal fruiting bodies (from heartwood rotting species, commonly bracket fungi)
- Pollard form or indications of past treatment as a pollard

Other features mentioned, such as the presence of epiphytes, sap runs and aesthetic interest, can apply to non-ancient oaks and are omitted here.

Each of these indicators can be present in oaks that are not ancient but in the late first or second stage of their life. Damage caused by storms or lightning, disease (pathogens) or weakening by drought or a combination of these factors can accelerate wood decay and make the tree look ancient when it is not. Girth can only denote age when assessed in combination with the other indicators. An oak that has a large, full crown and grew most of its life as a free-standing tree on a good site can attain 6.00 m girth and more in slightly more than 200 years (Mitchell, 1966). Such a tree, of which Mitchell gives the Knightwood Oak in the New Forest as an example, is unlikely to be ancient. In 1965, this tree was measured with a girth of 7.03 m and Mitchell, considering likely growth rates and site conditions, came to an estimate of 300 years, "but it could be less than 250." In 2007, this oak was measured with a girth of 7.38 m (ATI data), so in 42 years it added 35 cm circumference to its trunk. According to the Ancient Tree Forum (www.ancienttreeforum.co.uk), this tree was (once?) pollarded when 200 years old and last received this treatment about 150 years ago; they give it an age of more than 500 years but without rationale. Mitchell's estimate, brought up to the present, would be 300–350 years. We shall get into this girth–age equation in more detail later; here, this tree is merely cited as an example of a large (in terms of

ancienttrees) based on recommendations in Read (1999). The term 'ancient tree' encompasses:

- Trees of interest biologically, aesthetically or culturally because of their great age
- Trees in the ancient or third and final stage of their life
- Trees that are old relative to others of the same species

Apart from the element of subjectivity, it will be obvious that 'ancient' is a relative quality, much depending on the

Figure 2-3. A veteran oak with 6.70 m girth in Elmham Park, Norfolk

events of insect defoliation, late frost damage and other stochastic events, but on average, after an often rapid initial growth, the current annual increment of new wood (CAI) remains of the same width during this stage. This is so because although each year more wood is added than in the previous year, it has to be laid out over a larger surface.

After the optimum crown size is reached, barring intervening damage (which can be repaired to an extent), a more or less stable stage begins, the middle age of the tree. During this stage, the crown and root system do not expand, and if damage is not severe, replacements of lost foliage and branches by reiterative growth maintain the extent of the crown. A stationary crown size, and thus stable photosynthetic productivity, means increasingly narrower annual rings because the same amount of new wood is spread over an increasing area of stem and branches. Consequently, the rate at which the circumferences of the trunk and large branches increase slows down. In oaks, this 'mature' stage can be at least as long as the formative stage and is often much longer.

Finally, the trees enter the third stage. When conditions become right (by aeration through gaps for instance), the mostly endophytic wood rot fungi enter a mycelial mode and start colonizing the tree, expanding the area of wood rot, mostly but not exclusively in the heartwood (Figure 2-2). The relatively dense and decay-resistant wood of oaks (compared to that of beech or birch, for example) and a good capacity for reiteration and damage 'repair' will cause the ancient stage to be quite prolonged, adding to the potential age of the oaks. If this stage is attained without excessive damage in earlier stages, and the tree has reached a circumference ≥6.00 m, it is indeed likely to be ancient.

circumference and height) oak that is unlikely to be ancient. It is still in its second stage of life.

Which are these stages? White (1995, 1997, 1998) distinguished three stages of growth through which trees progress: a formative stage, the middle age or the 'mature state', and senescence. Because oaks at this third stage are normally quite healthy, I shall use the term 'ancient' in its place. The first stage is dominated by growth, in particular crown expansion. The total leaf surface, which is crucial for growth, expands every year and the annual increment of wood on the stem and branches is likewise increasing. Fluctuations occur due to differences in precipitation,

What are veteran oaks? The guidelines for the Woodland Trust ATI give the following definition:

"A **'veteran tree'** is usually in the second or mature stage of its life and has important wildlife and habitat features including: hollowing or associated decay fungi, holes, wounds and large dead branches. It will generally include old trees but also younger, middle-aged trees where premature ageing characteristics are present."

This definition is vague, it introduces indicators that are also used to identify ancient trees, and if it includes old trees, one wonders how to determine that the tree is still in its second, mature and stable stage as defined by White (1995, 1997, 1998). In his 1997 paper, White substituted the term 'veteran' for 'ancient', applying essentially the same criteria to define a veteran tree that are used by the ATI to identify an 'ancient' tree. Indeed, Read (1999), on whose guide the Woodland Trust's definition of ancient tree is based, used the term 'veteran tree' for the same. Fay (2007)

noted the confusion of these terms, but recommended that they be used separately, with the designation 'ancient' reserved for trees that are "old for its species". "While all chronologically ancient trees have veteran features sufficient to qualify them as veteran trees, all veteran trees will not necessarily have entered into the ancient age class." Veteran features may occur as a result of environmental impacts rather than of the natural ageing process. In practice, given that such damage may have occurred long ago, it can be difficult to distinguish accident from ancient. If we want to keep this distinction between ancient and veteran for oak trees, we will have to be able to estimate their age and agree on an age beyond which the tree is called ancient. We cannot obtain ages of ancient or veteran oaks by counting annual increment rings, because most of these no longer exist. If the tree is still alive, we will only have a number of rings that were laid down in the later stages of its life, often only the most recent years up to the present year. We need other indicators of age. The most commonly used proxy for annual rings is the circumference of the trunk, recalculated to the radius (or half the average diameter), with an inferred

Figure 2-4. Veteran oaks in High Park, Blenheim Park, Oxfordshire

Figure 2-5. The Marton Oak in the village of Marton, Cheshire

relation between radius and number of annual increment rings. Can this be a reliable proxy for the lost annual rings of an ancient oak? This question we will investigate next. The problem must be approached in two stages: first the accuracy of the measurement of circumference and its relation to radius, and second the validity of the relationship between this measurement and the age of the tree.

Measurements of circumference

The trunk of an average tree above buttresses or root bases and below its first branches only slightly tapers upwards and is more or less circular in diameter. In order to avoid the irregularities, measurements of circumference are usually taken at *c.* 1.30 m above the ground ('breast height') and can be expressed as diameter at breast height (dbh), which is calculated by dividing the circumference by 3.14. Such trees fall in category A of the Tree Register of the British Isles (TROBI) and are easy to measure. The radius, presumably proportional to the number of annual rings, is half of the diameter. The ATI project guidelines recommend that the circumference of ancient trees is measured at a height above ground of 1.50 m, and provide a number of further tips on how to avoid situations that deviate from the 'normal', such as trees on slopes, leaning trees, trees with branches or burrs at 1.50 m above ground, and other

irregular situations (www.ancient-tree-hunt.org.uk), for example by seeking the smallest circumference below 1.50 m if that area has burrs or large limbs or is broken. The main purpose of these recommendations is to encourage comparability between measurements made by different recorders. By following the guidelines and by noting what has been done in cases where 'normal' measuring at 1.50 m height could not be executed without undue bias, it is hoped that measurements between trees, or on the same tree at time intervals, will be comparable on an equal footing. However, while these guidelines are undoubtedly useful and may avoid exaggerated measurements in many cases, they are not sufficient to overcome the major problems encountered with many ancient oaks. I discuss some of these below, with examples.

For many ancient oaks, large sections of the trunk are missing from the ground up to 1.50 m or above, often because the tree is hollow. The trunk may have one or sometimes more gaps, or a once-entire trunk may have come apart into two or more sections. In the latter case, it may seem at first that there is more than one tree, but closer inspection usually reveals that the parts belong to a single tree and are probably still connected underground by the root system. In lapsed pollards, the parts often lean outwards, probably because the increasing weight of heavy uncut branches outspreads the trunk sections and widens the gaps between them. Ancient oaks of this shape may resemble category C trees as defined by TROBI, and their girth will be the smallest measurement recorded between ground level and 1.50 m above ground level, provided there are no substantial gaps.

The Queen Elizabeth I Oak in Cowdray Park, Sussex (Figure 1-12) has a circumference of 12.67 m measured in December 2007, or 12.50–12.80 m as recorded by David Alderman in June 2008. This tree is a very squat and entirely hollow natural pollard and a huge section of the trunk has been missing for many years. One can only measure air there. Yet it is stated to be "the third biggest sessile oak ever recorded" with the two other contenders being the Pontfadog Oak in Wales (now blown over and dead) at 12.85 m and the Marton Oak in Cheshire at 13.38 m.

The Marton Oak is often mentioned as the biggest ancient oak now alive (Figure 2-5), but it is made up of four completely separate sections above ground, three living and one dead. The gaps are wide enough for two or even three people to walk through side-by-side. The sections do not lean out below 2.50 m above ground, so it is possible to form a picture in the mind of the probable circumference at approximately the narrowest point when the trunk was still one, which was a long time ago. A drawing from 1810 reproduced in Miles (2013) shows a divided tree similar to the present situation. Of course, if it has been centuries since the trunk was entire, it would have been much smaller at the time when it began to split. There is no way that you can measure the circumference of this tree with an accuracy of 50–100 cm without a great deal of conjecture. In fact, this oak no longer has a circumference.

The Gospel Oak in Herefordshire has a recorded circumference of 12.80 m (Figures 1-14 and 5-31). This ancient pollard consists of four parts separated (in two groups) to such an extent that some have doubted that it is one tree (Archie Miles, who is not a doubter, pers. comm., June 2013, at the tree). Some of the sections lean outward almost from the ground and there are numerous burrs. The smallest 'girth' would have to be measured just above the ground, filling in the gaps with conjecture.

King Offa's Oak in Windsor Forest, Berkshire is recorded with a girth of 11.18 m (Figure 1-7). This measurement appears almost miraculously accurate, because this tree is falling apart despite props under the remaining limbs. One long-dead section lies prostrate having fallen from a gap that is at least 2.50 m wide, and other gaps are wide enough to walk through with ease. Other sections lean outward. This situation has more or less existed for at least a century (Menzies, 1904). One website about Windsor Castle and Park (www.theroyallandscape.co.uk/about-us/history/) gives this tree a growing start in the year 710.

A fragmentary ancient pollard oak in Danny Park near Hurstpierpoint in Sussex (Figure 2-6) was given a circumference of 10.50 m by the ATI recorders. This tree consists of one large section and two smaller sections, one of which is now dead. The two subsidiary sections are remote from the main section and lean outward. The parts of the original tree that are missing are so large that the former shape of the bolling when intact can only be guessed at. The recorded girth must be a conjecture of the tree as once was; it cannot be a measurement of any of the existing parts unless the tape stretched across much empty space. In defence of the recorders, there is a convention or recommendation that the tape is spanned straight across gaps, adding that distance to the girth.

Many large ancient oaks have squat trunks that taper rapidly, usually from near the ground up to well above 1.50 m. Branches often begin just above the ground and can be big or even massive, causing an abrupt narrowing above them. Measuring a few decimetres lower or higher will make a substantial difference, but the guideline "measure the trunk at the narrowest point below a big branch or a burr" cannot apply to such trees. Any level at which the tape is placed is arbitrary, but near ground level would seem to be the most logical choice. Measurements for such a tree would then not be comparable with trees measured at 1.50 m above ground. Some oaks of this shape could fall into category B of TROBI if they can be measured below 1.50 m avoiding the greatest irregularities.

An ancient pollard oak that I visited in April 2011 in a privately owned meadow (Salcey Little Lawn, Figure 5-10) in Salcey Forest not only has a large gap from the ground into the crown, but its trunk strongly tapers, with a narrowest point (with gap) under the main branches about 3.50 m above the ground. Its trunk is also full of burrs. No true measurement of its circumference is possible, and my estimate of 9.00 m at 1.50 m above the ground is as good

Figure 2-7. The Billy Wilkins Oak in Melbury Park, Dorset

or bad as the recorded girth of 10.62 m at 1.00 m, which was made with a measuring tape in December 2013.

The Billy Wilkins Oak in Melbury Park, Dorset (Figure 2-7) measured with a circumference of 11.88 m, tapers from the ground up. It also has large areas of burr, which can hardly be avoided unless one measures the trunk above 2.50–3.00 m, where it is substantially thinner than at ground level or at 1.50 m. It is possible to obtain different figures of circumference for this ancient oak from near ground level to 2.50 m, and none of these can be demonstrated to be more accurate than any other.

Burrs (burls in American English) are very common on ancient oaks (Figure 2-8), especially on the largest pollards. These irregular growth structures can form localized bulges, sometimes of considerable size, or they may spread over much of, or even the entire, trunk. They are an extreme form of reiteration. Normal reiteration occurs in the form of successive and ever-smaller branching beyond the basic architecture of the tree (Hallé, 2002) from new buds axillary

to leaves on terminal branches. It forms, expands and maintains the foliage crown of the tree. In burrs, new buds are formed in close proximity in the cambium under the bark of the trunk and sometimes on large branches. A few of these grow out to small leafy twigs (epicormic shoots) but most die shortly after appearance. Most of the buds in burrs form shoots that fail to penetrate the bark (Tyler, 2008). New bark and wood form in an irregular way in burr areas because normal and more regular incremental growth is disrupted and deformed at the cellular level by the emerging shoots. If the burr is caused by environmental trauma, the cambium can become hyper-stimulated, resulting in rapid expansion. In other cases, there may be moderate stimulation of epicormic growth, resulting in similar cambial growth but not forming the typical bulging burrs. If the burrs cover large areas of the trunk, they will contribute to the girth of the tree, but not in the regular way of annual increment rings of wood. There is some evidence that the trunks of ancient oaks that have extensive burr growth, such as the Lydham Manor Oak and the Billy Wilkins Oak, are expanding faster than they would if growing normally. Wherever this type of growth on the trunk is extensive,

Figure 2-8. A burred oak in Hatfield Park, Hertfordshire

measurements of circumference lose their meaning in relation to attempts at ageing the tree, because even if we could establish a rate of girth increase for burred trunks, we do not know how long the burrs have been present.

The largest of two oak pollards at West Lodge in the former Forest of Bere, Hampshire, measured as having a girth of 9.55 m, has, apart from a large cavity that threatens to rip the tree apart, large burrs all around the trunk (Figure 5-2). These combined factors stand in the way of accurate measurement of the Bere Oak.

A broad squat pollard at the edge of a wood and lawn in Spye Park, Wiltshire (locally known as the Pot Belly Oak, Figure 2-9) is nearly all burrs below and partly beyond the branches. Its trunk tapers strongly and there is a large cavity from the ground to *c.* 1.20 m above ground level, as well as a smaller one. Any attempts at measuring this tree would be spurious from the point of view of making an estimate of its age.

The Quarry Oak or Croft Castle Oak at Croft Castle, Herefordshire has burrs all around its trunk from the ground upwards. Its trunk tapers markedly, and it has a huge limb from *c.* 1.00 m to 1.50 m off the ground (Figure 4-25). Despite these insuperable difficulties, it has been given a measurement of 11.22 m "over one side limb" in the Tree Register of the British Isles (TROBI).

The Lydham Manor Oak in Shropshire, with a measured circumference of 12.53 m, is one of the most densely burred ancient pollard oaks that I know (Figure 2-10). These burrs occur all around the trunk and extend from the ground to the six or seven great limbs. This tree has no visible large cavities and it may be that the burrs prevent these from developing, or that they have closed them up. There is a series of relatively recent measurements of this tree that appear to imply a faster growth, due to burr development, than would be expected in a 'normal' bolling of this size class. No age can be estimated from this growth rate because we do not know when burr growth first enveloped the trunk. It may have developed late in the life of this undoubtedly ancient pollard.

The Old Man of Calke in Calke Park, Derbyshire, measured as 10.01 m in circumference, is a squat pollard full of burrs, roughly one half now being dead wood and the other half still with strips of living cambium and sound bark (Figure 9-9). Cavities and other irregularities abound, too. There is, in my opinion, no way to lay a tape around

Figure 2-10. The Lydham Manor Oak near
Bishop's Castle, Shropshire

this tree without measuring burrs and
empty space.

Wyndham's Oak near Silton in
Dorset, with a measured girth of
9.79 m, is a completely burred squat
pollard, with mostly dead wood from
which the bark has long been stripped
(Figure 5-6). Bark only occurs in a
few broad strips that keep the tree
growing and on the living branches.
Measurements near ground level on
this tree are as valid (or invalid) as
at any other height below the few
remaining branches.

It would be easy to expand these
examples with descriptions of many
more ancient oaks; in fact, among the
large trees listed in Chapter 5 (>9.00 m
circumference) only a few are not beset
with major irregularities and could be
measured with reasonable accuracy.
Most of the measurements given can
only be considered as approximations.
Measuring the girth of truly ancient
oaks is often impossible within an
accuracy of 1.00 m or more. To
assume that the trunk was round when
still intact, and not beset with burrs
or other irregularities, is not justified.
I often measure ancient oak trees and
usually photograph them. Often, in
going around the big trunk, I notice
how they are markedly narrower from
one direction of observation than from another, even when
largely intact. These oaks are not 'normal' growing trees.
Dead wood does not grow, and many trunks have had or
still have large exposed sections of dead wood. New wood
is only added where bark is still alive. In such trees, there is
no annual increase of girth in a conventional sense. Annual
rings are added on one side, but not on another side. Where
gaps have formed, there is no wood at all. Burr wood, if
present more or less all around the trunk, adds to its girth
but not in the same way as annual ring increments, and
this growth also occurs irregularly and at different times on
various parts of the trunk. All these factors cause the trunk's
growth to deviate markedly from the 'normal' circular

annual increment of a tree, and it is very likely that this has
been the case for many ancient oaks for centuries.

Why is this important? It is important because
circumference, from which radius is calculated, is used as a
proxy for the number of annual increment rings and thus
for the age of the tree. When the calculated radius is false,
or irrelevant due to irregular growth, calculations of age
become spurious. It is that calculation of age that we must
now address.

Calculating age from trunk circumference
All living trees add some wood to roots, trunk and
branches each year. It is the wood added to the trunk

a mature stage and an ancient stage. Here, I argue for a fourth stage that is relevant to English native oaks.

In the formative stage, during which the tree increases the size of the root system, the crown and the total leaf surface, there is rapid and steady increment of wood around the trunk. Subject to fluctuations due to climate, defoliations by insects or storm damage, the crown expands and more wood is formed each year; but as the circumference of the trunk is also increasing, each year's new wood is divided over an area that increases in size each year, and so the annual rings tend to remain of similar width during this stage (White, 1998).

Once the optimum crown size is attained, the second or mature stage begins. The amount of new wood added each year remains more or less constant, but since it has to be laid out over a still increasing surface of trunk, main branches and roots, the annual increment rings become narrower. For many mature oaks that have been cut and are still sound, with no heart wood rot or hollowing, a date at which this transition from constant annual increment rings to narrowing annual increment rings took place can be established within a few years. In maiden, free-standing oaks, this transition is often observed to arrive after about 100 years (White, 1998; the author, personal observations).

In theory, the current annual increment (CAI) produced by a constant volume of new wood should produce ever narrower rings in a thickening trunk, but in practice, this trend is often much less visible. Variation in CAI depends on locality within the stem and, above all, on fluctuating conditions in the local environment, such as dry or wet years, early or late spring, competition from neighbouring trees or their disappearance. Annual rings in the mature stage are narrower, sometimes much narrower, than those in the formative stage, but they do not narrow gradually. One reason for this is the physiological constraint of the ring-porous wood of oaks: vessels cannot operate properly for conduction below a certain functional diameter. Yet a crown that has reached maximum total leaf surface still has the capacity to increase the existing total wood volume, and it is possible that this works towards an average equal ring width. In this stage, there is also a minimum ring width of c. 0.5 mm, below which survival of the tree would become unlikely (White, 1998). The sapwood serves to transport water upwards through capillary action, which is brought about by evaporation through the leaves. Too narrow vessels in this wood will impede this flow.

In oaks, at least, it seems useful to further distinguish a 'late maturity' or 'veteran' stage. This stage is marked by an increase in dead wood in the crown in the form of 'stag horn' branches. The narrower rings possibly slow

that we are interested in here, and when there is seasonal distinction between growth in spring (early wood) and summer (late wood), with a virtual cessation of growth in winter, the tree produces annual increment rings (Figure 2-11). In native oaks these are usually clearly visible in a cross section of the trunk, branches and roots. There are many factors that determine the width of the annual rings in the trunk of an oak tree, resulting in numerous variables. Some of these are intrinsic and can be related to the life stages of a tree; other factors are extrinsic and are related to the environment, stochastic events or management. Earlier, we discussed how three stages in the life of a tree are usually recognized: a formative stage,

down water uptake, and consequently, water fails to reach the branches furthest from the roots and highest in the tree because the weight of the water column becomes a limitation. Blank (1996), in a careful study of the ancient oaks at Ivenack in Germany, found that the ring width of early wood remained fairly constant in veteran trees and that the decrease in ring width was mostly in late wood; in other words, a shift towards a higher proportion of early wood volume occurs with increasing age of the tree. This could mean that physiological functions are maintained at the cost of structural strength at this stage. It has also been suggested that reduction of the total surface area of the branches and foliage may restore a balance between maximum root extension and the uptake of water and nutrients necessary to maintain and grow the amount of wood in the crown (Read, 1999). Oak trees that have had their root systems reduced by ploughing around them in arable fields are often seen to become stag headed, even when their apparent age does not seem to be beyond the early mature stage. Following crown retrenchment, dormant buds are usually activated and lammas (late summer) shoots develop on the remaining living branches; these may grow out to form new branches and the tree will continue to add wood produced by a smaller crown of foliage. During the late maturity stage, there is also cumulative damage from stochastic events; the chance that the tree will have

suffered a number of these events obviously increases with age.

The last stage in the life of a tree is the ancient stage. By this stage, or even before if damage has been severe, decay of wood has become substantial. The heartwood (dead material in any tree) has been partly consumed by fungi, in oaks mainly those species that cause brown rot, as evidenced by fruiting bodies that emerge from crevices, cavities or often at the root bases of the trunk. Annual increment of wood still occurs, but it is no longer distributed evenly around the trunk because the bark is missing over crevices and/or elsewhere. Eventually, substantial parts of the outer trunk and its bark are involved in the die-back. Where there is no living bark, there is no increment of wood. The dead wood functions in supporting the tree, so that the trunk and branches can still expand where there is living bark. At this stage, growth is very slow and may merely compensate for losses elsewhere in the tree, including the trunk, so that effectively there is no change in the overall girth of the tree. This final stage, in oaks, can last a long time, but as experience shows, death can also occur quite suddenly. The intricate balance between root system, living bark and living foliage, and the restorative powers of the tree against increasing damage from droughts, storms, defoliators, wood decay, root damage and so on is easily disturbed. When the ancient stage has apparently lasted

Figure 2-11. Micrograph of a cross-section of the wood of an oak (*Quercus robur*) showing two annual rings. Photograph: Peter Gasson/ RBG Kew

for long, Read (1999) has suggested that it should be divided into three sub-stages:

- Early ancient – when, over a period of years, there is a trend for the amount of dieback to exceed growth.
- Mid-ancient — when the annual rings cannot form all the way around the stem and some discontinuities start.
- Ancient – the terminal decline of the tree, leading to death.

I call the variation in annual ring width that is inherent in these life stages of an oak tree 'intrinsic variation'. It would be observed in all trees that have grown to the ancient stage if there was a more or less complete ring sequence. Before the last two sub-stages of the ancient stage, there is a more or less linear relationship between the radius of a tree trunk and the number of annual rings, and taking the variables inherent in the life stages into account, this relationship can be determined. Mitchell (1966, 1974) gave a formula for ageing oak trees which has been represented in a graph and a table in Harris *et al.* (2003), in which trees in an 'open-grown situation', in an 'avenue' or in 'woodland' are separated. At the same age, their girth would be successively smaller under these three conditions. According to the formula, a 7.50 m girth open-grown oak would be around 525 years old; an oak of the same age grown in a wood would have attained only 3.76 m girth. A general formula seems to me to be too simple to represent this relationship; a full-grown oak that is part of the uppermost canopy in a forest is not much inhibited, as many vigorous oaks in tall forests in countries such as France testify (England does not have many such forests). CAI rates for such forest trees would be influenced by many factors, such as soil type, precipitation or access to ground water, defoliation by insects, and indeed competition for light and water. Unless all these variables and their effects on growth are known for a particular tree, we may miscalculate its age easily if we use a simple formula. What is valid is to assume that the greater the total surface of leaves, the greater the stem increment will be.

Some oaks grown in open fields have huge crowns and may indeed increase in girth faster than oaks in closed-canopy tall forest, provided they keep their crown intact. White (1998) has provided a stepwise assessment protocol with values depending on variables such as species and site characteristics. This involves first the calculation of core size, i.e. the size the trunk is likely to have reached at the end of the first life stage. Because many samples of cut trees at this stage were available, White (1998) was able to apply reliable statistics to give core size calculations in a table. An open-grown pedunculate oak on a site with 'average'

climate and soil conditions would attain mature age at 100 years and have an average ring width during this time of 3.5 mm. The core radius will therefore be 35 cm, the basal area calculated for this tree will be the square of the radius × 3.14159 = 3,848 cm², the basal area calculated for this tree will be the square of the radius × 3.14159 = 3,848 cm². The next step is to subtract one ring width from the core radius and calculate a new basal area for the core. The difference between the two core basal areas represents the current annual increment (CAI), which here comes to 76.6 cm². The basal area of the whole tree, minus that of the core (3,848 cm²), gives the basal area beyond the mature stage. This is then divided by the mature state CAI (the area of the outer ring on the core, here 76.6 cm²) to give the age of the outer section, which added to that of the core gives the age of the tree. After crown decline, annual rings will usually be around the physiological minimum needed to maintain live parts of 0.5 mm wide, and for very large trunks, it is likely that the ordinary mature stage will also produce such narrow rings. In such situations, the calculation procedure recommended will produce a correct estimate of the age of the tree (White, 1998).

The upshot of this is that, provided we can obtain an accurate measurement of the circumference of a tree with an intact trunk that is laying down annual increment rings around the tree, we can calculate its age, of course within a margin of error. However, Mitchell (1966) estimated from recorded growth rates that the diameter of "vigorous, large crowned trees" would be larger at the mature stage than the sizes given by White. If this is true, calculations of age would be influenced significantly, as we will see below. White's protocol also assumes that there is no extrinsic variation, caused by factors from outside the tree that affect growth, which substantially influenced the pattern and rate of growth even into advanced age. What about intrinsic variation? Do oaks grow at the same rate under the same external circumstances? There are three planted oaks (*Q. robur*) in Windsor Great Park just west of The Village, along the road to Cranbourne Gate, near each other in open grassland. All three have full crowns (but of different size) and were healthy trees at the time of writing. They have plaques commemorating their planting, with dates 30th December 1887, 9th December 1897 and 9th August 1902. When measured on 28th December 2013, the oldest tree had the smallest girth at 1.30 m (254 cm), the youngest tree had the largest girth (392 cm), and the tree planted in 1897 was of intermediate size (290 cm). Here are three oaks growing on the same site, in similar soil and other conditions, yet varying in growth rate by almost 100%. As Muir (2005: 174) reminds us, in almost any avenue of trees planted at the same time in the same environment, the girths of the individual trees vary. With such variation possible,

Figure 2-12. The Major Oak in Sherwood Forest; anonymous drawing (*c.* 1888)

extrapolation from girth to age becomes rather uncertain, even if these differences might phase out through crown reduction either by pollarding or by retrenchment when the trees become old.

The effect of pollarding on the annual increment of the trunk has been assessed differently. Mitchell (1966) argued that a pollard of the same age as a maiden tree would have a larger girth because it would, once past the initial shock of pollarding, produce a deeper and wider crown than a maiden. The implication is that a pollard would have faster growth than a maiden oak. Rackham (1976) in contrast to this notion, pointed out that pollards have a very small crown for much of the time, and therefore "tend to be older than maidens of the same size", with a very slowly expanding bolling. White (1998) did not explicitly consider the effect of pollarding, even though some of his examples from Moccas Park and Staverton Park are surely pollard oaks. Read (1999) did imply an effect on growth rate when she stated that pollarding results in severe periodic crown reduction and hence a reduction in photosynthetic activity. Only when the crown is more or less fully restored can the annual increment of wood resume at pre-cut levels.

Pollards may live longer than maiden trees because the veteran stage, involving retrenchment and the onset of decay, is delayed. The process of decay may be accelerated in pollards because repeated damage to the crown would allow access to water, fungal spores and invertebrates (Kirby *et al.*, 1995). This would cause the trunks of oak pollards to rot inside and become hollow, but the frame of the bolling is not necessarily compromised by this. Presumably, although we have little evidence apart from depictions in art, oaks were first pollarded at a fairly young stage. For most pollards, cutting was carried out repeatedly, as is evident from the architecture of many ancient pollards which have a massive short trunk (bolling) and many, now often large, branches emerging at more or less the same height; these trees lack a primary leader. It is possible that, at least in oaks, not all branches on the bolling were cut at the same time, allowing the tree to function and grow wood continuously. Other pollard oaks may have been cut only once, after which a few branches developed into massive limbs, giving the tree a large crown with possibly more foliage than would be present if the tree had been left uncut; such trees may confirm Mitchell's notion of faster growth in a pollard. Crown reduction, whether through pollarding or retrenchment, does appear to result in slower growth rates. Tyler (2008) asserted that very large oaks (9.00 m circumference) would still add 1.25 cm to girth annually, but if pollarded, this would decrease to 0.625 cm; however, he gave no reference for these statistics.

A field study of growth rates that I carried out in Richmond Park in March 1997 on maiden oaks and pollard oaks seemed to indicate a substantially slower growth rate in a middle section of the large branches of pollards compared to those of maidens, with an average of 54 rings per 100 mm in maidens and 85 rings per 100

Figure 2-13. The Major Oak in Sherwood Forest, Nottinghamshire

mm in pollards (AF field note book XVIII, unpublished). The maiden trees were all in the second or mature stage but three were showing dieback in the crown, while the pollards were in the veteran or early ancient stage. The branches of the pollards had been growing since the last cutting; two of these branches were found to be 235 and 285 years old. The sample —13 branches, five of maidens and eight of pollards — was small, dependent on branches that were sawn off presumably for public safety reasons and excluding undue variables related to compression growth or curvature, so conclusions must be drawn with caution. Blank (1996) found that, in the ancient oaks at Ivenack in Germany, the annual rings in the branches are smaller than those in the stem and have no direct relevance to the ageing of the tree.

Rackham (1976) found in small oak pollards in Epping Forest of only 1.25 m girth ages of at least 350 years (average ring width 0.4 mm); this confirmed a (much) slower rate of increment in pollards, at around the physiological minimum of increment on the bolling. However, as Rackham also pointed out, when pollarding ceased and the oak regained a substantial crown, growth accelerated and was comparable to that of maiden trees of similar size and age. Most pollarding was discontinued around 200 years ago, or even earlier as exemplified in the two branch ages mentioned above. Although it is very likely that pollard oaks have grown slower than maiden oaks, to make an assessment of their age, we would need to know: at what size and age pollarding began, how extensive and how often the cutting was done, at what size the cutting was carried out, and how long ago it stopped. Most, if not all, of these variables are unknown for most of the living ancient pollard oaks.

Evidence from a large dataset of oaks of at least 6.00 m in girth obtained via the Woodland Trust's ATI shows that pollard oaks tend to have slightly larger girths than maiden oaks, with the average girth of pollards at 6.96 m and that

Figure 2-14. The Signing Oak in Windsor Great Park, Berkshire

of maidens at 6.76 m. Is this because they also tend to be older, or does it support Mitchell's hypothesis that pollard oaks have grown faster than maiden oaks? The difference in numbers of pollard and maiden oaks recorded is most notable in the lower size classes, from 6.00 to 8.00 m, where maiden oaks are more numerous. While the total sample sizes differ in favour of maiden oaks, of those above 8.00 m in girth (size class 5 or above out of 10 classes) there are about as many pollards as there are maidens, which means that, of the recorded oaks, proportionally more pollards than maidens have survived to reach that size (Figure 1-13).

- Maidens of size class 5 or larger (8.00–12.18 m), n=126, average girth 8.83 m.
- Pollards of size class 5 or larger (8.00–13.38 m), n=146, average girth 9.17 m.

Mortality is greater in maiden oaks of smaller girths, probably because they are more likely to have been felled by the wind or the axe than pollards. Pollards with girths of 8.00 m or larger are on average slightly larger than maidens (p=0.0056).

When large ancient oaks retain living bark around all, or nearly all, of the trunk as well as a crown of foliage sufficient to meet the commitment of laying down more wood around the trunk and branches, they will continue to expand, even though the crown and the root system have long ago attained their maximum and have since been reduced. An example of such a tree is the Major Oak in Sherwood Forest, Nottinghamshire (Figures 2-12 and 2-13). This celebrated ancient oak pollard has been pictured in photographs since at least 1885, showing relatively little change since then. A series of circumference measurements is also available. In 1790, the circumference was 8.33 m, in 1906 it was 9.30 m, in 1965 it was 10.21 m, in 1990 it was 10.59 m, in 2004 it was 10.62 m and in 2006 it was 10.66 m. We cannot be sure that the accuracy of these measurements on a rough and unequal bolling is within a 1% margin of error, but the tree has no significant burrs and only two relatively narrow gaps at the height of its smallest girth, just above the root buttresses, where it is measured. In 116 years (1790–1906), its circumference increased by 97 cm, then in 59 years (1906–1965) it gained 91 cm, in the next 25 years (1965–1990) 38 cm, and in the final 16 years (1990–2006) it added only 7 cm. From the earliest measurement by Major Hayman Rooke (1723–1806) in

1790 to the most recent one by the ATI recorders in 2006 (216 years), this tree's circumference has increased by 2.33 m. If the measurements are correct, within the same 1% margin of error, the tree has not grown steadily; the period 1906–1965 saw faster growth (1.54 cm/year) than the period before (0.84 cm/year) or after (1.10 cm/year). An ancient oak at Ivenack in Germany (girth 10.96 m at 1.30 m in 2015) has a similar sequence of comparable measurements made between 1804 and 1996. Its annual girth increase has varied between 1.16 cm and 3.00 cm (Ullrich *et al.*, 2009).

The Major Oak is a pollard, whereas the Ivenacker Oak is a maiden. Another maiden oak, in Holme Lacy Deer Park in Herefordshire, is locally known as The Monarch (Miles, 2013). In 1867, this tree was recorded by the Woolhope Club as 6.65 m (21 ft 10 inches) in girth at 1.50 m above ground; in 2007, it measured 9.25 m at the same height. In 140 years, it had increased 2.60 m in circumference, or an average of 1.86 cm per year. This is more than the average for the Major Oak (1.08 cm per year), as might be expected if pollards grow slower. Major Rooke's drawing from 1790 of 'his' oak is not nearly as accurate as the photos of the tree taken since 1885, but it looks as if its crown had been greatly reduced with only smaller branches left with foliage. It may have taken many decades before the crown approached the size it has in the late nineteenth century photographs, hence its slower growth. This sequence of measurements and images of a well-recorded ancient pollard oak indicates that the relationship between girth and age is not necessarily linear. The Major Oak is relatively unscathed, apart from possible crown reduction at a time after regular pollarding had stopped towards the end of the eighteenth century. Most other large ancient oak pollards are much more damaged or reduced. The possible consequences for attempts to age these trees are discussed next.

Annual increment of wood occurs only where there is active cambium, and this has to be protected by bark. The cambium can die off, and this often happens in sections on the trunk while it remains active elsewhere. In the dead section, growth of both wood and bark stop and decay sets in, loosening the bark and exposing dead wood. These areas of dead wood are often associated with the breakage of limbs or dissections of the already hollow trunk. Oaks have a remarkably potent capacity for damage repair, during which cambium and living bark slowly grow over the damaged area and cover the dead wood underneath with new wood. Nevertheless, this sideways expansion of cambium and bark is limited, and large gaps and/or areas of dead wood will not be closed. In ancient oak pollards, therefore, we usually find a hollow inside, often with large gaps in the shell-like bolling, as well as large sections of exposed dead wood where all growth has ceased. In extreme cases, such as the Crowleasowes Oak in Shropshire (Figure 5-33), the Old Man of Calke (Figure 9-9), the Signing Oak in Windsor Great Park (Figure 2-14), Wyndham's Oak (Figure 5-6) and the Queen Elizabeth I Oak in Cowdray Park (Figure 1-12), one third to one half of the bolling is dead wood without bark and often with gaps to the hollow interior. In other cases, such as the Gospel Oak in Herefordshire (Figure 1-14) and King Offa's Oak in Windsor Forest (Figure 1-7), even less of the bolling is left to grow but the tree is still alive. These are only some examples of named oak trees; numerous unnamed oaks can be added. This is the common condition of large ancient oaks, not the exception.

Oak trees can live for a long time in a 'terminal' condition, the truly ancient stage, as demonstrated by both old photographs and anecdotal evidence. When I visited the ancient oaks in the Old Park of Chatsworth with the Duchess of Devonshire in April 1997, she pointed out a tree she had written about in a letter to the Royal Botanic Gardens, Kew, which had landed on my desk. From a massive, mostly bare, broken trunk just one living branch came forth. The Duchess, who was in her seventies when I visited, remembered this tree from childhood, with the same single branch; it had not noticeably grown or changed. This ancient natural pollard had lived for at least 50 years without adding to its girth other than some wood under a strip of bark. Using John White's methodology to estimate the ages of five oaks with an average of 6.12 m girth in the Old Park at Chatsworth, Geoff Machin reported to the Duchess an average age of 501 years. Another oak with 7.10 m girth was estimated at 653 years, while the single branched oak that had experienced "50 years of standing still" came out as 654 years old. The oaks at Chatsworth are on a "poor site, open grown, with some exposure" and these estimates seem to me to be within the bounds of probability.

Ancillary methods for estimating age

All of this evidence on the effects of pollarding and of extreme retrenchment caused by accidents and by the death of large parts of the bolling points to the likelihood of much slower growth in ancient oak pollards (or maidens) than in trees with a full crown and an intact covering of bark. It is clear that extrinsic factors such as pollarding need to be taken into account when estimating the age of ancient oaks. The difficulty is that most of the relevant variables cannot be quantified. Add to this the serious uncertainties inherent in measuring the circumference of ancient oaks, and the estimates of age may become little more than conjecture for many ancient oaks. It seems that we are in need of other evidence of age. Does it exist?

A potentially valid method is radiocarbon dating. While the wood was living material with growing cells, it absorbed radioactive carbon (^{14}C) together with non-radioactive isotopes (^{12}C and ^{13}C). When the wood cells died, which would have been some years after they were formed, they stopped absorbing carbon. While the initial numbers of non-radioactive isotopes of carbon remain constant, those of ^{14}C fall and are not replenished as the result of radioactive decay. The rate of this decay of ^{14}C is known, or rather, can be calculated; it is not entirely constant over longer periods and needs correction for this using calibration curves. The 'half life' of ^{14}C, the time it takes for the amount of ^{14}C to be halved, is 5,730 ± 30 years; the amount of ^{14}C actually found in an old wood sample results in an age expressed in 'radiocarbon years' before present (^{14}C yr BP, with present = 1950). Correction of this age with the appropriate calibration results in a calibrated or calendar date (cal yr BP). Wood, especially its cellulose, is an excellent material for radiocarbon dating, provided it is not contaminated by more recent carbon from sources such as fungal hyphae, plant rootlets or insect remains. The difficulty is to find a piece of wood from the earliest period in the life of an ancient oak which is hollow. Brown or white rot wood and mould will not do, as these are undoubtedly contaminated by additional carbon. As will be evident from previous descriptions of the condition of the trunks of ancient oaks, such pieces of wood are unlikely to exist. They have to be *in situ* or show, in cross section, the centre or initial growth of the tree. There are commercial laboratories which offer ^{14}C dating to anyone who wishes to send a sample. To my knowledge, no radiocarbon dating of living ancient oaks in England has been attempted, although there is plenty of dated sub-fossil oak wood ('bog oaks'), as well as dated timbers in medieval buildings.

Dendrochronology is a scientific method of dating that is based on a chronology of tree ring patterns. This chronology, with tree rings datable to an exact year, is the main tool for ^{14}C yr BP calibration to give cal yr BP. Owing to the influence of environmental factors, mainly climatic, the width of annual rings in a given tree varies from year to year. By comparing many trees in the same region, patterns can be identified that repeat themselves in most trees, and a 'smoothed average' of tree ring widths can be used to eliminate individual variation. By computerized comparison of these patterns, based on dated samples, a chronology of tree ring patterns can be constructed. Overlap of patterns between living trees and old wood, including samples

from ancient buildings, extends the chronology back in time. There is now a dendrochronology going back several thousand years in many regions in which seasonal tree ring growth occurs. This means that if a fragment of wood with a sufficient number of annual rings is preserved, it can be compared with the tree ring chronology of the same region and cross-dated, often to an exact year BP. Evidence presented by Fletcher (in Morris & Perring, eds., 1974) makes the dating of English oak wood back to Anglo-Saxon time feasible. This seems a marvellous method for the accurate dating of ancient oaks, but as with ^{14}C dating, it depends on finding a piece of wood that represents the earliest growth of the tree. I have often looked for such a piece, but have never found it. The older the oak, the less of its central wood still exists.

Perhaps it is possible to obtain sufficient tree rings in increment cores taken from ancient oaks, living or dead, to give an estimate of average ring width and to correlate with the dendrochronology obtained from oak timbers in medieval buildings. Watkins *et al.* (2003) conducted a tree ring survey of veteran oaks in the Buck Gates section of Bilhaugh, Sherwood Forest, dating samples either directly from ring counts or through dendrochronological comparison with data from the East Midland Tree-Ring Chronology. They found that ring widths vary considerably among trees and within trees, and that there is only a rough correlation between the length of a core and the number of rings. The oldest piece of wood sampled was from near the centre of a trunk and was dated to 1415; this was an exceptional case of a near solid dead and fallen tree. It was apparently a maiden tree that had grown most of its life among other trees, with an average yearly increase in girth between 1492 and 1903 of 7 mm or average ring widths of 1.12 mm, considerably lower than some estimates for open grown trees. This tree, estimated to be 530 years old when it died, reached a circumference of only 4.56 m (dbh 1.45 m) and is probably not representative of most ancient oaks which are growing on less poor soils.

Another example of estimating the age of an oak (partially) using dendrochronology was presented in a lecture to the Friends of Greenwich Park by Jane Sidell (on 1st February 2014) and was carried out on the remains of the Queen Elizabeth Oak. A sequence from 1568–1801 that excluded decayed heartwood and sapwood and that contained 233 rings was found; an additional sample of sapwood taken when the oak fell after 1878 provided an additional sequence of 26 years. The growth rate of a 100-year-old oak from nearby in Greenwich Park provided some evidence of the early growth rates of the ancient oak, but see my example of the three oaks in Windsor Great

Figure 2-15. The Bowthorpe Oak, engraving by B. Howlet after I. C. Nattes, publ. 1805

Park described above. Estimates of the life span of the 6.50 m girth (maiden) Queen Elizabeth Oak came to 550 ± 30 years (David Alderman, pers. comm.).

Historical evidence may be present to give us an estimate of age. In this field, anecdotes and folklore abound. Robin Hood, a legendary figure anyway but who is supposed to have lived in the thirteenth or fourteenth century, is said to have gathered with his merry bunch of outlaws under the Major Oak in Sherwood Forest. He is even said to have hidden inside from the Sheriff of Nottingham and his men. Such stories give the oaks eternal life. Other ancient oaks are said to have been planted by medieval kings. Neither the Conqueror's Oak nor King Offa's Oak, both in Windsor Forest just outside the medieval park pale of Windsor Park, have any historical connection with these medieval kings.

Popular books on ancient trees in Britain (such as Hight (2011) and Miles (2013)) relate many of these anecdotes, which are for the most part unhistorical. Medieval kings did not plant oaks in Royal Forests or anywhere else; there were plenty of such trees growing naturally and the notion of forestry planting evolved centuries later (Buis, 1985), with the earliest recorded tree planting (other than fruit trees) in England around 1500 by some religious houses (Rackham, 1986).

Historical records of trees can sometimes give an indication of a minimal age. The Queen Elizabeth's Oak on the village green of Northiam, East Sussex, was apparently the scene of a banquet given to the Queen on 11th August 1573 (Hight, 2011) of which there is some kind of a record. No other ancient oaks existed on this green, so that gives us a minimum age of 418 years as the tree was certainly dead by 1991 (Harris, Harris & James, 2003). Applying White's protocol of calculations to an oak in parkland on an average site, with a recorded circumference of the

Figure 2-16. The Bowthorpe Oak, Lincolnshire

dead stump of 7.35 m, this tree could have been 688 years old, or 270 years old when the Queen visited its location. However, the stump is much burred, which exaggerates its circumference for the purpose of ageing. The evidence is therefore equivocal, and a 100-year-old oak would also have been large enough to shelter a Queen for a banquet.

What we really must look for are records of individual oak trees in charters or other documents that can be linked to still-living ancient oaks. Boundary oaks, mentioned in Chapter 5, could be candidates for such an assessment. Medieval descriptions of perambulations, set down in charters since Anglo-Saxon times, often mentioned trees as landmarks, but more often these were 'thorns' rather than oaks (Hooke, 2010). Only careful reconstruction of these boundaries in the present rural landscape can establish the location of an oak tree mentioned in the charter. But even

if this is possible, it gives only partial evidence of the age of the tree. The Vicar's Oak in Norwood near Croydon, Surrey (now Greater London), stood where four parishes met. It is mentioned from 1583 to 1704 in perambulations (www.norwoodsociety.co.uk/articles/87-norwood-and-the-vicars-oak.html) and may have existed until it was last mentioned in an enclosure act of 1808. To be a boundary oak, it had to be a large tree, especially if it stood in a wood, so it may have been old in 1583. Neither its early life nor its demise is recorded and this is to be expected.

The name Gospel Oak can refer to an extant tree, such as that near Grendon Bishop in Herefordshire, or to a location where an ancient oak once stood, such as the suburban area of Gospel Oak at the southern end of Hampstead Heath, Greater London (en.wikipedia.org/wiki/Gospel_Oak). The Hampstead boundary oak was last recorded on a map of 1801 and vanished sometime later in the nineteenth century. There is no explicit medieval record of either of these trees, even though the name appears to apply to a boundary oak used in perambulations. The likelihood that historical records of trees will provide evidence covering more than a relatively short period in an individual tree's life is very small indeed.

So, it seems we are stuck. No large ancient oaks with entire sequences of annual rings in the trunk exist, thanks to the bracket fungi that inhabit them. The few old oaks found with apparently sound boles, and thus a complete set of annual rings, are much smaller in circumference than most of the ancient oaks we are interested in here. Examples are from Spessart in Germany with 588 rings, a stump in Sherwood Forest with 512 rings, a living oak also in Sherwood Forest that had 470 rings in 1981, two living oaks in northwest Poland with 441 and 433 rings, and a dead oak in Switzerland with 453 rings (Jean-Luc Dupouey, INRA-Nancy University, Forest Ecology and Ecophysiology Unit, pers. comm.). The proxies for annual rings are usually unreliable as it is difficult to measure the circumference of the trees, to estimate the growth rate of the trunk, or to assess the virtual lack of expansion and how long it has lasted while the tree lives on. The same absence of wood in the core of the trunk that has robbed us of annual rings prevents the use of ^{14}C or dendrochronological dating methods. The historical record is patchy and it would appear that no ancient oak alive today can be linked unequivocally to a tree in the medieval records.

Circumference revisited

We must return once more to the proxy of circumference and see what it can tell us. Let us assume that an ancient oak with a genuine circumference of 10.00 m exists. If we apply White's protocol of calculations to that girth, and we take the oak to have grown in a parkland situation on an average site, its core radius at 100 years of age would have been 35 cm. If the bolling of a pollarded tree is still circular, its present radius is 1,000 cm/3.14159 x 0.5 = 159.15 cm. Its core basal area is 3,848 cm² and its CAI is 76.6 cm², so the recalculated core basal area is 3,771.4 cm². The total basal area of this tree is [dbh/2]² x 3.14159 = 159.15² x 3.14159 = 79,572 cm². The basal area of the mature state (total basal area minus core basal area) = 75,724 cm². The age of this outer section is basal area divided by CAI = 75,724/76.6 = 988 years. The age of the tree is 988 + 100 = 1,088 years.

The hypothetical 10.00 m oak is a pollard, and we have seen that pollards in all probability grow slower in girth than maiden trees. White acknowledges an ancient stage, but states that a tree must keep adding wood at a physiological minimum rate of around 0.5 mm ring width, or die. We have seen that many oaks still live while this ring width is hardly achieved, as there are only certain limited areas of growth on the bolling or trunk. All of these considerations add to the age of the tree as calculated by White's method, and so we could arrive at an age of between 1,200 and 1,400 years for our 10.00 m girth oak. This seems somewhat implausible and hence something could be wrong with White's calculation for very large ancient oaks.

Mitchell (1966) estimated that in trees over 6.00 m in girth, the annual increase would be nearly 2 cm ("three quarters of an inch") but less than 1.3 cm on less good sites. Using this estimate, and the several measurements over time of the Major Oak (a 10.21 m tree in 1965), he came to an age of about 400 years, and a maximum of 640 years (this would now be about 690 years). This seems to be an underestimate, as we have seen that over a period of 216 years, at the start of which the tree was already over 8.00 m in girth, its annual increase in girth varied between 0.84 cm and 1.54 cm. On a more or less round trunk, these are very narrow annual rings of 1.3–2.4 mm wide, but much wider than the c. 0.5 mm estimated by White for the veteran to ancient stages of large oaks.

If we assume average growth for a young oak up to 100 years old as given in White's tables, with a girth of 2.20 m at the end of that stage and an average annual increase thereafter similar to that for the 216 years actually measured (1.08 cm), we can make the following calculation. Increase to 2.20 m girth took 100 years, then further increase to the 10.66 m girth measured in 2006 took 846/1.08 = 783 years, so the Major Oak was 100+783+9 = 892 years old in 2015.

For the Major Oak with a girth of 10.66 m, we therefore have three different estimates. White's protocol would give it 1,230 years, Mitchell's maximum estimate is 690 years, and my calculation comes to 892 years, all in the year 2015. The Major Oak is an easy case when compared to most other ancient pollard oaks in its size class. We have data, albeit incomplete and imperfect. These seem to show that the likely age of this tree is around 900 years and that White's method overestimates its age.

One of the ancient oaks at Ivenack in Germany, of 9.45 m girth, was estimated to be 600–800 years old based on similar calculations (Ullrich et al., 2009). A more elaborate set of measurements and calculations was carried out by Blank (1996) for six of the ancient oaks at Ivenack. This researcher calculated the radius minus c. 5 cm for bark and compared the outer annual rings of these oaks from increment borings, finding individual differences in average widths ranging from 1.43 mm to 1.87 mm. The widths of the inner rings, which are absent from these hollow trees, were estimated from analogues of younger and sound oaks growing in the same locality. Under some assumptions about the onset of the 'age trend' and its values, Blank estimated the age of the largest oak, which had a radius without bark of 1.66 m (= girth with bark 10.74 m), to be 826 years. This is reasonably close to my estimate for the similarly sized Major Oak, which unlike the Ivenack oak is a pollard. However, as we have seen, there are gnarled old oaks that may have hung on to life even after growing somewhat bigger in circumference than the Major Oak, such as the Bowthorpe Oak (Figures 2-15 and 2-16) and the Queen Elizabeth I Oak in Cowdray Park (Figure 1-12). These and others like them may not really grow in girth at all anymore, as growth in one area is compensating for loss in another. Both have smaller crowns than the Major Oak. A maximum age of around 1,000 years for these two ancient pollards seems not implausible. This, I think, is as far as we can take it within the constraints of science.

David Alderman, Director of TROBI, wrote in a comment on an earlier draft of this chapter on 2nd December 2013: "Unlike the yew and Scots pine, a ring-porous timber such as oak has ring width limited by the size of the vessels, ensuring that old trees inevitably have to fragment to survive. Growing large quickly is clearly not conducive to being able to sustain a large girth for many years. There must be an optimum growth rate for longevity, but as this is linked to crown size and conditions which are influenced by so many factors, the mathematical calculations are probably beyond any workable model. It does appear that there must be exceptional circumstances for an oak to exceed 600 years of age with any kind of measurable trunk." I believe such circumstances do exist. They are indeed found in the fragmentation that ensures survival. Oaks can sustain greater fragmentation without dying than most other trees (Chapter 1), relieving them from the obligation to expand total circumference and thus allowing them to maintain minimal functioning ring width only where it counts to keep parts of the tree alive. This stage can last for a long time, perhaps with luck a few centuries, thus adding some to the 600 years that might be deduced from circumference alone.

What is important for the main subject of this book, ancient oaks in the English landscape, is that we can be quite certain that the largest ancient oaks, certainly those of 8.00 m and perhaps 7.00 m in circumference upwards, are relics of the medieval landscape. They originated in the Middle Ages, before Cantor's (1983) cut-off date of 1485 for his gazetteer of medieval deer parks. Under certain circumstances, such as poor soil conditions or low average precipitation, even smaller ancient oak pollards may date back to these dates. Statistically, it is likely that an oak with 6.00 m girth dates from before 1600, that is, from before the end of the Tudor period. They are therefore relics of the unplanted forest, undoubtedly used for wood pasture and for cutting timber and other wood, but ultimately connected to the 'wildwood' of prehistoric times.

3

General distribution of ancient and veteran oaks

From data collected for the Ancient Tree Inventory (ATI) a database managed by the Woodland Trust, the Tree Register of the British Isles (TROBI) and the Ancient Tree Forum of ancient and veteran trees in the UK (Butler et al., 2015), I have compiled and edited a dataset of living native oaks in England with girths ≥6.00 m. These are the recordings by volunteers ('citizen science'), NGOs such as the National Trust and other organisations and represent the currently known distribution of oaks in these categories. The inventory, though comprehensive, is not complete and some sites with ancient and veteran oaks remain under-recorded or unrecorded. [As an example, ancient oaks in some parks in the north of Suffolk that remained unrecorded for the ATI have only come to my attention in October/ November 2016, after the database used in this book had been closed.] Recording has been more comprehensive in some counties than in others. 'Batch uploads' have, in some instances, introduced trees into the ATI that are neither veteran nor ancient. With the latter potential weakness in mind, I have applied a minimum girth of 6.00 m to the ATI database to filter out a subset of oaks that are likely to be at least 400 years old.

Fig. 3-1 An ancient pollard oak in Cowdray Park, Sussex

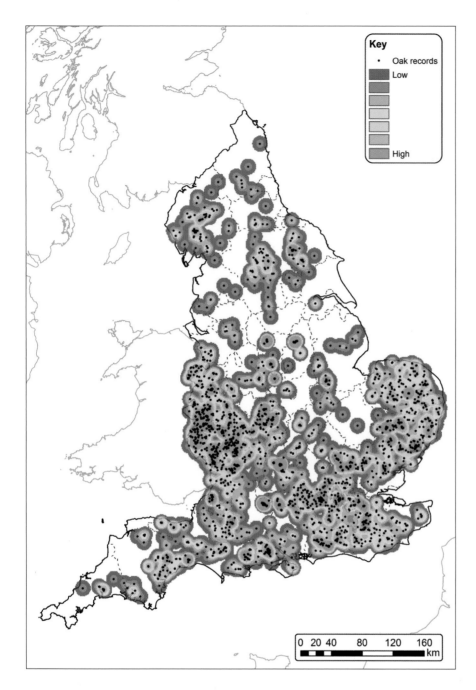

Figure 3-2. Distribution and density map of 3,320 oaks in England with ≥6.00 m girth, each of which is represented by a black dot

The terms 'ancient' and 'veteran' as applied by the ATI guidelines (Woodland Trust, 2008) refer to the life stages of the individual tree, not to its actual age in years. A birch is likely to be 'ancient' when 100 years old, but an oak of that age will have just reached 'maturity' (see Chapters 1 and 2). Many smaller oaks have been recorded as veteran or ancient, but because of a general correlation between girth and age in years, the likelihood that these are in fact relatively young increases with diminishing girth. Here, we are interested in the actual age of trees in a historical context; we are to look at trees that were in existence before the end of Tudor reign in England, 1603. After this date, oaks were increasingly planted and their distribution began to spread beyond the ancient pasture woodland with which the truly ancient oaks were mostly connected.

The correlation between the age and girth of a tree is not perfect. Oaks smaller than 6.00 m may be older than 400 years and oaks larger than 6.00 m may be younger, depending on factors discussed in Chapter 2. Perhaps oaks grow slower in the north of England and on the poor sands

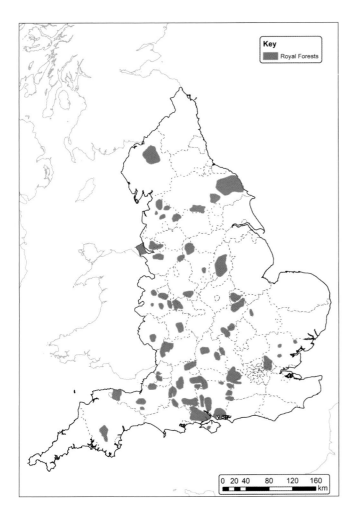

Figure 3-3. Royal Forests of England in the first half of the fourteenth century. (Originally compiled for Nelson, N. (1940). The Forests. In: J. F. Willard & W. A. Morris (eds), The English Government at Work, 1327–1336, vol. 1, map V. Medieval Academy of America, Cambridge, Mass.). Adapted for this book by Justin Moat, RBG Kew.

Chapters 4 and 5, we look in more detail at the distribution in an attempt to explain why ancient oaks occur in the places where we find them today.

Distribution of oaks in England

Where in England do the native oaks older than 400 years occur? The two species involved, pedunculate oak and sessile oak, are among the most common trees in Britain and are almost ubiquitous. Going west and north, sessile oaks become more common and pedunculate oaks less so, but taken together, as they are here, these species can be said to grow in virtually every hectare of native woodland, in every 100 metres of hedgerow with trees, and as uncountable solitary trees in fields up and down the country. By contrast, the distribution of truly old oaks is much less evenly spread. To demonstrate this, we have produced a map that shows the distribution of all 3,320 oaks recorded in England that have a stem circumference ≥6.00 m (Figure 3-2). Owing to the scale of the map, many of the dots that represent trees overlap, but the pattern is visible regardless. To show where the concentrations really are, a so-called heat map is overlaid over the dots. This heat map shows the areas of England where concentrations of ancient oaks occur using coloured fields. The colours change from low (green) to high (red) density, while the area of the coloured fields gives an additional indication of oak numbers. The densities of ancient oaks differ from a single tree on a farm to many trees in an ancient deer park that is too small to allow each tree to be seen as a separate dot on the map.

Now a clear pattern emerges. Ancient oaks are concentrated in the Welsh Marches (in Shropshire, Herefordshire and parts of Gloucestershire) and in the counties to the east (in Worcestershire, Warwickshire and, more isolated, in Oxfordshire and Wiltshire). In the southeast, concentrations are present in Berkshire, Buckinghamshire, north and southwest Hampshire, Dorset, Hertfordshire, north Greater London (former Middlesex), Surrey, West Sussex, East Sussex and western Kent. In eastern England, Essex, Suffolk and Norfolk are rich in ancient oaks. In the north of England, smaller concentrations are present in North Yorkshire, in Cumbria, and further south in Derbyshire and Nottinghamshire. Oaks ≥6.00 m in girth are notably scarce in the far southwest (west Devon and Cornwall), in Cambridgeshire and Lincolnshire (except for the southwest corner of the latter county), in an east-west belt from the East Riding of Yorkshire to Cheshire, in the Pennines and in Durham and Northumberland.

Incidentally, the map of oaks ≥6.00 m (Figure 3-2) is remarkably similar to the map of data hotspots for all

of Norfolk and Suffolk, whereas they might grow faster in the wetter West Country. Varying site factors also have effects on growth rates (White, 1998), and even on a single site, individual oaks may perform differently (see Chapter 2). However, there is no documented quantitative evidence of these factors that would allow me to correct the dataset accordingly, and so I have taken a girth of 6.00 m as a cut-off minimum size throughout England. I shall consider all oaks with girths ≥ 6.00 m to be ancient in the historical sense used here. On 1 April 2016, this dataset contained 3,320 trees, all georeferenced so that they can be plotted on maps. In this chapter, the general distribution patterns of these oaks across England are shown and discussed; in

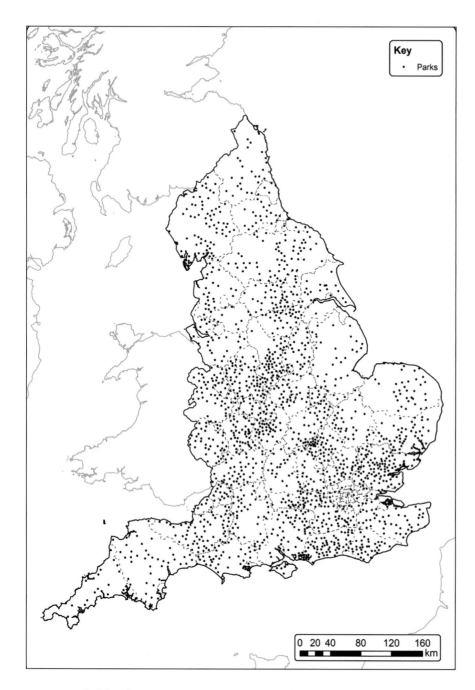

Figure 3-4. Medieval deer parks of England. (From Cantor, L. (1983). The Medieval Parks of England, a Gazetteer. Department of Education, Loughborough University of Technology, Loughborough.). Adapted for this book by Justin Moat, RBG Kew.

Key

• Parks

0 20 40   80   120   160
km

trees recorded for the ATI up to 2013 given in Butler *et al.* (2015). This probably means that oaks are the dominant component of the ancient and veteran treescape in England (and possibly the UK).

Maps of historical landscape features in England

The general distribution of ancient oaks with ≥6.00 m girth in England can be compared against several published maps of historical landscape features that could be relevant to the present distribution of these oaks.

Maps of medieval forests and chases (royal and private)

Various maps indicating the areas (Young, 1979; Grant, 1991; Hooke, 2010) or localities (Young, 1979; Rackham, 1976, 2003b) of (royal) forests have been published at various times from about the ninth to the fourteenth centuries. A comprehensive inventory with maps of forests and chases can be found at info.sjc.ox.ac.uk/forests/Index. html. Forests in the Middle Ages were areas reserved for hunting by royals and/or their magnates. The landscape would have been largely uncultivated but only partly

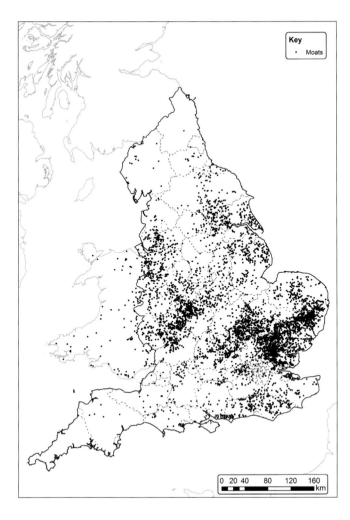

0 20 40   80   120 160
km

Figure 3-5. Moated sites in England and Wales. (From Aberg, F. A. (ed.) (1978). Medieval Moated Sites. Council for British Archaeology Research Report No. 17, Council for British Archaeology, York.) Adapted for this book by Justin Moat, RBG Kew.

the Middle Ages (Rackham, 1976; Vera, 2000). Only the Crown could bring about disafforestation, which meant excluding (parts of) the Royal Forest from Forest Law. A map of Royal Forests in the first half of the fourteenth century is given in Figure 3-3.

Maps of ancient woodland dating back to (at least) 1086 ('Domesday Book')

Rackham (1976) mapped areas in Domesday Book where "every settlement had woodland", indicating the areas in England covered in that eleventh-century record that were more or less wooded. For the same period, another map in Rackham (1976) calculates relative areas covered in woodland during this period by 10-km National Grid squares, or by county in east and southeast England. Hill (1981) mapped the distribution of Anglo-Saxon woodland in England and Wales; whereas Hooke (2010) recorded the distribution of named Anglo-Saxon forests on another map. Comparison of these two maps demonstrates that there is only partial congruence between the 'woodland' and 'forest' categories. Most woodland was managed during the Middle Ages and well into Early Modern time in a way that did not allow oaks to grow to become large or old, except for occasional boundary pollards that marked the edge of a wood.

Maps of medieval and Tudor deer parks

During the period from the Norman Conquest to the Battle of Bosworth (1066–1485) possibly more than 3,000 deer parks were created. About 1,900 of those in England were mapped by Cantor (1983), and Rackham (1976) published a map for Great Britain that includes Cantor's data. Other recent mapping of deer parks has been regional, and includes that for Hertfordshire (Rowe in Liddiard, ed. 2007), Dorset (Pollard & Brawn, 2009) and the Yorkshire Dales (Muir, 2005). From the late sixteenth century onwards, maps of the English counties began to depict deer parks with sufficient accuracy to locate them on modern maps. Most were medieval but those created during the reign of the Tudors can also have relevance to ancient oaks. Deer parks were usually established in wooded areas and needed trees, usually oaks, to shelter and feed the deer. They were often relatively small, park-like landscapes surrounded by an earth bank with a wooden fence or pale, and usually contained solitary oak trees as well as woodland patches and open grassy areas. Royal and some ducal parks could be much larger and were usually situated in Royal Forests or very near these. Besides providing deer, these parks could be used as pasture for domestic animals. The trees could be utilized but were then retained through

covered by trees. In the twelfth and thirteenth centuries, nearly one quarter of England was Royal Forest, that is, subject to Forest Law (Young, 1979; Grant, 1991). All of the trees in these forests would have been non-planted native trees, and oaks would have been common almost everywhere. Forest Law forbade "trespasses of the vert", which mostly amounted to prohibition of the cutting of trees (particularly oaks) unless licensed and outlawed the making of 'assarts' or clearings of forest for agriculture (Manwood, 1598). Afforestation and disafforestation were used according to their medieval meaning; unlike at present, these were legal actions and had nothing whatsoever to do with planting or felling of trees. There was no planting of trees other than fruit trees in gardens or orchards in

pollarding or sometimes shredding. A map of medieval deer parks is shown in Figure 3-4.

Map of moated houses or homesteads
Many manor houses, granges, freehold homesteads and hunting lodges were moated, depending on topography and soil. Moats were usually meant to be wet (filled with water) and provided some protection against banditry and raiding. Depending on the local situation, manor houses in the Middle Ages were usually either moated, fortified ('crenellated') or both. Manors can have relevance to the distribution of ancient oaks because on a manor often all elements in the landscape for self-sufficient existence would be present including woodlands for pasture which could have preserved ancient oaks (see Chapter 5). Granges were outlying farms belonging to a monastery and exploited under the manorial system, whereas freehold homesteads, which became more numerous after

Figure 3-6. Ancient lapsed pollard oak in Alfoxton Park, Somerset

provided a map of medieval moated sites in England and Wales (Figure 3-5).

Map of 'Ancient Countryside' and 'Planned Countryside'

Rackham (1986, 2003a) presented a map which divided England into 'Ancient Countryside' (AC) and 'Planned Countryside' (PC). PC is that part of England which was already mostly turned over to large-scale agriculture in Roman times. This land was subject to 'parliamentary enclosures' in the eighteenth century and again to large-scale agriculture from the nineteenth century to the present. By the eleventh century, woodland in PC already covered only 8% of the land, as against about 19% in AC; woods were absent or few and large in PC, where pollard trees other than willows mostly occurred in villages and rarely in landscapes away from habitation; by contrast, pollarded trees and woods in AC often occurred far from villages and towns.

Historical landscape features and the distribution of ancient oaks

There is a general congruence with all of the features in the historical maps described above and the distribution of ≥6.00 m native oaks in England. Royal Forests and chases were present in nearly all areas in which concentrations of ancient oaks are found today; they were absent in Norfolk and Suffolk, where however there was much woodland in Anglo-Saxon times. Some Royal Forests had little relevance because they were mostly treeless moors, especially in the north and in the southwest. Deer parks were ubiquitous in most of lowland England, but most abundant in the larger areas with concentrations of ancient oaks: in the counties surrounding London and in western Kent, Sussex, Essex, Suffolk and in the West Midlands. Areas with abundant moated sites, such as north Gloucestershire, Warwickshire, Worcestershire, Hertfordshire, Essex and Suffolk, also often have high densities of ancient oaks, but here the congruence is less consistent, with many oaks but fewer moats in the Welsh Marshes (too hilly for moats?) and many moats but fewer oaks in Northamptonshire, Bedfordshire and southwest Cambridgeshire. The correlation between the distribution of ancient woodland and that of ancient oaks is relatively weak because the term 'woodland' includes coppice woods as well as pasture woodland. Areas that had much woodland in 1086 presumably also had pasture woodland in which oaks could grow large. Their value may be measured in Domesday Book in terms of the numbers of pigs allowed in the woodlands for pannage, but this was not recorded consistently across England (Rackham, 1976), and we do not know whether coppice with oak standards was really separated for this use from pasture woodland.

the Middle Ages when the feudal manor system gradually became obsolete, could nevertheless have a similar division of land and its uses. Hunting lodges in forests and in or near parks were often moated to protect the forester or parker, who had to face armed poachers and venison thieves. Many medieval deer parks had a moated house nearby or inside. Even when the dwelling has long ago disappeared, the moat, unless filled in, can often be seen from the air or found in the terrain. Aberg (1978)

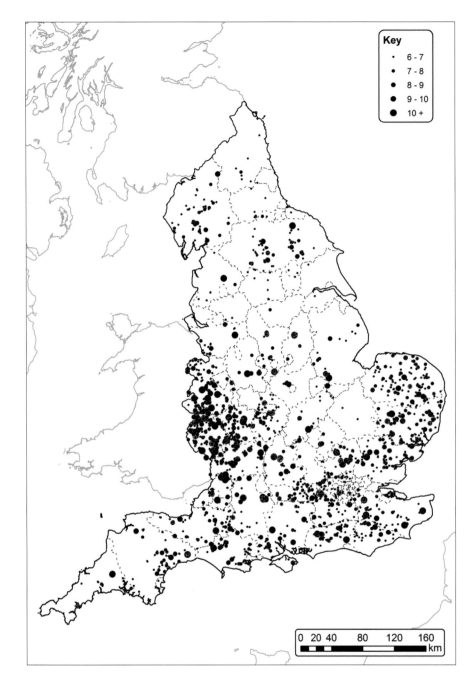

Figure 3-7. Distribution of 3,320 oaks in England according to five size classes (ranging from 6.00 m to ≥10.00 m)

Key

·   6 - 7
·   7 - 8
●   8 - 9
●   9 - 10
●   10 +

0   20   40   80   120   160   km

(I suspect not, depending on the availability of woodland with oaks.) Correlation between historical areas of 'Ancient Countryside' (AC) and 'Planned Countryside' (PC) and the presence of extant ancient oaks is clearly present but on a very coarse scale.

The relationship between absence of these historical landscape features and the absence of ancient oaks is almost a better match than that for the presence of these variables. West Devon and Cornwall have few ancient oaks and had few medieval forests, deer parks, moats and, to a lesser extent, ancient woodland. The zone of England that Rackham mapped as 'Planned Countryside' as opposed to 'Ancient Countryside', which runs from the Dorset coast in a widening belt northeast to Cambridgeshire and Lincolnshire, is also relatively poor in ancient oaks. In the eighteenth century, common land in this belt was mostly enclosed and taken into more intensive cultivation. There is, however, less correlation between the presence of lowland AC and the presence of ancient oaks: several lowland AC areas, such as Kent and Lancashire, have relatively few

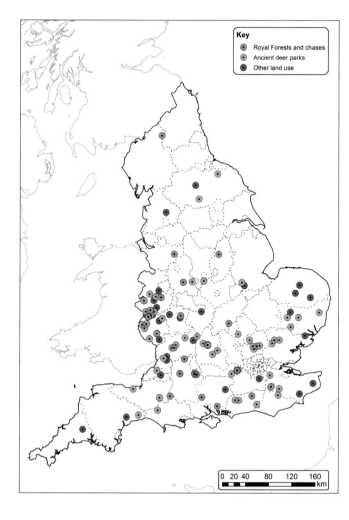

Figure 3-8. Distribution of 115 oaks with >9.00 m girth in England, with historical land use indicated.

ancient and veteran oaks, while others are richer. In Kent, AC in the more hilly parts is often replaced by PC on more level ground, so that actually there is a mosaic of the two categories of countryside in parts of this county. Ancient oaks are found in a small concentration within the zone of PC between Corby and Peterborough, which is rich in woodland and formerly a part of Rockingham Forest. This area largely escaped modernisation. Medieval deer parks had a high concentration in north Buckinghamshire and adjacent parts of Bedfordshire that more or less coincides with present locations of ancient oaks.

Despite exceptions like these, the negative influence of 'Planned Countryside' seems convincing and especially appears to explain the lack of congruence with the medieval presence of deer parks in this belt. Most of these deer parks probably disappeared in the zone of PC

(see also Table 4-1 in Chapter 4); they were also lost or were much altered elsewhere, but by other causes such as urbanisation or by landscaping in the style and on the scale of Lancelot ('Capability') Brown and Humphrey Repton in the eighteenth and early nineteenth centuries (Rotherham, 2007b). Counties such as Lincolnshire and Cambridgeshire had few medieval deer parks and moats and have few ancient oaks. The one notable exception in Lincolnshire, Grimsthorpe Park, was a Tudor deer park. The few medieval deer parks and fewer moats in the far north of England coincide with the paucity of ancient oaks there.

Finally, the upland areas of England (the Pennine Chain, Cheviot Hills, Yorkshire Moors, Dartmoor and Exmoor, but not the Lake District) are poor in ancient oaks. Most of these upland areas were already extensively deforested in prehistoric times and grazing by livestock has kept them mostly treeless ever since. Harsher winter conditions on the moors limited the number of deer parks in these upland areas, especially during the Middle Ages when fallow deer (*Dama dama*) were the most common species in these parks.

In conclusion, there appears to be a general match between the distribution of ancient native oaks and the past distribution of Royal Forests and chases, of medieval and Tudor deer parks, and to a lesser extent, of moated houses. In the next two chapters, we investigate these and other possible historical associations in much more detail. This may lead us to a greater understanding of the presence of ancient oaks in the English landscape.

Are there relationships between the size and the distribution of ancient oaks?

We can also investigate whether there is an additional distribution pattern according to size. Are trees of a similar size distributed evenly or unevenly? We can divide the measured girths of all 3,320 oaks into five size classes (the range is 6.00 m to ≥10.00 m) and present these in another map of the distribution of ancient oaks (Figure 3-7). Broadly described, the two maps are similar. The most striking difference is found between East Anglia and the counties bordering with Wales: ancient oaks are significantly smaller in the east than in the west. The north of England also has relatively few very large oaks. The largest oaks, here considered to be oaks with a girth of more than 9.00 m, also have an uneven distribution (Figure 3-8). Nearly all occur south of a line from Shrewsbury in Shropshire east to Norwich in Norfolk; only five are found north of that line. Herefordshire has 17 such great oaks and Shropshire, Herefordshire and Gloucestershire together have 34 out of a total of 115. The Welsh Marches are thus particularly rich in the greatest oaks. High Park in Blenheim Park, Oxfordshire,

Figure 3-9. Ancient oak with 9.68 m girth at Porter's Hall near Stebbing, Essex

has four living oaks >9.00 m in girth, the greatest number at any one site. The great oaks occur mostly in ancient deer parks (or what may have become of these) where 67 out of 115 (58%) are found; Royal Forests and chases account for only 20 (17.5%). In the category 'other land uses', 16 oaks of >9.00 m girth have been assigned to 'manors'. Deer parks usually belonged to manors, in this case, the 'manors' category includes only those sites where there is evidence of a manor before 1600 but not of a deer park (see Chapter 5 for a rationale and more details). Apart from three great oaks (probably) associated with wooded commons, there are seven great oaks in the category 'other land uses' of which the historical association with any of the land uses defined here is unclear or unknown (see Table 5.3).

The geographical patterns correlated with girth in both maps could be explained by climatic differences across lowland England. Climate maps in *The Times Atlas of Britain* (Riches *et al.*, 2010) show lowest annual rainfall in a widening belt from York to London and lowest average July temperatures in the north. Both lower temperature and lower rainfall are likely to have negative effects on the growth rate of oaks. In the north, a reduction in average July temperature of about 4 °C results in a shorter growing season, especially in a later coming into leaf (White, 1998). Lower rainfall in the east results in slower rates of wood growth, which leads to narrower annual increment rings (White, 1998). If the maximum attainable age of native oaks were equal across the country, the oldest oaks in the east and north would be smaller in girth than those in the south and west, consistent with the data.

Distributions of maiden and pollard oaks

Rackham (1986, 1996) associated medieval deer parks with pollard trees, and his opinion has been repeated by many other authors. In the next chapter, I will demonstrate that this was not the case everywhere, and that across the whole of England there are in fact more maiden oaks than pollards with girths ≥6.00 m in deer parks. Is there perhaps a geographical bias? Figure 3-10 confirms this, with pollard oaks more common in exactly the area Rackham knew best, East Anglia. Pollard oaks dominate in Essex, Suffolk and Norfolk and not only in (former) deer parks. In the rest of the country, there is either a balance between maidens and pollards or maidens dominate, as in High Park (Blenheim Park), Birklands (Sherwood Forest), and especially in later Tudor deer parks.

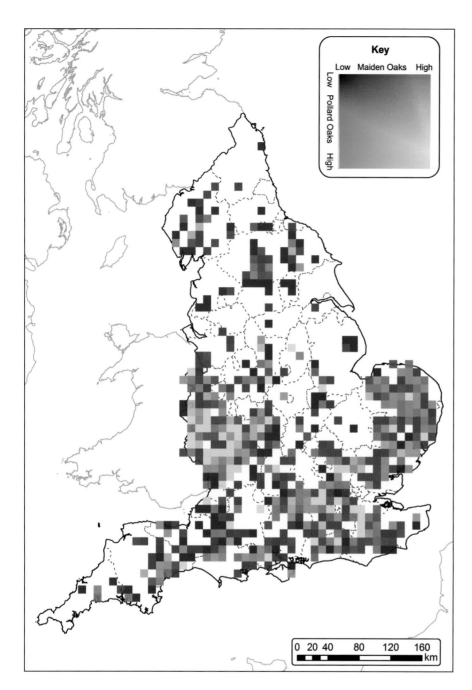

Fig. 3-10. Distribution of pollard oaks (green) and maiden oaks (red) per monad (10×10 km) calculated per 1km²; yellow = ± equal numbers and hues towards red have increasing proportions of maidens. Dark monads have low numbers in both categories.

The activity of pollarding oaks ceased nearly 200 years ago, but a lapsed oak pollard, now measuring ≥6.00 m in girth, would have been a 'managed pollard' around 1800. Only managed pollards are counted here, trees that have become a 'pollard' as a result of storm damage ('pollard form, natural' in the database) are included here with the maiden oaks. It is assumed that some recorders in the early days of ATI recording were not aware of this distinction — the category 'pollard form, natural' was introduced in a later phase of the project — and that there may therefore

be a bias towards pollards in the database. I have corrected this where possible from images or after a revisit, but not all trees have (good) images, so some bias may still be present. It does not in my opinion influence the general pattern: ancient pollard oaks dominate in East Anglia but, with local exceptions, not in other parts of England.

Historical land uses and the distribution of ancient oaks
The ancient oaks can be associated with certain forms of land use, which is discussed in detail in the next two

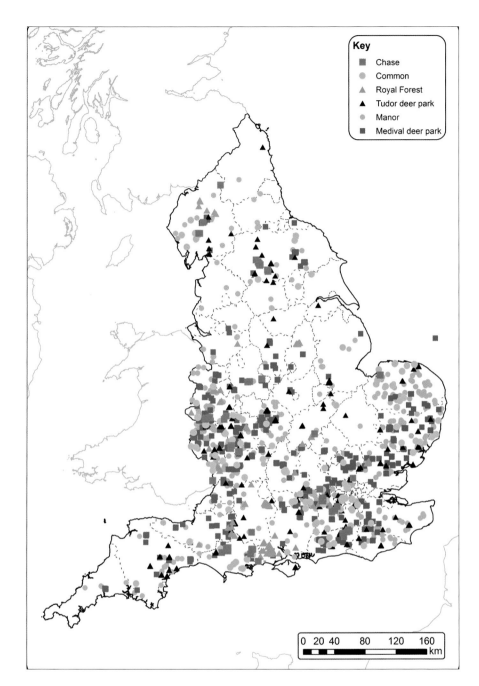

Figure 3-11. Distribution of ancient and veteran oaks with girths ≥6.00 m according to different categories of land use prior to 1600 CE. CHA = chase, COM = wooded common, MDP = medieval deer park, RF = Royal Forest, TDP = Tudor deer park, and manor without evidence of a park

Key

■ Chase
● Common
▲ Royal Forest
▲ Tudor deer park
● Manor
■ Medival deer park

0 20 40   80   120   160
km

chapters. The types of land use I have found to be relevant are: medieval deer parks (MDP), Tudor deer parks (TDP), Royal Forests (RF), chases (CHA), wooded commons (COM) and, to a lesser extent, manors (when not having deer parks). This land use had to be in place before 1600, that is, more than 400 years ago. Of the total of 3,320 oaks in the dataset, 482 could not be allocated to any of these categories (see Table 5.5) and they are not shown on the map in Figure 3-11. This map shows a few patterns; for example, in north Norfolk, commons and manors without

deer parks dominate and the only deer parks relevant to ancient oaks are from the Tudor period. In Kent, medieval deer parks with ancient oaks are in the west, nearer to London; elsewhere, the few relevant parks are Tudor and the remaining oaks are associated with a manor without evidence of a park. In Cumbria, ancient oaks are mostly associated either with Royal Forests or with Tudor deer parks. The New Forest in Hampshire stands out as a Royal Forest because it is such a large area, but it had two medieval parks inside it (Tubbs, 1986). Otherwise, the

Figure 3-12. Pollard oak in the snow in Richmond Park, Surrey

categories in which ancient oaks are found are more or less evenly distributed, with commons and Royal Forests having ancient oaks more thinly spread than the other categories.

In summary, some general patterns, possibly inter-related, emerge. Where ancient oaks ≥6.00 m in girth are abundant in the eastern counties they tend to be smaller, more often pollards, not or rarely associated with Royal Forests, and more often associated with parks or indeed with manors without parks. In the west, Royal Forests are added into the mix of land uses, but here the ancient oaks tend to be larger and less often pollards. In other parts of England where the densities of ancient oaks are high, these patterns are only evident more locally and they sometimes seem inconsistent. Great oaks with girths >9.00 m are almost limited to the southern half of England. Climate could be influencing sizes, with both temperature and precipitation determinant factors. Why pollards tend to be more abundant in East Anglia than in the rest of England (with local exceptions) is not easy to explain. If the oaks would grow slower in the drier climate of this region, could it be that the land owners abandoned hope of growing large timber oaks and allowed their tenants to use them as pollards instead? Elsewhere, we will see that the privacy of deer parks tended to favour maiden oaks. In the next two chapters, we shall look at these questions in more detail.

4

Distribution of ancient and veteran oaks in England explained: Deer parks

The history of localities with ancient oaks can give clues to the reasons why these oaks are (still) present. In this and the next chapter, these historical connections are investigated. A very substantial proportion of oaks with circumferences of ≥6.00 m, and therefore likely to be older than 400 years, are in localities that have been deer parks. We can divide these deer parks into medieval deer parks and Tudor deer parks. Medieval deer parks date from around the time of the Norman Conquest (1066) to the Battle of Bosworth (1485), in which the Tudors gained the throne of England. Tudor deer parks date from the period between this battle and the death of Elizabeth I (1603), which ended the reign of the Tudors. After around 1600 fashions changed. Parks increasingly became pleasure grounds surrounding a country house belonging to a member of the landed gentry, who were no longer exclusively members of the nobility. Large ornamental gardens were laid out in which deer had no place; later, these gardens were 'landscaped' and extended so that deer could once again be accommodated in parts of the park. The great royal and ducal parks especially continued to have deer as a symbol of status. For every ancient and

Figure 4-1. Remains of the medieval park pale in Woodend Park (Shute Park), Devon

veteran oak in the database I have tried to establish whether it occurs in a locality that was once a medieval or a Tudor deer park. If it was a medieval deer park, it does not matter if the large oaks now present are, according to their size, not or not all dating from before 1485, as long as they are likely to date from before 1600. The rationale is that such oaks (≥6.00 m in girth) are unlikely to have been planted and more likely to be the survivors of spontaneously grown populations of oaks that do go back to the Middle Ages and possibly beyond.

Medieval deer parks in twelve counties

The hunting park was a very prominent feature of the medieval landscape (Cantor & Hatherley, 1979; Muir, 2005; Mileson, 2009). Every county in England had several or many deer parks (Cantor, 1983); there is documentary evidence for at least 2,000 (and an estimated maximum of 3,000–3,200; Rackham, 1986) in England. Many of these have left various traces, while some have persisted to the present in a condition that resembles their medieval state; examples include High Park (part of Blenheim Park) in Oxfordshire, Bradgate Park in Leicestershire, Bear's Rails (part of Windsor Great Park) in Berkshire, Moccas Park in Herefordshire, Old Park (part of Chatsworth Park in Derbyshire) and Staverton Park in Suffolk. From the beginning, general congruence existed with wooded areas; most counties with a high proportion of woodland in Domesday Book (Rackham, 1976) subsequently experienced much imparkment. Disafforestation of parts of Royal Forests often led to imparkment, as the terrain and partly wooded landscape of these uncultivated areas were ideal for deer parks, which could create revenue for the Crown. From the early thirteenth century onwards, those who wished to impark land had to obtain a royal licence. Early parks (licensed before 1325) were mostly created within Royal Forests, but when these had become much smaller and fragmented (from 1326 to 1500), new parks appeared mostly outside (Mileson, 2009). Within counties, the density of medieval parks was usually concentrated in the more wooded areas, although there were exceptions, such as in Hertfordshire (Rowe in Liddiard, ed. 2007). Counties with a high density of parks were (part of) Derbyshire, Staffordshire, Worcestershire, Warwickshire, Gloucestershire, Buckinghamshire, Berkshire, Hertfordshire, Essex, (part of) Suffolk, Surrey and Sussex. Fewer parks were present in Cumbria, Northumberland, Durham, North Yorkshire and East Riding, Lancashire, Lincolnshire, Cambridgeshire, Norfolk, Devon and Cornwall. Muir's maps of parks in the Yorkshire Dales include medieval and Tudor parks. There were very few parks in Great Britain outside England.

In most parks, a surrounding earth bank, topped by a wooden paling fence, and an inside ditch prevented the deer escaping; a few parks were surrounded by a stone wall. The park was part of the demesne lands of the lord of the manor and was usually situated in an uncultivated area at the edge of the manor, which most often included some woodland to provide shelter ('covert') for the deer. This landscape remained mostly 'unimproved', trees were not planted; the emphasis on visual, ornamental aspects ('landscaping') that determined or modified the character of the later parks around country houses was not truly considered in the Middle Ages. The aim was the production of venison, and usually some entertainment in obtaining it. The enclosure of the deer park stood in contrast with the open Royal Forest and chase, but the landscape of a deer park was not unlike that of the Royals Forest, albeit on a smaller scale.

There is evidence of 'deer folds' in Anglo-Saxon time, some of which may have been precursors to post-Conquest parks (Muir, 2005) but the medieval deer park is essentially a Norman creation that came to England with the Norman Conquest of 1066. The park deer *par excellence*, fallow deer, were introduced by the Normans to stock these parks (Fletcher, 2011), probably from Sicily or Apulia in the beginning of the twelfth century (Rackham, 1986). In the thirteenth and fourteenth centuries, the deer parks reached their peak in numbers and importance in the expanding agrarian economy (Cantor & Hatherley, 1979). The high period of imparkment was between 1200 and 1350, when the labour needed to construct the park pale was easily obtained through the feudal manorial system. After the demographic collapse of the Black Death (1348–1350), labour became more expensive and many park pales fell into disrepair, the deer escaped and the number of newly created parks fell sharply (Chapter 7, Figure 7-2).

According to Rackham (1976), there were two types of park, 'wood pasture' park and 'compartmental' park. In the latter type, further enclosures protected coppice woods from the deer, whereas in the former, the entire park was open to them. In both, however, the deer must have had access to some wooded part for the necessary shelter. Not all parks in the Middle Ages were used for deer, or for deer only; domestic animals may have been kept in some parks, and in these, there were not necessarily many trees. Parks differed greatly in size, with the largest up to 400–500 ha (rarely to 1,000 ha) belonging to the Crown and to a few great magnates, secular or ecclesiastical. Most deer parks were much smaller, between 40 and 80 ha in the thirteenth century (Cantor & Hatherley, 1979). After 1350, when much land fell into disuse, parks were often enlarged and a few became much larger than any in the previous centuries;

Figure 4-2. An aerial view of Staverton Park in Suffolk (©Suffolk County Council, 1999)

oak in grazed areas, especially on limestone substrates.

Not only utilitarian, but also aesthetic considerations often determined the divisions between open 'launds', parkland with trees, and woods, as well as the location of a hunting lodge (often moated) or a fishpond (Muir, 2005). Moats are often indicators of vanished parks; the distribution of medieval moats and medieval deer parks is strongly correlated, especially in southern East Anglia, Hertfordshire, Buckinghamshire, Gloucestershire and Warwickshire, where both attain greatest densities (Cantor, 1982; see also Chapter 3). Where parks remained for deer or for the grazing of domestic animals, new oaks were often planted from the late sixteenth or seventeenth century to the present and the grassland was often improved to give higher yields of venison, beef or mutton. After the middle of the seventeenth century, these developments ushered in the early modern landscape park. From the eighteenth century onward, deer parks were often substantially redesigned and planted with trees other than native oaks. Where they still survived, the ancient trees, mostly oaks, were sometimes incorporated into the landscaped park as relics of a venerable past (Rotherham in Liddiard, ed. 2007). They may still be there, even after the park's use changed in modern times to become, for instance, a golf course, such as Lullingstone Park in Kent and Stoneleigh Park in Warwickshire. Farming could be more destructive if trees stood in the way of arable fields, but they may have been preserved on pasture and sometimes incorporated into hedgerows. Most destructive has been urbanisation, which has swallowed a good number of medieval deer parks in Birmingham, London (including the royal parks of Eltham), Milton Keynes and elsewhere (Cantor & Hatherley, 1979).

First Epoch Ordnance Survey Maps, hereafter referred to as nineteenth-century OS maps, were issued *c.* 1843–93, a period when railways became established but the motorcar had yet to arrive. Comparison of these maps with those of the twenty-first century is indeed a shocking experience from a countryside point of view. In preserved parks,

however, only the very rich and powerful could afford to do this. Chalk downland, elevated moorland, fens and areas of intensive agriculture were usually avoided.

Disparkment accelerated after 1500 and in the seventeenth century many deer parks were converted for the first time to more intensive agriculture, including arable farming (Fletcher, 2011). However, many stayed in private landownership or became private after the dissolution of the monasteries during the 1530s. New deer parks were still created during the reign of the Tudors; examples of such later (and large) parks are Grimsthorpe Park in Lincolnshire, Packington Park in Warwickshire, Parham Park in Sussex and Bushy Park near Hampton Court in Middlesex, the latter imparking arable land (hence the scarcity of ancient oaks there). Evidence from several regional studies appears to show that, in terms of numbers of parks, disparkment and new imparkment balanced each other out during the Middle Ages (Robert Liddiard lecturing on 24 March 2015). A century after Bushy Park was imparked, nearby Richmond Park in Surrey was enclosed by Charles I (see Chapter 9). Domestic animals were increasingly grazed within the parks that remained, leading to a gradual decrease of woodland and increase of grass pasture. This practice may also have favoured oaks among the remaining trees, as these are somewhat more persistent under grazing pressure than most other native trees (Rackham, 2003b). Ash is much more palatable than oak and may therefore have gradually been replaced by

with or without deer, the presence of ancient and veteran oaks is obviously related to their past. If land use and the landscape have drastically changed, only traces of the park may remain to indicate the relationship between ancient oaks and a medieval deer park. We could also assume this relationship if no other park traces but the ancient oaks are left and the locality of the trees is congruent with a former deer park. Many ancient oaks stand on farms, often solitary with no other examples anywhere near.

I have investigated the relationship between medieval deer parks and the presence of ancient oaks still alive in the second decade of the twenty-first century. This has been done by county for the twelve counties that had most parks in the Middle Ages, using the information given by Cantor (1983), indicators on current OS maps and on nineteenth-century OS maps, aerial views from Google Earth (Figure 4-2), distribution data for oaks on the Interactive Map of the ATI (www.ancient-tree-hunt. org.uk/discoveries/interactivemap/) and the author's field observations of ancient oaks. Since the inventory of medieval deer parks by Cantor (1983), more parks of the period have come to light. In Hertfordshire, Rowe (in Liddiard, ed. 2007) counted 64 such parks against 45 in Cantor's list; in Sussex, Mileson (2009) found 130 against 108 in Cantor's list. However, I do not have additional information on medieval deer parks for all twelve counties, and those in the two cited publications are presented on a small county map but not named and listed as in Cantor (1983), making identification and comparison 'park for park' impossible. I have therefore not included these additional parks. Medieval deer parks with ancient oaks that still exist are mentioned, and deer parks that are still parks, either with or without ancient oaks, are also noted; but what we really want to know is whether ancient oaks that are now in a different setting may be there because they once belonged to a medieval deer park. The names of these vanished parks are in italic text. In only a few cases are vanished parks without ancient or veteran oaks mentioned; the results of the inventory are given below and are summarized in Table 4-1.

Berkshire
Aldermaston has been mentioned as a deer park since 1202, and a small part of it still exist as a park around Aldermaston Court (now Aldermaston Manor, a Victorian stately home). The existing park is partly landscaped with plantings of non-native trees, but an unimproved section of ancient pasture woodland still contains 14 ancient oaks, ranging from 5.10 m to 8.56 m in circumference, that have been recorded, mapped and photographed for the Ancient

Tree Inventory (ATI). The park is overgrown and nine oaks were seen on a visit in July 2015 to be very fragmented and hemmed in by secondary tree growth. The property stood empty and a developer was at the time of writing seeking planning permission for a housing estate that would destroy this last remnant of a medieval deer park with ancient oaks.

Benham Park near Speen has been mentioned since 1349. It is now a landscaped park south of the A4 that contains woods, lawns and fields with scattered trees, as well as a house and adjacent commercial buildings. There are no ancient oaks.

Figure 4-3. Ancient oaks in Staverton Park, Suffolk

Easthampstead just south of Bracknell has been mentioned as a deer park since 1365. It is now a golf course, but it retains one ancient maiden oak with a girth of 9.05 m (Figure 4-4), which stands on the course near buildings at the end of Old Oak Court.

Foliejohn has been mentioned as a deer park since 1317; there are remnants of the old park pale, but Foliejohn is now a much smaller landscape park around a country house and it has no ancient or veteran trees.

Hamstead Marshall Park (or Hampstead Park) has been mentioned since 1229 and is still mostly extant,

with the old park pale visible in many places. Remains of three castle mottes exist in the northern perimeter and near St. Mary's Church. The fish ponds in the park are also medieval. The River Kennet (and now also a canal) borders the park on the north side. The park was enlarged in the Tudor period and was subsequently landscaped, with planted wood copses and many trees in the lawns and meadows, although most of the avenues of the seventeenth to eighteenth century landscape park have now disappeared. The present house inside the park dates from the eighteenth century when it was expanded from an earlier hunting lodge. There are 21 ancient and veteran oaks in the park with 6.01–8.35 m girths, but four of these are standing outside the medieval park pale boundary; these oaks could date from the time of the Tudor enlargement when a mansion was built near the church. A large number of smaller oaks has been recorded throughout the park, many as 'ancient' although most of these are likely to date from the seventeenth century or later. A more detailed description of this park is given in Chapter 9.

Kingston Lisle has been mentioned since 1336. On the nineteenth century OS map, it is larger than now, with a

section called North Park and another known as Green Park. North Park is now arable fields with hedgerows; Green Park has been landscaped, with the inclusion of some of the copses. There are no ancient oaks.

Radley (now in Oxfordshire) has been mentioned from 1262 onwards and was held by the Abbot of Abingdon. This park extends from Radley Lodge to Park Farm on the nineteenth-century OS map and, apart from a double

Figure 4-4. An ancient oak in Easthampstead Park, Berkshire

avenue and cricket grounds, was still very much a park at this time, although compartmented with hedgerows in the western half. Today it is mostly arable fields, a golf course and sports grounds. Near the end of the avenue close to the old fish pond stands an ancient oak with a circumference of 9.40 m and there are two veteran oaks nearby.

Remenham was a deer park on the Thames near Henley-on-Thames first mentioned around 1250. Park Place and Parkplace Farm are now on this site, along with a landscaped park and farmland. There are no ancient oaks. A veteran oak with a girth of 7.70 m is nearby at Remenham Place.

Swallowfield has been mentioned as a park from 1232 onwards. On the nineteenth-century OS map it is bounded by the River Loddon, the Great Wood, a hedge from that wood to All Saint's Church, and The Broadwater. These features still form the boundary of Swallowfield Park, which comprises a country house surrounded by a landscaped park consisting of open fields with solitary trees and denser tree plantings. There are no ancient oaks.

Ufton Park was first mentioned in 1338. On the nineteenth-century OS map its rounded shape can be seen to be intact in the southern half. At the southern end is the toponym Old Park, but in the fourteenth century, Ufton Park must have been larger than shown on the nineteenth-century map, as Ralph de Restwold was licensed to impark 100 acres (40 ha). There is a 'park piece' in the midst of woods on the nineteenth-century OS map, but the park is now all planted woods and a few small fields. There is one ancient oak with a girth of 7.55 m near the end of the drive to Ufton Court, just outside the former deer park.

Windsor Great Park has been known since 1132 and still contains many ancient and veteran oaks; it is one of the primary sites for these trees in England and indeed the world (Green, 2001; 2010b). Whichmere (first mentioned in 1359) was in Windsor Park and is here included with it. There are long sections of park pale remains indicating the park's medieval extent. The present deer park is in the northern section and east of the A332. There are areas with open parkland reminiscent of a medieval park, but also elements of landscaping such as the Long Walk, enclosed plantations and ponds. The Village and other houses are just outside the present deer park. There are also farms with arable and pasture fields and forestry plantations. In the southern part of Windsor Great Park, there are large polo fields, and in those areas where the terrain is dissected by valleys and hollows, the Savill and Valley Gardens are woodland gardens planted with exotic trees and rhododendrons (*Rhododendron* spp.). Virginia Water, a large artificial lake, is situated at the southern

end of the park. The entire park is *c.* 2,000 ha, but here I have excluded the area around Cranbourne Tower which historically was part of Windsor Forest, later Cranbourne Chase. Ancient and veteran oaks are concentrated in the northern half of the park and are almost absent from the woodland gardens and farms. In the present deer park stand some of the largest pollard oaks in the Windsor area; other ancient oaks are situated around Cumberland Lodge and along the road from there to Sandpit Gate. This park is described in more detail in Chapter 9.

Yattendon was imparked by Philip de la Beche in 1335. There is a small park 'Attendon' on the nineteenth-century OS map, with Clack's Copse on its east side and a fish pond. There are now two large houses, some woodland and a small landscaped park. There are no ancient oaks.

Buckinghamshire

Bulstrode was a deer park of unknown date, owned by the Abbot of Bisham. In the nineteenth century, the park included the Iron Age fort known as Crab Hill; this is now surrounded by suburban development. Some parts of the park are now wooded or landscaped, but much remains open grazed fields. There remain several ancient and veteran oaks, mostly in the southern and central sections of the park as it still was in the nineteenth century. Two of these exceed 8.00 m in girth and are now on separate private property. The ring wall of the fort, also known as The Camp, has several smaller veteran or ancient oaks, some dying or dead. The large meadows of Bulstrode Park are occupied by herds of horses that are damaging the bases of the oaks there, as observed in January 2015.

Burnham was a deer park from 1226 onwards; it is now Hitcham Park, shown on the nineteenth-century OS map as divided into two parks by a road and some landscaping. This is also the present situation, but there are now more planted trees and wooded areas with buildings. There are no ancient oaks.

Chenies was mentioned as a deer park in 1335. The manor house dates from the fifteenth and sixteenth centuries and was owned by the Earls of Bedford. The house is surrounded by formal gardens and in the parterre (former paddock?), beyond these is the Chenies Oak (Figure 4-5), an ancient maiden (see Burgess, 1827) with 6.75 m girth. It is difficult to see where the deer park once was; now, there is a wood (Placehouse Copse), farm buildings, an old church and a newer country house with a small landscaped park. The shape of that small park and the adjacent wood suggest that this was the location of the medieval deer park, but the ancient oak is on the other side of Chenies Manor.

Figure 4-5. The Chenies Oak at Chenies Manor House, Hertfordshire

Ditton was a deer park from 1335 onwards; it is situated between modern Slough and the M4, which curves around it. Since the nineteenth-century OS map was made, this area has become more wooded and some major development has occurred in the southwest corner, whereas the northwest section is now arable fields. There are still park elements remaining, but now landscaped. There are two ancient or veteran oaks recorded with girths of 5.26 m and 6.10 m.

Gayhurst has been mentioned as a deer park since 1229 and still exists as a small park adjacent to Gayhurst House; today, it has the same shape and size as on the nineteenth-century OS map except for recent farm buildings in the northern end. It consists of meadows grazed by sheep or cattle, with solitary trees. More landscape development

has taken place nearer the house and two ponds have been created on the southeast side. There are no ancient oaks.

Great Hampden was licensed for the imparkment of 600 acres (240 ha) in 1447. This deer park was situated on the chalk of the Chiltern Hills, an unusual location. The park of Great Hampden around Hampden House was much smaller in the nineteenth century; at present the northern and eastern parts are woods and a small landscaped park with hay meadows and tree copses extends south of the house. There are no ancient oaks.

Ibstone was mentioned as a deer park in 1281 and is situated in the Chiltern Hills on chalk. Its more or less oval outline is recognizable on the OS maps as well as on aerial photos of the landscape. Ibstone House is on the southwest edge and the park is now partly landscaped, partly arable and grazing fields and partly wood, much of it recently planted. There are no ancient oaks.

Langley Marish (Langley Park) has been mentioned as a park since 1202. It is a landscaped park on the nineteenth-century OS map, on which some sections had already become farmland. Some of these sections became urbanised with the expansion of Slough. The park is now a country park for recreation, but it retains elements of the ancient park, among which is one recorded ancient oak and one veteran oak, both with girths >6.00 m.

Lathbury was a small deer park first mentioned in 1230. On both the nineteenth-century OS map and the current OS map, it is still named Lathbury Park and it is surrounded by wooded hedgerows or narrow woods, with a country house and church in the northwest corner near the River Great Ouse. It consists of a grazed field with a few solitary trees and has no ancient oaks.

Penn was mentioned as a deer park in 1325. On the nineteenth-century OS map there is a park-like field south of Penn House, surrounded by a band of wood which could have been a woodbank. A group and a line of trees in the field are also marked. Some of these features survive, but not the trees in the field, and it now appears to be an arable field in which new tree copses and an avenue were planted. There are no ancient oaks.

Stowe was imparked in 1257 by Sir John Chastillon. In the eighteenth century, it was substantially landscaped and included a lake and numerous follies, so that here nothing remains of the medieval park. Stowe is renowned for its landscape and park design and architecture, but the intensive remodelling has apparently been the reason why there are no ancient oaks inside the landscaped park. An ancient oak nearby, with a girth of 7.14 m, was probably in the medieval deer park; another ancient oak with a girth of 8.83 m was recorded in a remnant of pasture woodland with several veteran oaks to the north of the landscaped park, which may also have belonged to the deer park.

Waddesdon was a deer park in 1374, at which time it was owned by the Earl of Devon. It was extensively landscaped in the nineteenth century and Waddesdon Manor was built in a Neo-Renaissance French chateau style by Baron Ferdinand de Rothschild. The house and gardens are now owned by the National Trust. There are no ancient oaks.

Widdington (in West Wycombe) has been mentioned as a deer park since 1350. This is probably now West Wycombe Park. On the nineteenth-century OS map much of it is still open park, but with some landscaping, including a lake in the northern half and arable fields in the southwest. At present, the northern half is landscaped and the southern half converted to agriculture and a wood. There are no ancient oaks.

Derbyshire

Alderwasley was mentioned as a deer park in 1360, owned by the Duke of Lancaster. The park is still largely intact, with Alderwasley Hall at its northern end and little evidence of landscaping. In the northern section are some farm fields, but in the southern half there is rough grass invaded by bracken fern (*Pteridium aquilinum*) and trees from the adjacent wood. There are solitary trees in the fields but no ancient oaks.

Belper has been mentioned as a park since 1297. On the nineteenth-century OS map 'The Park' is situated just south of the small town, with a 'Manor Farm' and fields around two woods separated by a broad ride. The present park is smaller, more wooded and surrounded by suburban development and sports grounds. There are no ancient oaks.

Bolsover was first mentioned as a deer park in 1297. The park is now known as Scarcliffe Park and is situated *c*. 1.5 km east of Bolsover. It has extensive remains of the medieval park pale, showing that its medieval extent and shape were similar to the present. However, on the nineteenth-century OS map the park is shown as a wood and it remains as such, with planted conifers and broadleaved trees. There are no ancient oaks.

Bretby was mentioned as a deer park as early as 1295, and on the nineteenth-century OS map it is still marked as a deer park. The park was landscaped with a string of small lakes by damming a stream, but otherwise retained much of its old parkland and some woods, with Bretby Hall in the centre; Bretby Castle was formerly on the northern edge of the deer park. Much of the park is now farmland and there are some additional planted woods. There are no ancient oaks.

Calke Abbey was an Augustinian priory founded between 1115 and 1120 by the Second Earl of Chester. The canons moved to Repton Priory in 1172 and made Calke a subsidiary cell. It is likely that Calke was managed as an agricultural estate rather than as a religious house during the fourteenth and fifteenth centuries, but documentary evidence is lacking. It is possible that a deer park was established in this period, but none is listed by Cantor (1983) for this location. The outline of the deer park is still clearly visible, marked by hedges and lanes, but no park pale remains are indicated on the current OS map. Much of Calke may have been disparked by the end of the sixteenth century. The outline of the park on the nineteenth-century OS map is very similar to the present, more or less a heart shape with the point in the south and Calke Abbey in its centre. There are many ancient and veteran oaks in the remnants of pasture woodland, among which is the 10.01

m girth 'Old Man of Calke'. A detailed description of this park is given in Chapter 9.

The manor of **Chatsworth** was acquired by the Leche family in the fifteenth century; they already owned property nearby. They enclosed the first park during their tenure, perhaps in the late fifteenth century. The present *c.* 400 ha landscaped park contains a substantial number of ancient oaks, some in excess of 8.00 m circumference. Most of these are in unimproved parkland areas to the north and south of Chatsworth House (the latter known as Old Park), and these areas may coincide with the first deer park. This park was probably too recent to be listed by Cantor (1983), who mentions few fifteenth-century parks, and explicitly none later than 1485. Chatsworth's many ancient oaks are most likely associated with this late medieval deer park. See a detailed description of the Old Park in Chapter 9.

Codnor was a large deer park imparked in 1308 by Henry de Grey, who built a stone castle on the site of a Norman earth and timber castle. On Speed's map of 1610 (Nicolson & Hawkyard, 1988), Codnor is the largest park in Derbyshire and it includes Butterley Park. In the nineteenth century, the former deer park was all farmland but railways traversed it, in part allowing access to coal and ironstone mines. The area is now urbanised and industrialised. Dismantled railways are replaced by roads, remaining farm fields are much enlarged, and plantation woods occupy other fields or mining spoils. The only reminder of this once extensive deer park is the ruins of Codnor Castle. There are no ancient oaks.

Derby refers to **Markeaton Park**, now a city park for recreation activity. It had been a deer park, probably since 1250, but certainly from 1333 when the Earl of Lancaster owned it. It was surrounded by a wooded margin both in the nineteenth century and at present. The park is now mostly golf course, but with a landscaped eastern part with a lake and a meadow in the western corner. In the landscaped part are two veteran and ancient oaks (with 5.62 and 6.38 m girth).

Eckington has been known as a deer park since 1356; this park is now called Renishaw Park. In the nineteenth century, it was mostly parkland, with Renishaw Hall in its centre and a few woods; its nearly circular outline was also evident. The present situation is very different, with a golf course occupying the northern half, arable fields and a limited landscaped and partly wooded park. There are no ancient oaks.

Longford was first mentioned as a deer park in 1258. Formal gardens and, beyond these, a park surrounded Longford Hall in the nineteenth century. There is now a modern farm and most of the park is now farmland, except

for a small park south of the house. An ancient oak with 6.18 m girth stands in the corner of a field that was once in the deer park.

Mansell has been mentioned as a deer park since 1329. Little remains of this park but there are remnants of the park pale. An active sand pit is swallowing it from the west side, and there are large ponds resulting from earlier sand extractions. In the nineteenth century, apart from sections of the wooded park pale, Mansell was all farm fields, but on Speed's map of Derbyshire of 1610 (Nicolson & Hawkyard, 1988), it is one of four deer parks in close proximity in this area. There are no ancient oaks in the entire area of these four parks, two of which were created in the Tudor period.

Quarndon is also known as Kedleston Park, a large deer park west of the village of Quarndon; it was imparked for deer in *c.* 1400, but the Curzon family have owned the estate since at least 1297. The present landscaped park, which was designed by Robert Adam, is 328 ha, with Hay Wood in the northern part; it is owned by the National Trust. A large eastern part is made into a golf course. This park is now grazed by sheep and cattle and has no deer. Some parts are still unimproved grassland or parkland. There is evidence of ridge and furrow in a National Trust section just north of the golf course and possibly elsewhere, on which stand some of the 6.00–7.00+ m oaks, indicating that the park was created on or expanded from farmed land, and giving a maximum age of about 500 years for these oaks. The park retains a substantial number of ancient and veteran oaks, with 44 having a girth of ≥5.00 m, 15 of which have been measured with girths from 6.16 m to 8.37 m (Figure 4-6). Whitaker (1892) mentioned this park together with Windsor and Cornbury as "pre-eminent for magnificent oaks". However, landscaping and the golf course, as well as management and use of the grassland and parkland, have substantially altered this late medieval park and most of the oaks now present were young or nonexistent in the eighteenth century.

Ravensdale is mentioned as a deer park from 1297. It was adjacent to Mansell and is also being destroyed by sand extraction. Apart from a small section of remains of the park pale in the north, this park was farm fields and two woods in the nineteenth century. There are no ancient oaks.

Snelston was imparked in 1428. There is still a landscaped park, a few small woods and some arable fields; Snelston Hall is at the western end. Parkfield Lane and Windmill Lane mark two sections of the boundary. There are no ancient oaks.

Walton upon Trent was imparked in 1316. The site of the Old Hall on the OS map marks the beginning of an old wood bank that formed the southern boundary of the deer park. A new house, with a small landscaped park and a village expansion occupy the northern part, the rest is arable fields and a small marginal wood. There are no ancient oaks.

Essex

Coggeshall is presumably the same as Marks Hall Park, 2–3 km north of Coggeshall; it was imparked in 1203 by the Abbot of Coggeshall. This park has mostly been transformed into pine plantations and farm fields, but a landscaped core remains and has more recently been developed and planted as an arboretum. There are two ancient oaks, the largest locally known as the Honywood Oak and measured with 8.61 m girth by the recorders of the ATI.

Danbury was imparked in 1282 by William de St. Clare; St. Clere's Hall is just north of the present park. On the west side, the park (now Danbury Country Park) has been converted to agriculture and the remainder is a wood with two ponds or 'lakes' and partly landscaped parkland. There are a few ancient oaks, one with 7.09 m circumference. A >10.00 m girth oak known as Big Ben stood here before 1900 but was 'decayed' by 1907.

Gosfield was first mentioned as a park in 1314. Gosfield Park was a park with a large lake, with Gosfield Hall in the centre and with two woods, one at the end of the lake. The lake has partially silted up and the park has been converted to a golf course, woods and arable fields. There is one ancient oak (with 8.20 m girth) on the edge of a wood and arable field, which is all that remains of the medieval deer park.

Great Braxted was imparked in 1342. The park outline is still well visible on the OS map and from the air, formed by lanes, wood banks and hedgerows. The park was landscaped with a lake and small woods, but most of it is now either a golf course or agricultural fields, both pasture and arable. Three ancient oaks >6.00 m and several veteran oaks <6.00 m are recorded within the former park boundary of Braxted Park.

Great Hallingbury has been mentioned as a deer park since 1238. Hallingbury Park is situated just west of Hatfield Forest and was still a substantial park in the nineteenth century. It is now fragmented into arable fields, some woods and some remaining parkland, some of it landscaped around houses. There are no ancient oaks.

Great Waltham has been mentioned as a deer park since 1283. In the eighteenth century it was known as Langleys Park, and on the nineteenth-century OS map it was still a substantial park, but arable cultivation has since claimed large parts. There is a landscaped park near the house and a grassland park with scattered trees (of 67.6 ha) extending north along the River Chelmer and east and south along a stream. There are no ancient oaks.

Havering was first mentioned as a deer park in 1159 and was imparked within Havering Forest, itself a section of Waltham Forest and owned by the Crown. It was a large park of *c.* 440 ha, as was common for a royal deer park. In the nineteenth century, the park was much reduced, with most of the western half turned to agricultural fields. Some of these fields have been abandoned and are turning to rough grass and to shrubs and trees that are encroaching from the nineteenth-century part, which is now much more wooded and constitutes Havering Country Park. There are no ancient oaks.

'**Haselden**', also known as Hazelton, was first mentioned as a deer park in 1300. Nothing of it remains except (perhaps) for Hazelton Wood at the western end and an ancient oak (with 7.21 m girth) at the site of a medieval hall in White Notley at the presumed eastern end.

High Ongar was imparked in 1355 by Joan de Walden. In the nineteenth century it was still mostly intact and mostly park, with some landscaping

Figure 4-6. An ancient oak with 8.37 m girth in Kedleston Park, Derbyshire

around Forest Hall, now a farmhouse. Almost the entire park is now arable fields and a few small woods, but East Park Lodge and West Park Lodge still indicate its former width. Only near Forest Lodge at the southern end a small section of park remains. There are no ancient oaks. Nearby Chipping Ongar with motte and bailey ruins had a park (Ongar Great Park) from *c*. 1045, but nothing of this remains except a section of the park boundary, in part with bank and ditch.

Lawford was first mentioned as a deer park in 1279. On the nineteenth-century OS map it is a small rectangular park between a wood called The Rookery north of Lawford Hall and the village of Lawford as it was then. There are now more woods but much of the parkland with solitary trees remains. One of these is an ancient oak with 7.89 m girth.

Layer Marney has been mentioned as a deer park since 1264. There is little left of the park, but its southern boundary follows country lanes and some parts of Layer Wood are actually park-like. Near the house, Layer Marney Tower, is some landscaped parkland and a small wood. There are no ancient oaks.

Leyton was imparked in 1253; it is now situated in the London Borough of Redbridge. In the nineteenth-century Eagle Pond was dug in the northeast corner, and there is now an additional 'natural' pond with islands called Hollow Pond. The park is partly wooded and partly rough grassland. There are no ancient oaks.

Little Easton had two deer parks in 1302. The nineteenth-century OS map shows Easton Park to be a large park, nearly 3 km long and 2 km wide. This has all disappeared, converted to arable farms and a wooded garden around Easton Lodge. However, there is one large ancient oak with a circumference of 9.86 m near a farm *c.* 400 m southwest of Easton Lodge (Figure 4-7). In the nineteenth century, this tree stood in the park not far from a pond which still exists. Another remaining ancient oak on the farmland is 7.44 m in girth and a third measures 6.80 m. These two trees now stand on the edge of an area of sand and gravel extraction.

Little Parndon was first mentioned as a deer park in 1358. On the nineteenth-century OS map half of the original (?) park, south of Parndon Hall, still existed. This and the agricultural northern half have now been engulfed by Harlow's suburban development, including a hospital. Two or three small woods were not present in the nineteenth century. A single ancient oak (with 6.27 m girth) is all that is left of this deer park.

Little Waltham was first mentioned as a park in 1263. On the nineteenth-century OS map it appears to be divided into two small parks, Little Waltham Hall and Little Waltham Lodge, with farm fields in between. Little Waltham Lodge retains some park character, but elsewhere the village has expanded while fields have turned to rough grass, shrubs and some woodland. There are no ancient oaks.

Rivenhall was first mentioned as a park in 1296. On the nineteenth-century OS map the park is mostly intact except for a large field in the northwest corner, and it has a lake and some other landscaping. Since then, more lakes and also compartmented fields surrounded by woods have been created. There are no ancient oaks.

Saffron Walden was a park owned by the Crown in 1323; Walden Abbey came under the Duke of Lancaster and then the Crown later in the fourteenth century. On the nineteenth-century OS map there is still a large park, called Audley Park. It was substantially landscaped after the building of Audley End in the seventeenth century. The northern part is now a golf course and there are some large arable fields within the boundary of the former park. There are no ancient oaks.

Stansted Montfichet has been mentioned as a deer park since 1185; in the fourteenth century it was owned by Gilbert de Montfichet, Earl of Oxford. In the nineteenth century the park was still mostly intact, but a railway was laid across the northern part. Now the M11 also crosses the park and the remaining Stansted Park is limited by the railway, the motorway and a country lane along the southern boundary. The park is partly landscaped and there are some arable fields. There is one recorded ancient oak along the drive to Stansted Hall.

Stapleford Tawney has had a deer park since 1258; it is now the park of Hill Hall. Still largely intact in the nineteenth century, the M25 has now cut off the southern third, which has become part of a golf course. The remainder is partly farmland and partly park, with only light landscaping and scattered trees and copses in unimproved grassland. There are no ancient oaks.

Terling was first mentioned as a park in 1230. Much of it was still a park in the nineteenth century, but subsequently, most of it has been converted to agriculture and some woods. A small park remains in front of the present house, separated from it by a semi-circular haha. At the back of the house are landscaped and wooded gardens. There are no ancient oaks.

West Horndon has been mentioned as a deer park since 1363, when a licence was granted to impark 400 acres (160 ha). This park is not the same as Thorndon Park, and in the nineteenth century was still a large but partly wooded park. The oldest park was "more than a mile away" from the present Thorndon North Country Park, once a wood pasture common (Rackham, 2003b), in what is now Thorndon South Country Park, where there are remains of the old hall. The most extensive part of a later park, the Great Park, is now a golf course; in fact, all true parkland has been so converted except for a small landscaped part near Thorndon Hall. The North Country Park and The Forest are plantations and only a small central part is still more or less original pasture woodland, now overgrown with trees. In this section, three ancient or veteran oaks (with *c*. 6.00 m girth) have been recorded; a survey carried out in 1774 recorded 2,080 oak pollards.

Wivenhoe Park dates from 1427 and belonged to the Earl of Oxford. It lies east of the River Colne on the rising land beyond Colchester. On the nineteenth-century OS map it is mostly open parkland with a few coverts and park trees, either in rows or scattered. A small lake was created by damming a small stream. Little remains of this park, which is now the site of the University of Essex. There are, however, two oaks with girths of 6.14 m and 6.97 m near the mansion in what remains of the park.

Gloucestershire

Badminton has been mentioned as a deer park since 1236. In the nineteenth century it was a large park, with geometrical landscaping (Worcester Avenue) from an earlier period through extensive parkland, much of which remains. The park is surrounded by a stone wall and still has deer, but no ancient oaks. Two great oaks of around 10.50–11.00 m girth were recorded here in 1948 but are long since dead and gone.

Barnsley has been mentioned as a deer park since 1197. The present park is approximately the same as it was in the nineteenth century, with the house in its centre. In the northwest, a 'field system' is indicated on the OS map, this may imply that imparkment took in some earlier medieval ridge and furrow ploughland. The park has some landscaping and small woods, but much open parkland remains, albeit compartmented. There are no ancient oaks recorded in the present park, but nearby Barnsley Wold has some (Keith Alexander, pers. comm.) that remained unrecorded on 1 April 2016. It is unclear if this location was part of the medieval park; the nineteenth-century OS map indicates common pasture land in the area.

Brockworth has been mentioned as a deer park since 1243. On the nineteenth-century OS map there is no clear boundary of a park but Castle Hill, a motte hill, could have been part of the park. The land is all compartmented by hedges into separate farm fields and some woods. On either side of Painswick Road near Green Street is some remaining parkland, used as pasture. It has an ancient oak. Three ancient oaks are located near Abbotswood Farm, one of these has a circumference of 9.10 m and the second largest oak is 8.60 m. A fifth ancient oak stands in a hedged field to the southwest of Castle Hill; in all, 10 oaks with girths >6.00 m have been recorded at Brockworth (Figure 4-8).

Frampton-on-Severn has been mentioned as a park since 1296. About half of it was farmland in the nineteenth century. At present, this is secondary woodland. Much of the middle part of the park was dug out for sand or gravel

and is now landscaped with islands and trees; grassland remains nearer Frampton Court. There are no ancient oaks.

Great Barrington has had a deer park since 1412. Unusually, this park has a rectangular shape with almost equal sides. It is parkland except for the rectangular Grosvenor Plantation; there were a few plantations like this in the nineteenth century, now gone. There are no ancient oaks.

Ham should be **Whitcliff Park** (also known as Whitley Park) to the southwest of the village, which was imparked in 1250–70. It extends from near Ham to Upper Hill and is still a deer park. It has a few wood copses and light landscaping but is mostly open parkland. A brick wall was built around the enlarged park, now 120 ha, in the eighteenth century. There is one ancient oak with 6.80 m girth and only a few veteran oaks (some now dead).

Hardwicke dates from the late twelfth century. In the nineteenth century it was a small park between Hardwicke Court and the Roman Road (A38). Its southern half is now arable fields and a small wood and north of a small stream is lightly landscaped parkland. There are no ancient oaks.

Haresfield has been mentioned as a deer park since 1251. In the nineteenth century most of it had been converted to farm fields and orchards. The arable fields remain but the orchards have gone, although a few old apple and pear trees remain. There is the ruin of a motte between Mount Farm and the medieval church, indicating a fortified house. Although the park has almost entirely gone, there are five ancient oaks recorded in what remains of the park between the church and Haresfield Court, the largest of which has a 9.77 m girth and is probably a high pollard (Figure 4-9). An 8.50 m girth maiden oak stands along the road to Haresfield. The former boundaries of the park are obliterated and this oak may not have stood within them.

Hill was imparked in 1309 by Nicholas de Hill. Little of the park remains except for a small piece of landscaped park near Hill Court and a wood in which are the remains of a moat. The curved Wooded Lane probably marks part of the boundary, with possibly some remains of the ditch and bank alongside. There are no ancient oaks.

Hinton is Dyrham and Hinton, where a deer park west of Dyrham was imparked in 1368. This park is substantially landscaped with avenues, wood copses and trees, and a few ponds. It has no ancient oaks.

Miserden has been mentioned as a park since 1297. Many of its parts on the steeper slopes were already wooded in the nineteenth century. The park is partly landscaped, partly wooded and partly converted to arable fields. It contains the remains of a motte and bailey castle but no ancient oaks.

Figure 4-8. An ancient oak with 7.34 m girth in Brockworth Park, Gloucestershire

Northwick was first mentioned as a deer park in 1339, owned by the Bishop of Winchester. It was still mostly intact on the nineteenth-century OS map, with a country house and other buildings in its centre, a lake, and a wood in the northern part. A business centre has now been inserted between the lake and the wood in what was parkland with many trees. Some further landscaping has taken place, but much parkland remains. One oak with a girth of 7.19 m stands probably just outside the original deer park, another with 7.26 m girth is inside.

Painswick has been mentioned as a deer park since 1283. Just a small park remained in the nineteenth century but it had been much larger. Its outline is partly indicated by remnants of a wood bank and by Edge Lane. A small part near Painswick House remains as parkland today, but the rest is farm fields. An ancient pollard oak with a circumference of 9.78 m in a field between Edge Farm and Packhurst Farm may have been within the deer park.

Sapperton was imparked by William de Lisle in 1308. It was a large park, but in the nineteenth century it was substantially altered by the laying out of the Broad Ride and further rides radiating from a central point. The resulting compartments were planted to create Oakley Wood. While substantial lengths of the ancient park pale (i.e. ditch and bank) remain, there is at present little else left, with arable fields predominating and only a very small portion left as park. There are no ancient oaks.

Sevenhampton is first mentioned as a park in 1262, when it was owned by the Bishop of Hereford. Only a small section of parkland, now situated amidst meadows and arable fields, is all that has remained since at least the second half of the nineteenth century. There are no ancient oaks.

Sudeley is listed by Cantor (1983) as a "possible" deer park because there are no written medieval records of it. Nevertheless, extensive park earthworks provide archaeological evidence of a park that is "almost certainly medieval". The park is now mostly farm fields, but at the edge of a hedged lane stands an ancient oak pollard that I measured in May 2015 as having a 9.40 m circumference under a huge burr. Three other ancient oaks >7.00 m remain elsewhere in the former park and there are many veteran oaks, often in hedgerows.

Tewkesbury has been mentioned as a park since 1201, and a small part of it was still a park in the nineteenth century. There is now a golf course, agricultural fields, sewage works and a hotel. There is a large ancient pollard oak of 8.68 m girth, which may be a remnant of the medieval deer park.

Thornbury was a large civil parish (CP) in the nineteenth century and the park listed for it in Cantor (1983) is most likely Eastwood Park, now *c.* 3 km to the northeast of Thornbury. The history of the deer parks of the Earl of Gloucester has been described by Franklin (1989). On the nineteenth-century OS map this park is still extensive, but with some compartmented fields and a few small woods, and the original boundaries are difficult to detect. The present park is much the same except for a prison in its northern part. There are two ancient oaks in a meadow on the west side of the park, one with a circumference of 11.28 m (Figure 11-9).

Tormarton is now Dodington Park, first mentioned as a park in 1336. It is now bounded by roads, the A46 in the west and Catchpot Lane in the east. The park is extensively landscaped and there are no ancient oaks.

Woodchester was a deer park in 1308. It is situated in a short side valley of Nailsworth Valley, bounded by Dark Wood and Dingle Wood on the steep slopes and Woodchester on the west side. The park is partly landscaped and partly compartmented into pasture fields by hedges. There are no ancient oaks.

Hertfordshire

Amblecote was first mentioned as a deer park in 1292. In the nineteenth century, a small park around a house called The Hill remained, with Amblecote Hall on the other side of a railway. The urbanisation of greater Birmingham has now surrounded it, with a modern hospital on the site of The Hill. A tiny park remains as a municipal nature reserve. There are no ancient oaks.

Ardeley was first mentioned as a deer park in 1222. In the nineteenth century, as now, it was a small park called Ardeley Bury. It is partly landscaped and surrounded by wood banks; the one along the northern perimeter appears to be newly planted. There are no ancient oaks.

Ashridge has been mentioned as a deer park since 1286. The nineteenth-century OS map shows an elongated park adjacent to commons and woods, which was partly landscaped. This park is now a golf course and other sports fields, partly laid out in the woods. A single ancient oak with 6.85 m circumference remains in the landscaped grounds of Ashridge College, and a 6.33 m oak near Old Park Lodge also once stood in the park. The National Trust has recorded an oak coppice stool with *c.* 10.00 m circumference on the Ashridge Estate, which includes former coppice and secondary woodland belonging to Aldbury and Berkhamstead Commons. There is a herd of free roaming fallow deer at this location.

On the nineteenth-century OS map, **Brocket Park** is a landscaped park surrounding Brocket Hall that contains parkland, woods, copses

Figure 4-9. A great oak with 9.77 m girth in Haresfield Park, Gloucestershire

and a lake. This park has mostly been overlaid by a golf course, which has taken over all the parkland and several woods and copses, with new woodland strips planted. The shape of the park on the nineteenth-century OS map is reminiscent of a medieval deer park bounded by roads and streams, except along its northern limit where it was adjacent to farmland and a wood. The earliest mention of a manor house on this site is in 1239, when it was recorded as Watership or Durantshide, held variously of Hatfield Manor or the Bishop of Ely. Nevertheless, Brocket Park is not listed as a medieval deer park in Cantor (1983) or mapped as such by Rowe (in Liddiard, ed., 2007). In 1700 it was described as "the ancient seat of the Brockets situated upon a dry hill in a fair park, well wooded and greatly timber'd enclosed with a brick wall on the west side of the road [now the A1] for the length of a mile..." (Chauncy, cited in English Heritage listing report, November 2000). There are many recorded ancient and veteran oaks, the largest of which has a girth of 9.54 m, and several of these must date back to the Middle Ages.

'Broadfield' in Cassio was imparked in 1297 by John de Wengham. There is some uncertainty (Cantor, 1983) but this was presumably Cassiobury Park, which may have been extended later. It was a large park in the nineteenth century but the spread of urban Watford has consumed large parts, and much of what did not disappear under houses and roads is now a golf course and planted woods. A small section between Park Drive and Cassiobury Park Avenue is landscaped park. Three ancient oaks with girths from 6.00 m to 7.26 m have been recorded here.

Easneye was first mentioned as a deer park in 1332. Most of it was already wooded on the nineteenth-century OS map, with some parkland left south of the house. This house is now a college. There is only a small section of parkland left, in which there are no ancient oaks.

Great Gaddesden was recorded as having a deer park in 1241. This is the park of present day Gaddesden Place. In the nineteenth century it was about the same size as now, but the course of hedges in fields indicates that much of the southwestern part had been converted to farmland. There are several woods as well as open parkland, now with fewer trees than in the nineteenth century. There are no ancient oaks.

Great Munden refers to present day Hamels Park, now a golf course and farmland. There are no ancient oaks or remnants of a park pale, so this park no longer exists in any form. Just to the north there is Knights Hill / Coles Park, in Westmill CP. Of this park a core area of parkland still exists and there are three ancient or veteran oaks with between 6.00 m and 7.00 m circumference. This park is not listed in

Cantor (1983), but the nineteenth-century OS map and the ancient oaks make it very likely that it was a medieval park.

Hatfield had two parks, Great Park and Middle Park, which were imparked in 1277 and were owned by the Bishop of Ely. The nineteenth-century OS map shows Home Park and Hatfield Park, which were adjacent; the former was already mostly wooded and has a lake called The Broadwater. Hatfield Park is a landscaped park that now contains woods, (arable) fields and some parkland. There are several ancient and many veteran oaks, some of considerable size and one with 10.50 m circumference (see Chapter 9 for a detailed description).

Hertingfordbury Park was first mentioned in 1285 as a deer park. A small park known as Hertfordingbury Park still existed in the nineteenth century but this has been much reduced owing to suburban development, to the extent that all that is left are a few hedged fields and large landscaped wooded gardens. There is one large ancient oak in a field, now surrounded by gravel pits, with a girth of 9.44 m and three oaks with girths >7.00 m, all in Panshanger Park, which historically fell within Hertfordingbury Park. One of these three oaks, the Panshanger Oak, is well known locally and was measured repeatedly since the early nineteenth century.

Hoddesdon was imparked in 1227 by Stephen de Bassingburn. A house and small park called High Leigh are marked on the nineteenth-century OS map. Some landscaped parkland and rough grass fields still exist between the A10 and Park View, a residential road in Hoddesdon. There are no ancient oaks.

Hunsdon has been mentioned as a park since 1296. In the nineteenth century it was still a sizeable park, bordered by a plantation in the west, where across the road was Bonningtons Park. Hunsdon Park is now fragmented and contains more woods and arable fields. Some parkland remains in the western part. There are no ancient oaks.

Kingswalden Park was not listed by Cantor (1983) but is probably one of the 'other parks' mapped by Rowe (in Liddiard, ed., 2007) in the northwest of Hertfordshire. The present parkland away from the country house is grassland and contains planted native oaks of various ages to mature and veteran (under 6.00 m girth). It also contains the Great Oak of Kings Walden, a 10.70 m girth maiden pedunculate oak (Figure 4-10). This tree undoubtedly dates from the time of the medieval deer park. Along the country lane to Whitehall Farm, on a low bank inside the park, is a row of veteran oaks, this is perhaps a remnant of the park pale. A total of 17 oaks with girths ≥6.00 m has been recorded in the park, but

many of these are maidens with large crowns and may be younger than 400 years. The site is grazed with longhorn cattle (observed in May 2015).

Knebworth is a late park, first mentioned in 1472. The park was landscaped with avenues of which remnants survive. Today, the park is open grassland with few trees and a more densely planted area around Knebworth House. There are no ancient oaks.

Lichfield had a park in 1361 owned by the Bishop of Coventry and Lichfield. Much of this park had become farmland by the nineteenth century, but some parkland remained. Some of this, as well as the remains of a moat, are still present and some hedgerows mark the medieval park boundary. There are no ancient oaks.

Little Hadham is mentioned as a park from 1275 onwards. On the nineteenth-century OS map the eastern part is arable, truncating the park; it may or may not have once extended north of Standon Road (A120), which coincides with a Roman road in this location. At present, the park is even smaller, consisting of a few meadows with trees and a landscaped and wooded part around the house. There are no ancient oaks.

Maydencroft was first recorded as a deer park around 1320. Little of this park remains, but in the 19th century the northern part was still a park called Priory Park. Still less remains today as a landscaped park, the rest is arable fields and a section of Hitchin's urban expansion. Nevertheless, the outlines of the medieval park are mostly visible from the air and urbanisation has largely avoided it so far. There are no ancient oaks.

Moor Park (near Rickmansworth) was mentioned as a deer park in 1475 and was owned by the Crown. In the nineteenth century it was a triangular park with a house in the centre from which avenues radiated, but also with extensive parkland and a wood. It is now converted to golf courses. The only remnants of the deer park are an ancient oak with 7.40 m girth on the park boundary and some ancient and several veteran oaks of lesser size distributed mostly in the western half, where some small bits of unimproved parkland remain between the fairways. Only one ancient oak appears to be a pollard oak.

Much Hadham had a deer park from 1199. In the nineteenth century a small park remained around Moor Place and there was a farm. The farm now dominates the scene and there is only a landscaped park or garden behind it and near the house. There are no ancient oaks.

Pendley Manor was imparked in 1440 and is now known as Tring Park. Much of it was open parkland in the nineteenth century, with a wood on the western side and an avenue, but little other landscaping. This situation still exists, but the A41 now cuts the park in two. There are no ancient oaks.

Pishiobury (now in Essex) was first mentioned as a deer park in 1343 and was still a park in the nineteenth century. It has now virtually disappeared owing to the urbanisation of the long drive to the house and conversion to arable fields. Some parkland remains towards the River Stort. There are no ancient oaks.

Sawbridgeworth (now in Essex) has been mentioned as a park since 1237, but there was no longer a park on the nineteenth-century OS map. On modern street maps covering its possible locale, names like East Park and Park Way may be modern reminders of the medieval park that once existed. Near the location of Sheening Mill, now also gone, is some open land; this was hedged fields in the nineteenth century. Near a lock in the River Stort, in what is now a small modern city park, stands an ancient oak with 6.40 m girth, but it is not certain that it ever stood in this park.

Ware Park is a landscaped park west of the town of Ware. Its rounded boundary is indicative of a medieval deer park and the park is recorded in Domesday Book (1086) as property of Hugh de Grentmasnil, from whose family it passed on by marriage to the Earls of Lancaster in the fourteenth century (Cantor, 1983). A Tudor house was built by Thomas Fenshawe in the second half of the sixteenth century. There are now several arable fields and woods and only a small part is more or less parkland. The woods date from the twentieth century as they do not appear on the nineteenth-century OS map. There are nine recorded ancient oaks with girths >6.00 m, one of them with a girth of c. 9.50 m. It is possible that the famous Bed of Ware, made in the sixteenth century and now in the V&A Museum in London, was made of oaks that grew in Ware Park.

Weston has been mentioned as a deer park since 1231. The actual park appears much reduced on the nineteenth-century OS map, with much converted to farm fields and with some woods, among which is the still-existing Park Wood. The woods have since increased, but there is also more parkland, although much of this must be more recent. Near Park Lodge there is one recorded ancient oak that has a girth of 7.20 m.

Staffordshire

Alton has been mentioned as a deer park since 1313. In the nineteenth century there was still a park, but the area around Alton Towers was already wooded and partly developed. The park is now the site of a major 'theme park' with, among other development, huge car parks. Little parkland remains, and probably not for much longer. There are no ancient oaks.

Cheadle has been mentioned as a deer park since 1369. In the nineteenth century it was much reduced. Roads and streets such as Park Lane and Park Drive are indicators of its former extent. Much of it is now fields and urbanisation is encroaching. The remaining park has become much more wooded and little parkland is left. There are no ancient oaks.

Cheddleton was imparked in 1374. On the nineteenth-century OS map, it is marked as 'deer park' and is situated less than 1 km south of the village. It is now known as Ashcombe Park, a small park which includes a house and some woodland. Its irregular shape is undoubtedly due to the encroachment of farmland from different sides. There are no ancient oaks.

Chillington has been mentioned as a deer park since 1287. On the nineteenth-century OS map much of it is wooded (Big Wood) and this is still the case. A large lake and other landscaping were also present, so that not much ancient parkland remained. Some of this has later been turned to farmland. On the edge of Old Parkwood is a veteran oak which probably does not date back to the Middle Ages.

Leycett was first mentioned as a deer park in 1372. Much of it had been converted to farmland by the later nineteenth century and railways had been built through it. These railways are now dismantled. South of Scot Hay Road some of the park remains and is bordered by a wood bank. There are no ancient oaks.

Oakley was first mentioned as a park in 1322. Truncated and partly landscaped, the park on the nineteenth-century OS map is more or less the same size as at present. It is partly landscaped with a lake, now in dense woods. There are no ancient oaks.

Okeover has been mentioned as a deer park since the late twelfth century, when it was owned by Hugh de Okeover. There is still a park, of roughly of the same size and shape as on the nineteenth-century OS map, but a former path or track is now a paved road. There are remains of a medieval moat in Smythe's Plantation on the southeast edge of the park. The park is only lightly landscaped, but there are no ancient oaks.

Patshull has been mentioned since 1393 and was still a substantial park in the nineteenth century, albeit with the expansion of woods. The estate is now mostly woods and much of the remaining parkland has been converted into farm fields. A golf course has been laid out in the remaining parkland near Patshull Hall, leaving just a few patches of unaltered parkland. There are no ancient oaks.

Pattingham had a deer park in 1326. This park now belongs to Rudge Hall and was probably larger in the past. It is surrounded by arable fields. There is some landscaping along the margins that includes ponds and woods. There are no ancient oaks.

Stafford was imparked in 1285 by Nicholas de Stafford. On the nineteenth-century OS map the motte and bailey of Stafford Castle are surrounded by Castle Wood; the remainder of the park is farmland. Much of this farmland has now been taken by urbanisation and includes a golf course. Doxey Road and the A518 probably still mark the boundaries of the park. Just inside the A518 at Castle Bank there is an ancient oak with a girth of 7.78 m in a churchyard. The church is medieval, but its present churchyard may have expanded to include the oak which once stood in the park.

Stretton was mentioned as a park in 1175, and was probably imparked by Hervey de Stretton. In the nineteenth century, and at the present time, little remains of this park except a small piece of parkland and a lake surrounded by trees or woods. There are no ancient oaks.

Swynnerton was mentioned as a park in 1324. In the fifteenth century it was owned by Thomas Swynnerton. The present Swynnerton Park is just south of the village of that name, but *c.* 5 km to the north is a large wood called Swynnerton Old Park. This was partly heath and scrubland

Figure 4-11. A broken natural pollard oak with a girth of 9.55 m in former Yoxall Park

in the nineteenth century but is now all plantations. The smaller park near the village is now partly an arable field with trees and partly parkland. This is more likely to have been the medieval deer park. There are no ancient oaks.

Weston-under-Lizard has had a deer park since 1346. On the nineteenth-century OS map it is a large park, partly landscaped with woods, copses, a plantation and a lake, but with much remaining parkland. Woods, plantations (including conifer plantations) and farm fields now occupy much of Weston Park. The remaining parkland is dissected by roads and poorly preserved. There are, however, a number of ancient and veteran oaks, the largest of which has a circumference of 8.30 m. Most of the ancient oaks in this park are maidens, including the largest one.

Wootton was first mentioned as a park in 1316. It was a much reduced park in the nineteenth century, with fields and a wood taken from it. The situation at present is similar, with some landscaping around the house and a few clumps of trees remaining in rough grassland. There are no ancient oaks.

Wrottesley was first mentioned as a park in 1353. In the nineteenth century most of it was farm fields and two woods (one a new plantation); this reduction is now almost complete with only two small landscaped sections left. There are no ancient oaks.

Yoxall Park was first mentioned as a deer park in 1303, when it was owned by the Earl of Lancaster. In the nineteenth century a central open parkland area was surrounded by woods ('coverts'), with Lin Brook running through the park. Most of the woods are still there but the central area is now farmland. There are a few ancient and veteran oaks left, mainly in a remnant of unimproved parkland in the southeast corner known as Oakwood Pasture Nature Reserve. A ruinous ancient natural pollard oak here has a girth of 9.55 m (Figure 4-11). Some smaller ancient oaks in nearby farmland probably belong to the former deer park.

Suffolk

Aspal Close near Mildenhall is the remnant of a small and probably medieval park; it has the remains of a moat on its eastern edge. It was not listed by Cantor (1983). In the nineteenth century, there were only surrounding farm fields and the house, Aspall Hall, near the moat. The park is now completely surrounded by urban development, and the old hall has been replaced by smaller houses. There is a playing field within the boundaries of the park, but otherwise much of it remains in a medieval state with open acid grassland, partly overgrown with young oaks and thorn scrub, as well as more wooded parkland, with around 200 old pollard oaks. None of these are large and most are *c.* 4.00–6.00 m in circumference, but as they are growing on poor sand, they could still be (late) medieval. Owing to the surrounding housing estates, visitor pressure on the park is substantial. It is a local nature reserve managed by the Forest Heath District Council. A more detailed description is given in Chapter 9.

Benhall has been mentioned as a deer park since 1349. It is now Benhall Lodge Park, which has arable fields, woods and a landscaped park which is clearly a remnant. The house is now a farm. There are no ancient oaks.

Bentley was mentioned as a park in Domesday Book (1086) and is one of the earliest deer parks in England. Not much of it remains as parkland, but its extent and shape are partly indicated by roads and lanes around it. Engry Wood is probably an ancient coppice wood within the park. Some parkland remains around (new) Bentley Hall, the rest is farmland. There are no ancient oaks.

Bradfield St. George had a park from *c.* 1130, owned by the Abbey of Bury St. Edmunds. It was also known as Monk's Park and formed part of the Bradfield Woods. According to Rackham (2003b) it had many coppices as well as deer. In the nineteenth century parts of this park still existed, surrounded by arable fields. One of these parts was West Lodge, which must have once been at the western end of the park. Some remnants of parkland still exist here. Near the house is an ancient pollard oak with a girth of 7.30 m.

Chelsworth dates from *c.* 1130. In the nineteenth century, as now, Chelsworth Park was a lightly landscaped park with only a few parkland trees. There are no ancient oaks.

Earl Stonham was mentioned as a deer park, owned by the Earl of Norfolk, in 1283. Almost nothing of this park remains except a little park at Deerbolt Hall that has a lake and woods. There are no ancient oaks.

Elmsett has been mentioned as a deer park since *c.* 1180; it was also known as Long Melford Great Park. It was apparently an imparked wood (Rackham, 2003b). The property was owned by the monastery of Bury St. Edmunds before the dissolution of the monasteries in the sixteenth century gave it over into private hands. By this time the park had been much reduced in size. This park and some adjacent farmland and woods now belong to Melford Hall, a Tudor house owned by the National Trust. Only a small corner of Melford Park remains as ancient parkland, in which stand nine ancient pollard oaks (Figure 4-12), the largest with a recorded girth of 9.50 m.

Framlingham had a deer park in 1276, owned by the Earl of Norfolk. The famous castle is surrounded by two rings of dry moat and wooded banks and there are low lying wet meadows surrounding a lake nearby. There is not much left of the park, now mostly arable fields and a small golf course. There are no ancient oaks.

Great Bradley was first mentioned as a park in 1316. Almost nothing of it remains except for a tiny park with some landscaping features near The Hall, which also has the remains of a moat. There are no ancient oaks.

Haughley was imparked in 1262 and was owned by the Crown. There are several possibilities for the location of this medieval deer park, with a motte and bailey in Haughley, a moat at Haughley Green and the present Haughley Park west of Haughley. In Haughley Park, woodland is now planted at the western end, but a small park still remains. Near the mansion is an ancient oak with a reported circumference of 9.67 m (Figure 4-13). Near the moat of Haughley Green at New Bells Farm is a large pollard known as the Haughley Oak, which has a circumference of 11.22 m; this tree is now dead.

Holbrook has been mentioned as a park since 1300. Holbrook Gardens were still part parkland and included a lake in the nineteenth century. Today, a trace of the park pale remains north of Park House between arable fields, and the wood that shares the Holbrook name is outside the medieval park. There are no ancient oaks.

Huntingfield was imparked in 1313 by William de Huntingfield. In the nineteenth century there was still a park extending east from Huntingfield Hall, but it was broken up by farm fields and by 'Broomgreen Covert' and other planted woods. There were fish ponds along the southern margin beyond a road (now the B1117). Huntingfield Hall is now a farm and new trees have been planted in the park. An ancient oak, known as the Queen's Oak (with 7.20 m girth) stands near the farm in the sheep-grazed park.

Ickworth ('Ixworth' in Cantor, 1983) was first mentioned as a deer park in 1313. On the nineteenth-century OS map, it is a large park, with extensive parkland and several woods; in fact, it was much enlarged in the eighteenth

Figure 4-12. Ancient pollard oaks in Melford Park, Suffolk

century by imparking farmland, as is still evidenced by visible remnants of hedgerows. Its extent in the 19th century was greater than at present and parts of the parkland were later converted back to arable fields. The view expressed by Rackham (2003b) that the entire park dates from the eighteenth century is incorrect. From Google Earth imagery and the nineteenth-century OS map, it is possible to more or less reconstruct the outline of the 'old park', which may well have been the medieval park. There are many ancient or veteran oaks at Ickworth, nearly all in the 'old park' where the parkland landscape is best preserved, although 'improvement' of the grassland has meant that the landscape is somewhat sterile, with (mostly) oak trees on a green sward. The largest ancient oak at this location is known as the 'Tea Party Oak', now a living wreck of a pollard oak. The park is managed by the National Trust; see Chapter 9 for more details.

Kentwell was mentioned as a deer park, first owned by Aymer de Valence, in 1298. The remaining park is now landscaped, and includes a long avenue of lime trees (*Tilia*), but much of the ancient parkland has been turned into farm fields and there are several woods. One ancient oak stands near Kentwell Hall and there is another in a wood with a pond, both are more than 7.00 m in circumference, but the first tree was much reduced in the October 2013 storm. There are also a few smaller ancient and veteran oaks.

Lavenham Park has been mentioned since 1296, when it was owned by the Earl of Oxford. It was a sizeable park adjacent to Elmsett in Long Melford, but apart from some possible boundary features nothing remains of the parkland and it has all been turned to arable fields and farms. In Rackham (2003b, p. 194) a 'Manorhouse Moat' is indicated, but aerial photography (Google Earth) reveals no traces of this feature. There are no ancient or veteran oaks.

Laxfield had a park in *c.* 1300; in Cantor (1983), it is listed as a "possible" park but there are indications in the field that there has, in fact, been a deer park here. One is a large moat of which half remains. The other is an ancient pollard oak (with 7.50 m girth) recorded in a field at Old Oaks Farm (Oaks Farm on the nineteenth-century OS map).

According to Cantor (1983), **Staverton** has been listed as a deer park since 1306, but an earlier mention dates from the 1260s when it was seized by the Crown from Hugo le Bygot, Earl of Norfolk (Rackham, 2003b). It still is a medieval deer park, in private ownership, and one of the few unaltered or little altered parks left in England. Although on poor sand (Suffolk Sandlings), there is no podzol formation in the soil here, indicating more or less continuous tree cover since 'wildwood' times. Lanes, wood banks and what are most likely remnants of the park pale

indicate that arable fields especially have made substantial inroads into the park's original area (see map in Rackham, 2003b p. 293). Its present angular shape was already there in the nineteenth century, but it is not that of a medieval deer park. There are now pig farms on several of these fields, but some of these sections of the park were probably more or less open heath rather than woodland prior to ploughing. What is left is the core of 55 ha, a rather densely wooded park known as 'The Park' that has an estimated 500–750 oak pollards, many of them ancient and not cut for some 200 years. At its southern end, 'The Thicks' (26.5 ha) has massive holly trees. Deer are still held within a moveable fence, but there are also free roaming deer that come in from the surrounding forestry plantations. Staverton Park is one of the most important sites for ancient oaks in England and an SSSI for this reason. For a detailed description see Chapter 9.

Stoke-by-Nayland has had a park since 1300. In the nineteenth century this park was known as Tendring Hall Park and consisted of a partly landscaped park, woods and a plantation. This is mostly the situation at present except for the southern part, which became a farm field, and the house, which has gone. The Grove is a section of undeveloped parkland, in which stands an ancient oak. Another recorded ancient oak with 8.00 m girth is on the edge of the park along the B1087.

Wrentham had a deer park in 1339. The present park, just north of the village, is known as Benacre Park. This park is bordered on the west side by the A12 and on the east side by a wooded lane along which Benacre Hall is situated. The park has a few wood copses and scattered trees and is only lightly landscaped. There are no ancient oaks.

Surrey

Beddington was mentioned as a deer park in 1335. In the nineteenth century it was still a park, but much of it is now in use as sports grounds and only a small section remains as a landscaped park. There are no ancient oaks.

Betchworth was first mentioned in 1317 as a deer park. On the nineteenth-century OS map, as at the present time, there is a large meadow with some solitary trees between Betchworth House and the River Mole. This does not much differ from the surrounding land, but can be considered a park. There are no ancient oaks.

Cobham was imparked in or around 1123 by the Abbot of Chertsey. It is situated in a bend of the River Mole on the south side of Cobham (or the much smaller Church Cobham on the nineteenth-century OS map). The medieval extent of the park may have been greater than the park marked on the nineteenth- and twenty-first-century OS

Figure 4-13. The Haughley Park Oak in Suffolk

maps. In a field of grassland west of Plough Lane stand several ancient and veteran pollard oaks, one of which was recorded with an 8.15 m circumference. This field was a part of Hampton Court Chase, which included the earlier park. Farm fields and buildings now separate it from the remaining landscaped park.

Farnham was imparked in 1242 by the Bishop of Winchester. On the nineteenth-century OS map this park is mostly intact and situated north of Farnham Castle. This situation has changed relatively little despite extensive urbanisation on all but one side of the park. A small golf course has been established in the southeast corner. There is an avenue oriented on the castle, but otherwise not much landscaping. Elsewhere, much of the vegetation consists of grassland, scrub, more or less recent woodland and unmanaged hedgerows. There are some veteran oaks, the largest with a girth of 6.33 m, and many mature maiden oaks.

Gatton has been mentioned as a park since 1278 and was owned by Hamo de Gatton. It was a large park well into the nineteenth century, with some farm fields in the northern part. The present situation leaves a landscaped park, some woods, a cluster of buildings that forms a school, and a golf course. There are no ancient oaks.

Oxted was mentioned as a park in 1278, imparked by Roland de Oxted. A small landscaped park around Oxted Place ('The Rectory' on the nineteenth-century OS map) is all that remains. There are no ancient oaks.

Sheen was first mentioned as a deer park, owned by the Crown, in 1292. From the late fifteenth century, this park became known as Richmond Park; it was associated with Richmond Palace in Sheen (built 1501 demolished 1650–60). Another park of uncertain locality also existed along the Thames. These parks have either been destroyed, having been taken over by a golf course, sports fields, housing, roads and railways, or transformed to a city park with modern plantings. There is just one veteran oak with a girth of 6.14 m in what is left of the medieval park of Sheen, it now stands on the golf course. The present Richmond

Figure 4-14. Pollard oaks in Richmond Park, Surrey

Park is a large (*c.* 1,000 ha), partly landscaped deer park surrounded by a brick wall and was imparked by Charles I in 1634-37. It is described in Chapter 9.

Shere was known to have a deer park in 1303. This must be the present Albury Park, which contains an Anglo-Saxon church, while the estate ('Eldeberie') is mentioned in Domesday Book (1086). The sixteenth-century mansion was nearly rebuilt in the seventeenth century, with subsequent alterations. An earlier house or lodge may have stood on the same site, near the church. The park was extensively landscaped in the seventeenth century by John Evelyn, and fragments of avenues remain from this work. Much of the present park is wooded, including modern plantations, but patches of ancient parkland remain. In this parkland stand several ancient and veteran oaks, nine of which are recorded as having girths of between 6.00 m and 8.00 m (Figure 4-15).

Tandridge was first mentioned as a park in 1351. The park is now divided into fields in its southern half and into woods and some landscaped parkland near Tandridge Court. The former parkland to the north of this is now a golf course. There are no ancient oaks.

Vachery was first mentioned as a park in 1245. There is little evidence of a park on the nineteenth-century OS map, apart from Vachery Pond, a landscape feature created by damming a stream. Vachery House stands in a landscaped wooded park near this lake, but was not present on the nineteenth-century OS map. There are no ancient oaks.

Sussex

Arundel was imparked in 1299 by the Earl of Arundel and is still a large landscaped park north of the castle. In the nineteenth century there apparently was still some less developed parkland, but this has now also been heavily landscaped with plantations, copses and large lawns. There are few parkland trees and no ancient oaks.

Battle was imparked by the Abbot of Battle in 1397. It is the site of the famous Battle of Hastings that took place in October 1066, giving the English crown to the victorious William of Normandy; this in turn led to the introduction of deer parks with imported fallow deer. Battle Park is still the same size and shape as it was in the Middle Ages. It has become somewhat more wooded than it was in the nineteenth century and there are now sports grounds in its centre. The marshy ground in the hollow that featured in the battle became a pond with an island, the only landscaping that was carried out. There are no ancient oaks.

Buckhurst (in Withyham) was first mentioned as a deer park in 1274. On the nineteenth-century OS map it is a landscaped park with parkland, a lake and ponds, woods (including a large coppice enclosure) and parkland. The shape and size of woods, open parkland and the lake have changed since, and few trees remain in the central parkland part. There are no ancient oaks.

The park in **Barlavington** CP is now known as Burton Park and was first mentioned in 1296. It had woods and parkland in the nineteenth century and was flanked in the south by artificial lakes. A farm has been established in the northern part, the woods have been extended, and the area around the present mansion has been landscaped, but some unimproved parkland remains. There are seven recorded ancient and veteran oaks in a strip of unimproved parkland that is surrounded by a housing development at Lodge Green, one with an estimated 8.00 m in girth and another which may have been just as large before it broke up. The unimproved parkland also has several veteran or ancient sweet chestnuts (*Castanea sativa*).

Buxted was mentioned as a deer park in 1274. On the current OS map it is marked as a deer park, but on the nineteenth-century OS map it is called Buxted Park without this indication. A small stream, the River Uck, meanders through it. A large farm field has been taken out of the park in the northwest corner, but elsewhere it remains mostly intact. The park has been lightly landscaped and there is a lime avenue of which only one row of old lime trees still survives. This row has several ancient lime trees, but there are no ancient oaks.

Coates was first mentioned as a deer park in 1335. Some parts of it were farmland in the nineteenth century, and another part is marked on the nineteenth-century OS map as 'Warren'. Arable fields with hedgerows have expanded in the northern part, as has woodland in the centre near Coates House. The 'Warren' is now re-established parkland. There is one recorded ancient oak in the farmland area.

Cuckfield has been mentioned since 1241 and was first owned by the Earl of Warenne. In the nineteenth century there was a park west of the village of Cuckfield with the same name. It was an open, lightly landscaped park situated beyond the house and its formal gardens. At present, a plantation divides the parkland into two halves, but not much else has changed. There are no ancient oaks.

Eridge Park, with Saxonbury Hill to the south, is not listed as a medieval deer park in Cantor (1983) but is reputed to date from Norman times; it was inherited by the Nevilles in 1448. It lies on Tunbridge Wells Sandstone with steep slopes descending to a stream that has been dammed to create ponds. The northeastern part (Eridge Old Park)

retains much of the character of ancient pasture woodland with patches of heather (*Calluna vulgaris*), scrub and much bracken, in which are scattered mature oaks and (mostly) beeches, as well as old field maples (*Acer campestre*), ash and hawthorns (*Crataegus monogyna*) with many species of epiphytic lichens. There are two large oak pollards with girths of 7.66 m and 8.15 m (which I measured 19 March 2014). The ecological quality of the site suffers from invasive rhododendrons.

Findon was first mentioned as a park in 1333. This small park still existed in the nineteenth century, minus a large arable field taken off its northern part. There is now also arable farmland on the east side. A small wood and some parkland remain in the central and southern parts, respectively. There are no ancient oaks.

Firle Park was first mentioned in 1333. Its outline is clearly visible as a wood bank and/or country lanes on the nineteenth-century OS map. A large section in the western part of the park was already converted to farmland in the nineteenth century; these arable fields surrounded the wooded hill that is topped by Firle Tower, a landscape park folly. More farm fields are now present in the northern section of the park, but extensive parkland remains. There are no ancient oaks.

Glynde was mentioned as a deer park owned by the Archbishop of Canterbury in 1423. By the nineteenth century the northern third had been converted to farmland, while on the west side there were grazing fields and Decoy Wood. The present situation is similar, with much open grassland and only a small area of parkland. There are no ancient oaks.

Halnaker has been mentioned as a deer park since 1281. There are now a large field surrounded by coppice woods, plantations, and the remains of avenues and park borders with veteran sweet chestnut trees. Almost none of this can be described as true parkland. There are no ancient oaks.

Heathfield is first mentioned around 1341, when it was owned by Andrew Peverel. A large park still existed in the nineteenth century, designed and landscaped by Humphrey Repton including woodland along a valley stream that has ponds at its lower end. At this time, there were two fields on higher ground. The whole park has now become much more wooded with both native trees and planted exotics, so that it resembles an arboretum. There are no ancient oaks in this park.

Herstmonceux has been mentioned as a deer park since 1264, and was imparked by Waleran de Monceux. In the nineteenth century, it was a sizeable park but already had extensive woods. At present, the woods and farm fields dominate and there is little parkland left, in which grow a

Figure 4-15. Veteran pollard oak in Albury Park, Surrey

number of ancient sweet chestnuts but no ancient oaks.

Hurstpierpoint has had a deer park, first owned by Simon de Pierpoint, since the early thirteenth century. On the nineteenth-century OS map a few remnants of radiating avenues or lines of trees and some ponds are all that seem to have remained of a landscaped park called Danny Park. Near a country lane leading north from the former Danny Park, in an area called The Plantation, are two ancient oaks, both pollards, the largest of which has been measured with a girth of 10.50 m but is very fragmented. These oaks stand in what may be the only remnants of the deer park, a field of grassland with some scattered trees.

Isfield was first mentioned as a deer park in 1349. In the southwest corner are the remains of a motte and bailey and a medieval church. The later mansion, Isfield Place, is situated more centrally and a road runs north–south through the former park. The former boundaries of the park are not difficult to discern and all of the parkland has been converted to farm fields or copses of wood. There is an ancient oak with 6.20 m girth near the Old Rectory Farm.

Laughton has been mentioned as a park since 1293. In the nineteenth century the park was partly landscaped and partly compartmented into fields. Parkland existed mainly to the west and northeast of Laughton Hall. These areas too are now mostly fields, some with a scattering of park trees. There are no ancient oaks.

The park at East Lavington was mentioned as 'Woolavington Park' in 1329; it is now called **Lavington Park**. Although reduced by the nineteenth century, it still had a core of parkland at that time. This is now mostly sport grounds and a small golf course, with some hedged (compartmented) parkland to the north. There are two ancient oaks, one on either side of the road through the park.

There was a deer park in **Harting** CP from 1274 onwards. This presumably became Up Park in South Harting. In the nineteenth century this was still a large and partly landscaped park with a more or less circular shape, only encroached upon by farm fields in its southern part. Besides the near surroundings of the house, which include a landscaped garden and a wood, there is little parkland left and most of the park is now given over to farming. There are no ancient oaks.

Knepp is mentioned as a deer park from the late twelfth century onwards. This park appears to have been reduced by farm fields around it, but has a large lake as the most conspicuous landscaped feature. There is now a wood north of Knepp Castle and some remaining parkland to the south of this house. Near the house is an ancient oak with a girth of 7.07 m. This tree measured 6.00 m in 1981, 6.50 m in 1996 and 7.07 m in 2009, so it has been growing at a good pace and is probably a veteran rather than an ancient tree. Another oak in the park has a girth of 7.30 m but its growth rate is unknown.

Midhurst has had a deer park since 1285, imparked by John de Bohun. Since the fourteenth century it has been known as Cowdray Park. On the nineteenth-century OS map it is a large and partly landscaped park extending between Easebourne and the River Rother. Much of this park is now taken up by farming and plantations; there is also an 18-hole golf course. Some parkland remains in the northeast of the former park and here is the Queen Elizabeth Oak, an ancient and hollow pollard form (from natural causes) with a circumference of 12.67 m. There is another oak pollard nearby measuring 10.32 m girth (Figure 4-16) and there are several ancient oaks of lesser size (7.20–8.70 m; all maidens) on the golf course.

Petworth was imparked in 1296, but Henry Percy had three parks here around the turn of the century. Petworth is now a large park (of 283 ha) which was extensively landscaped by Lancelot ('Capability') Brown in the eighteenth century. It has a herd of deer, but its medieval character has virtually disappeared due to the major alteration of the park landscape. Vast open lawns, lakes and numerous planted trees, including non-natives, now dominate this deer park. The only remaining relics of the medieval park are four recorded ancient oaks of 6.57 m, 8.38 m, 8.85 m and 9.13 m in girth. It could of course be argued, aside from the fact that most of the trees are planted and several are non-native, that modern Petworth is not that far removed from a medieval deer park and that the younger oaks may grow old and 'ancient' in due course.

Plashett Park ('Plaskett Park' in Cantor, 1983) was mentioned in 1397 as a park owned by the Archbishop of Canterbury. It is now landscaped with a string of dammed lakes, and there are fields bordered by hedges and woods. A small part is still parkland. There are no ancient oaks.

Plotsbridge (in Framfield) was a park around 1400; this is now called Framfield Place. The former extent of this park is unknown but may have included fields to the west

and southwest bordered by woods. The present park near the house is small and landscaped, with just a small portion as parkland. There are no ancient oaks.

Poynings had a park in 1339. This park was 1–1.5 km north of the village near a moated castle. A house called Newtimber Place is now on that site, and near it is a small landscaped park. The rest of the park is now mostly woods and a few farm fields. There are no ancient oaks.

Shermanbury was first mentioned as a park in 1361. In the nineteenth century about half of the park remained, while the other half was converted to farmland and

Figure 4-16. An ancient pollard
oak in Cowdray Park, Sussex

woods. The woods are now larger and there is more arable farmland. A small section near the house, which stands within a medieval moat, is still park but with few trees. There are no ancient oaks.

Shoyswell was mentioned as a park in 1345. In the nineteenth century little parkland remained and, apart from a pond and its surroundings, this location was probably all farm fields and woods. At present, some of the fields have been abandoned, turning to rough grassland and scrubland. A small area of parkland has been created from Hogtrough Wood. There are no ancient oaks.

Slindon was imparked in 1317 by the Archbishop of Canterbury. In the nineteenth century, as now, it was a rather open small park surrounded by Slindon Wood, but with Butcher's Copse inside and compartmented by fences for grazing sheep. A small part near the college is used as sports grounds. It has substantial sections of remnant medieval park pale. There are no ancient oaks.

West Dean was mentioned as a park, owned by the Earl of Arundel, in 1327. On the nineteenth-century OS map, as at present, this is known as West Dean Park. Its shape is still visible from the air and is typical of a medieval park, with rounded margins that are now mostly roads. Traces of the park pale, a ditch, are visible along the southern edge. There are plantations and an arboretum, as well as some farm fields, but the central part is parkland. There is one recorded ancient oak in this park but it is smaller than 6.00 m in girth.

West Grinstead was mentioned as a park in 1326. It was still a large park in the nineteenth century. At present, almost all of it is farmland and woods, with only a few more or less park-like remains in strips along the western margin and in the centre. There are no ancient oaks.

Wiston was mentioned as a park in 1357. A lightly landscaped park with some small woods or copses, hedgerows and a pond existed in the nineteenth century. This park largely still exists, but its central area is almost tree-less and in part used as sports grounds. There are no ancient oaks.

Warwickshire

Arrow was first mentioned as a park in 1333, when Robert Burdet obtained a licence to impark it. The medieval park is now part of the Ragley Estate, and although its boundary on the south side is no longer discernible, it is likely that part of the more recent Ragley Park once belonged to it. Around 'The Garden House', which is surrounded by a rectangular brick wall, grow three recorded oaks that have girths >6.00 m, which probably belonged to this deer park. Perhaps an 8.44 m girth ancient oak 250 m southsoutheast of 'The Garden House' (Figure 4-17) was also in this park; it is now in the edge of a plantation of larch (*Larix*). Additional ancient and veteran oaks, but most of them smaller than 6.00 m in girth, were observed along a stream through the present park on 23 November 2015. There were also some gnarled hawthorns and an old field maple here, indicative of an ancient deer park. The seventeenth-century Palladian mansion was once surrounded by extensive formal gardens, and in the eighteenth century the park was landscaped by Lancelot ('Capability') Brown. The extent of this landscaping was such that little remains of the ancient deer park except for these few oldest oaks, which are separated by wide open spaces of grassland and planted patches of mostly non-native trees.

Astley has been mentioned as a deer park since 1296, when it was imparked by Andrew de Astley. The shapes of hedgerows on the nineteenth-century OS map suggest it was only a small park surrounding the moated castle. This situation prevails at present, with a pond or lake (Astley Pool) and a small park around the castle ruins and the medieval church in a corner. There are no ancient oaks.

Baginton was first mentioned as a deer park in 1285. A large park-like field to the west of Baginton Hall was still indicated on the nineteenth-century OS map, in it are the remains of a castle. The rest of what may have been the medieval park has mostly become farm fields; a water meadow is indicated along the River Sowe. The castle ruins are now barely present and the surrounding land is part waste (ruderal vegetation), part secondary woodland, part nurseries and part industrial development. A golf course occupies much of what may have been the rest of the park. Near the castle ruins is an ancient oak of 6.81 m in girth.

Beaudesart ('Beaudesert' on OS maps, but it did not mean 'beautiful desert') is mentioned as a deer park since 1265. Nothing is left except for some park-like undeveloped land and hedged fields around the remains of a motte. There are no ancient oaks.

Berkswell has been mentioned since 1322 as a deer park. On the nineteenth-century OS map outlines of the park are visible as wood banks and there is landscaping with copses and woods. At present, much of the western half of the park is farm fields, some with trees of the park remaining. Parkland still covers a large part of the park, but there are no ancient oaks.

Kenilworth had a deer park since 1165; it was first owned by the Crown but in the thirteenth century by the Earl of Lancaster. Nothing remains of this large park, which was situated west and south of the castle, but some sections of the park pale (remains are visible as a wood bank). These remains and Rouncil Lane outline the western and southern perimeter of the park, but the evidence of the park's extent has been lost in the north and east, mostly because of the development of the town of Kenilworth. There is an ancient oak at Oaks Farm with a girth of 7.53 m, and two oaks >6.00 m also stand within the perimeter of the former park. Chase Wood is a reminder of Kenilworth Chase, which extended beyond the park.

Middleton was first mentioned as a park in 1258. In the nineteenth century only a small park remained to the southwest of Middleton Hall and there was an artificial lake, Middleton Pool. The park remnant is now mostly farmland and woods, and the only park-like section is around Middleton Pool, with more formal landscaping around the house. There are no ancient oaks.

Nuthurst was first mentioned as a deer park in 1332. The park was established to the southwest of the small church that still stands alone in the fields. In the nineteenth century there was a large park called Umberslade Park, which still exists (now close to the M40) but with a local public road through it. The former parkland to the west of the road is now used as arable farmland. The piece that remains is a landscaped park with lakes or ponds, woods

Figure 4-17. This pollard oak with 8.44 m girth in Ragley Park may have grown in the medieval deer park of Arrow

and parkland. Near the northern wooded park boundary stands an ancient oak with a girth of 8.28 m.

Oversley has been mentioned as a park since 1283. The outlines of a park are still visible from the air as wooded banks and/or farm tracks. However, most of the park has been converted to farmland, with some landscaped parkland remaining around Oversley Castle. This property is shown on the nineteenth-century OS map as 'Upper Oversley Lodge' and is now converted to private dwellings. There are no ancient oaks.

Packwood has been mentioned as a park since 1279. Encroached on by farmland since at least the nineteenth century, Packwood is now a small park owned by the National Trust. It has been landscaped with crossing avenues and a lake. There is parkland with some large mature oaks but there are no ancient oaks.

Snitterfield was first mentioned as a park in 1427. It was already gone — converted to farmland — by the nineteenth century. Some scattered trees in an arable field north of the A46 are perhaps reminiscent of a park. In grassland near Park House stands an ancient oak of 8.03 m in circumference; two other oaks are >6.00 m in girth.

Sutton Coldfield has been mentioned as a park since 1298. It was a large park with an almost circular shape. In the nineteenth century, the eastern part of Sutton Park was broken up by farm fields and railway lines with adjacent development. There were also woods and landscaping with lakes and plantings, but the western half remained largely parkland. Now, urban spread has taken up the farm fields and there is a small golf course, but much of the woods and parkland remains. The park is now a National Nature Reserve, surrounded by urbanisation. There are no ancient oaks.

Worcestershire

Abberley has been mentioned as a deer park since 1280, when it was owned by the Earl of Warwick. On the nineteenth-century OS map, the park at Abberley Hall has the shape and size of a modest deer park. It was then already quite wooded, with only small areas of parkland. This parkland is now limited to a small area near Abberley Lodge and the hall is a school. There are no ancient oaks.

Beoley has been mentioned as a deer park since 1248. There are ancient earthworks south of the B4101 (Beoley Lane) marking the site of a castle, and on the nineteenth-century OS map there is some parkland north of this road. A small park, compartmented by hedges, still exists here. There are no ancient oaks.

Bushley Park has been mentioned as a deer park since 1296. It was already converted to mostly arable farmland in the nineteenth century. Until recently, there was one living ancient oak, but it is now a dead stump. This tree

Figure 4-18. Oaks in Elmley Castle Deer Park, Worcestershire

was near a field with ancient oaks across the border with Gloucestershire, which once formed part of Corse Chase, belonging to the Earls of Gloucester (see Chapter 5).

Elmley Castle has been mentioned as a deer park since 1234, when it was owned by the Earl of Warwick. It is a small park that is remarkably well preserved, with the remains of ramparts and a dry moat on a castle hill (part of Bredon Hill) a little south of its former centre. The most southerly parts are now hedged pastures. There is a small herd of fallow deer, but the park pale has long since disappeared and the deer have easy access to the wider countryside. There are two small ponds (one known as Castle Pool) and quite a few of the parkland trees are later plantings of species such as horse chestnut (*Aesculus hippocastanum*). Among the trees are two recorded ancient oaks, the larger measuring 8.80 m in circumference and probably once a pollard. There are about ten oaks with >5.00 m circumference in and around the deer park (Figure 4-18). The site is part of a larger SSSI that includes former pasture woodland on Bredon Hill.

Hallow was first mentioned as a park in 1309. It was situated on the west bank of the River Severn. No park remains, as it was either converted to woods and farm fields or subsumed in the expansion of the village. An ancient oak with a girth of 7.45 m stands in the boundary hedge between the churchyard and a field.

Hanbury was first mentioned as a park in 1339, owned by the Bishop of Worcester. On the nineteenth-century OS map, Hanbury Park is a substantial park that has avenues, copses, parkland and more formal gardens near Hanbury Hall in the middle of the park. The northern section of parkland is marked 'deer park' on this map. There are several ancient oaks in the parkland, but most are not of great size (or age); the largest (of 8.45 m girth) is on the other side of School Road in an arable field that is part of Beck's Farm, which may once have been part of the park.

Hanley Castle has been mentioned as a park since 1229. In the nineteenth century there were three avenues radiating from Severn End at the eastern margin of the park. The central one still has the old trees, but the other two have recently been replanted. Arable fields truncate the park in the west and south. There are no ancient oaks.

Ombersley was first mentioned as a park in 1358. In the nineteenth century it was a landscaped park with a lake, woods and extensive though rather open parkland. Large arable fields have now reduced the park, the lake has silted up and the wood around the lake has expanded. A strip of parkland with a wooded fringe extends from Ombersley Court southward meeting the A4133. There are no ancient oaks.

Shrawley was first mentioned as a park in 1304. It was landscaped in the nineteenth century, when dams created a series of small and large 'pools' and much of the park became a wood. Only a small section of landscaped park remains near nineteenth-century Shrawley Wood House. There are no ancient oaks.

Shrawnell (in Badsey) was a park in *c.* 1340, when it was owned by Abbot Chiriton of Evesham. No trace of a park is left in this village and its agricultural surrounds (with many nineteenth-century orchards in the 19th century). This park is mentioned here because Bond (2004: 174) stated that the abbot "beautified his newly enclosed park ... with oak and ash plantations", presumably because there were no trees here. If true, this would stand out as the earliest recorded instance of the planting of woodland trees in England. There are no ancient or veteran oaks in the area now.

Upper Arley had a park called 'Eymore Park' in 1305. A park with unclear boundaries is indicated north of Arley Castle on the nineteenth-century OS map. This area is now nearly all farmland except for a walled garden and a woodland garden near Arley House and a small landscaped park. There are no ancient oaks.

Analysis of the survey of twelve counties

The results of this survey of medieval deer parks are summarized in Table 4-1. The first observation to be made is that 577 or 75% of the medieval deer parks in these twelve counties, which once had so many of them, no longer exist as a park in any form. Without doubt, the greatest consumer of medieval deer parks has been agriculture. Even where parks still exist, the great majority

have been reduced, most often also by farming. The second, but more local, cause of disappearance is urbanisation, which here includes the creation of a golf course or other sports grounds. Parks have been sliced through by railways (in the nineteenth century) and especially by roads (in the twentieth century), and often these developments marked the onset of park demolition on the far side of the road from the country house. In 18 (2.3%) cases of park disappearance, one or more ancient oaks remain within the boundaries of the former park. These are here taken to be relics of the deer park, although they may not always be old enough to have stood in the park during medieval times (i.e. pre-1485).

In many cases, the OS maps show that there is a moat close to, or even in, the park, and sometimes there is a small church or chapel associated with the moated manor house. While the church usually still exists, the fortified lodge, manor house or castle has long since disappeared, and perhaps with it the park, or at least the deer and the park pale. From this survey, I cannot conclude that a church in a park is "an infallible sign of later emparking" (Rackham, 2003, p. 199).

Only two existing parks in these twelve counties (0.3%) still retain much of their medieval character, including the deer: Staverton Park in Suffolk and Windsor Deer Park in Berkshire. There are a few other extant deer parks like these in England, such as High Park (part of Blenheim Park) in Oxfordshire, Bradgate Park in Leicestershire and Moccas Park in Herefordshire, but it is clear that very few medieval deer parks have been left mostly untouched. The landscaping rage has sometimes destroyed the medieval character of former deer parks completely, as in Stowe in Buckinghamshire. In other cases, only small sections of the park have escaped major changes. It is therefore not surprising that the number of landscaped former medieval deer parks without ancient oaks, 112 (14.5%), is nearly twice that of those with ancient oaks, 61 (7.9%). Traces of the former park pale also appear to be rare, they were noted in just 17 (former) parks (2.2%). If not indicated on the modern OS map, sometimes traces of a former park pale could be suggested by the Google Earth image, but such indications must be 'ground-truthed' and archaeological field work may reveal quite a few more. Such features are not very relevant to the presence of ancient oaks, but could help to define the park boundary. Often, this boundary can also be deduced from aerial imagery or even from the OS maps. Deer parks in the Middle Ages had rounded outlines, not the angular outlines typical of most farm fields (Liddiard, ed., 2007; Rotherham, 2007b). The boundaries of deer parks are often preserved in hedges, wood banks

County	Deer parks with ancient oak(s)	Landscaped parks with ancient oak(s)	Landscaped parks without ancient oak(s)	Other land use with ancient oak(s)	Other land use without ancient oak(s)	Traces of park pale extant	Total of parks listed in Cantor (1983)
Berkshire	1	2	6	3	23	3	35*
Buckinghamshire	0	5	8	0	41	0	54
Derbyshire	0	3	6	0	41	4	50
Essex	0	8	9	4	81	1	102
Gloucestershire	0	6	12	2	41	3	61
Hertfordshire	0	9	11	2	25	1	47**
Staffordshire	0	4	11	1	73	0	89
Suffolk	1	8	8	1	47	2	65**
Surrey	0	3	8	0	39	0	50
Sussex	0	9	20	1	79	2	109**
Warwickshire	0	2	7	2	47	1	58
Worcestershire	0	2	6	2	40	0	50
Totals	**2**	**61**	**112**	**18**	**577**	**17**	**770**
Percentages	**0.3**	**7.9**	**14.5**	**2.3**	**75**	**2.2**	**100**

Table 4-1. The twelve English counties with the highest density of medieval deer parks and their status at present in relation to ancient and veteran oaks. The landscapes and treescapes of the deer parks included in this table are essentially medieval, and these parks contain several to many ancient oaks dating back to (late) medieval times. Landscaped parks have planted trees (native and usually also non-native), avenues, wood copses, lawns, lakes and other garden-like features. If there is grazing, it is usually not by deer but by domesticated animals such as cattle and sheep. Large or small parts of these parks may be converted to other uses. Other land use here means that there is essentially no park left; the park may have become a golf course or farmland, or have been given over to urban development or any other use. There may be traces of the medieval deer park, in particular the park pale; if present, these are noted for all categories.

* In the total count for Berkshire, I have omitted Whichmere, which was inside Windsor Great Park.

** I have included four parks that were not listed by Cantor (1983): in Hertfordshire, Brocket Park and Kingswalden Park; in Suffolk, Aspal Close; and in Sussex, Eridge Park.

and country lanes, but only recognisable remains of the park pale are listed here.

The 'harvest' of ancient oaks garnered by this survey seems not particularly rich. The twelve-county sample taken here is essentially random and is large enough to be representative of the whole of England; it seems unlikely that the situation would differ much in other counties. On the assumption that medieval deer parks in the twelfth to fifteenth centuries had trees, and that these trees would have been mostly native oaks, not planted but derived from natural woodland ('wildwood'?), we can see that relatively few of these oaks survived. Yet they are not a rarity in the English countryside, witness the ATI database on which this survey is mostly based. There must have been so many oaks to begin with that the destruction had to remain incomplete. For example, a 1624 survey of Sheriff Hutton Park in North Yorkshire described "4,000 decayde and decaying okes" (see Dennison, ed. 2005). In some counties, the majority of ancient oaks ≥6.00 m in girth now present are associated with medieval deer parks: 93% in Derbyshire, 82% in Oxfordshire and 79% in Hertfordshire (Table 5-5 in Chapter 5).

Other medieval deer parks

Medieval deer parks in English counties other than the twelve summarised in Table 4-1 that still have a minimum of five ancient living oaks with a girth ≥6.00 m, or if smaller a minimum of 25 oaks, are briefly described below.

Ashton Court Park in Somerset near Bristol is a landscaped park of 340 ha that is owned and managed by the City of Bristol. In 1392, it was imparked for deer by Thomas de Lyons. The estate was bought by a Bristol merchant in 1495 and Ashton Court, a Tudor mansion, was built near the centre of the park earlier in the fifteenth century and later enlarged. The medieval deer park, Ashton Lyon, was much smaller than the present park. The park had extensive areas of parkland and woodland well into the nineteenth century. Two golf courses have now been laid out, but since 1970 there have also been two enclosed deer parks on the estate, as well as woods and remnants of the unimproved parkland. There are several ancient pollard oaks, among which is the Domesday Oak (9.19 m girth), which is supported by props to prevent the heavy branches from further splitting the trunk. A more detailed description is given in Chapter 9.

Boconnoc in Cornwall has been mentioned as a deer park since 1381. On the nineteenth-century OS map, a 'Deer Park' at this location is surrounded by woodland; this is now almost tree-less open grassland. Landscaped parkland is present nearer the house and there is now a deer park near the southern end of the park, which is dominated by beeches but with some oak pollards. Most of the oaks recorded in the park by the ATI are maiden oaks and not ancient, but a few are pollards and the largest of these is recorded with a 7.71 m circumference. Boconnoc Park and Woods is an SSSI, mainly on the basis of the high diversity of lichen species associated with old or mature trees.

Bradgate Park in Leicestershire has been mentioned as a deer park since 1241, when it was owned by the Earl of Winchester. On the nineteenth-century OS map, parkland and wood copses are marked within a large and lightly landscaped park. This situation still exists in the park, now managed as Bradgate Country Park. This park also has deer, both fallow and red deer (*Cervus elaphus*), as well as unimproved acid grassland and the ruins of Bradgate House. It has several ancient pollard oaks, which are concentrated in some of the more wooded areas of the parkland. A visit in October 2011 established that there are more ancient and veteran oaks than had been recorded by the ATI, so on a second visit in May 2016, I added another 25 oaks from just one small area to the database. A more detailed description is given in Chapter 9.

Brampton Bryan Park in Herefordshire is situated on the Welsh border. It has been in the ownership of the Harley family since 1309 and was imparked by them, but it is not listed as a medieval deer park in Cantor (1983). The earliest mention of Brampton Bryan as a deer park is on Saxton's map of 1577, but it is likely that it was imparked much earlier. The ancient deer park is largely intact today except for some fields in its northern part and some modern forestry plantations. Much unimproved parkland remains, but lines of old sweet chestnuts indicate early landscape 'improvements', which were followed by the planting of non-native park trees and enclosed forestry plantations. There are several recorded ancient oaks and many veterans at this location (Figure 4-19); one of these has a girth of 9.20 m. Brampton Bryan is an SSSI because of its ancient trees and woodland. A more detailed description is given in Chapter 9.

Cobham Park in Kent between Gravesend and Rochester is a landscaped park. Cobham Hall is a Tudor period building, but there has been a manor house on this site since the twelfth century. On the nineteenth-century OS map the park has a rectangular outline, with Cobham Hall in its centre and has the designation 'deer park' in its eastern part. Because of its unusual shape for a medieval deer park, Cobham Park may date from the late Middle Ages or even later; it is not listed in Cantor's (1983) gazetteer of medieval deer parks. The park contains earthworks of a Roman villa. The deer park is now a golf course. Five ancient or veteran oaks are located in the park area around the mansion (now a school); the largest of these is 7.40 m in circumference.

Cornbury Park in Oxfordshire lay within the Forest of Wychwood and was first mentioned as a royal park in 1244 (Cantor, 1983). In 1337 it was more permanently imparked with a stone wall similar to that of Woodstock Park. A lodge in the park also existed in that year, and it was occupied by the keeper of both Cornbury Park and Wychwood Forest until the park was given to Edward Hyde, Earl of Clarendon, by Charles II in 1661 (Schumer, 1999). The present deer park, with the seventeenth century mansion at its centre, is slightly smaller than the medieval park, the boundary of which can be seen on the nineteenth-century OS map to extend closer to the River Evenlode. At the southern boundary of the present deer park, a series of dams were laid in a tributary stream to the Evenlode to create fish ponds. Beyond these ponds is an area called Little Park, originally the southern part of the medieval deer park, but separated from the present deer park since the creation of the ponds. The parkland consists of (mostly) improved grassland and many planted trees, including lime and beech avenues radiating from the mansion, among which remain a number of ancient oaks, one of these with a girth of 10.05 m. On two visits in 2014 and 2015, I established the presence of nine oaks with >6.00 m girth, of which five are >8.00 m. In some areas of the park with unimproved grassland several old field maples and many hawthorns were seen. The landscaping and planting of non-native trees, as well as mowing of the sward, have diminished the ancient character, and presumably the biodiversity, of this park, but not all is lost and especially the great oaks and the area with hawthorns are very valuable in this context. The park is used for annual events but is otherwise a strictly private domain.

Donington Park in Leicestershire belonged to Castle Donington. It was first mentioned in 1229 when the property was given to John de Lacy. This park extended west of the castle to the banks of the River Trent and was of considerable size well into the nineteenth century. Developments varying from the building of Donington Hall at the end of the eighteenth century, to landscaping,

Figure 4-19 (top). Ancient
oaks in Brampton Bryan Park,
Herefordshire
Figure 4-20 (bottom). Oaks on a
ridge in Far Moor Park (Duncombe
Park), North Yorkshire

agriculture, and the building of a motor racing circuit in the twentieth century have altered and reduced the medieval park considerably. Old maps of the park up to the nineteenth-century OS map provide evidence of the presence of many trees, presumably mostly oaks, in the now-developed parts. The western part of the park, towards the river, remains largely intact and a number of ancient oaks and other native trees remain on the slopes and ridge above the Trent. This area is now a deer park once more, sympathetically managed by the owner, John

Shields. Donington Park contains 15 oaks exceeding 6.00 m in girth, with the largest ancient oak measured at 8.85 m. Smaller veteran oaks are abundant, especially in the southwest part of the deer park. However, dead oaks, including standing trees as well as old stumps, appear to be more numerous than the national average of around 5% (see Chapter 11). An active programme to plant new oaks grown from acorns harvested in the deer park will eventually mitigate these losses. Numerous old hawthorns, as well as some field maples and old ash trees, complement the native treescape on unimproved acid grassland. The remains of crossing avenues of lime trees in the northernmost corner and a planting of sycamores around a small monument are the only landscaping features inside the present deer park. Two conifer plantations are being phased out and replanted with oaks to enlarge the deer park, which is an SSSI on account of its ancient oaks.

Eastnor Park in Herefordshire was a late-medieval deer park belonging to Castle Ditch, probably in existence from the middle of the fifteenth century onwards. A moated manor house existed here in the seventeenth century; the current large mock medieval castle dates from the nineteenth century. In the eighteenth and nineteenth centuries land was acquired from neighbouring estates and the park was much enlarged. It is now a mosaic of fields for pasture, woods and parkland, including a deer park. There are nine oaks with girths ≥6.00 m in the park and the largest oak measures 7.88 m in girth. Many other oaks have been planted since the nineteenth century and are now mature trees.

(Little) **Glemham Park** in Suffolk has a number of ancient and veteran oaks, two of which have been measured with girths of 7.90 m and 8.53 m. A house called Glemham Hall has been on the site since the thirteenth century, but a deer park at Little Glemham has been mentioned only since around 1560, when a new house was built. The park is surrounded by farmland and some woods, and is now divided by the A12 which runs through the northern eighteenth-century extension of the park. The ancient moated manor house was demolished and the moats filled in at the time of that extension, when the park was (lightly) landscaped. Only the two largest oaks are likely to date from before 1560; most of the oaks are between 5.50 m and 6.50 m in circumference. The park is laid to pasture grazed by cattle or sheep and is used for annual events.

Helmsley in North Yorkshire has been mentioned as a deer park since 1230. In the fourteenth century there were two parks, the 'Old' and 'New' Parks. With the castle situated on the west side of the town of Helmsley, the Old Park would have been congruent (more or less) with the present Duncombe Park on the left bank of the River Rye. This park has many oaks, but only about ten are large enough to be ancient or veteran trees and these are all pollards. On the slopes of the opposite side of Rye Dale is a wood called Beech Wood, and here there are about 20 ancient oaks. A more open area with farm fields is known as Far Moor Park; on its valley-side margin there are some more ancient oaks, and on a now wooded ridge are also smaller ancient oaks (Figure 4-20) and some large veteran small-leaved limes. The oaks on this side of

the dale are often larger than those in Duncombe Park, with girths of between 6.00 m and 7.00 m. Beech Wood is partly coniferised deciduous broadleaved woodland, but this location and Far Moor Park may have been part of the 'New Park' mentioned in the fourteenth century. The ancient oaks near Helmsley are therefore considered to be associated with medieval deer parks. About half of Duncombe Park is a National Nature Reserve (NNR), and this park and Beech Wood are both SSSIs. Beech Wood and Far Moor Park are leased by the Forestry Commission, which is now doing conservation work to remove planted conifers in areas that have ancient oaks (pers. obs. June 2015).

Henderskelfe in North Yorkshire was first mentioned as a deer park in 1359, when it was owned by the Barons of Greystock and situated in Bulmer Parish. It is now part of the Castle Howard estate, where the great Van Brugh house was built in the eighteenth century on the site of the ruined medieval castle. Farming and landscaping on a grand scale have destroyed much of the deer park. Nevertheless, a section of the medieval park pale can still be seen as a bank and ditch, now followed by a later stone park wall presently forming the southern boundary of the newly established Yorkshire Arboretum. A park is shown near 'Hilderskil' on John Speed's map of the North and East Ridings of Yorkshire (1610), and according to Whitaker (1892), there was still a deer park here at the end of the nineteenth century. Its outline beyond the remaining sections of pale is now difficult to establish, but four areas with ancient and veteran oaks may give an indication of the park's former extent. One group of such oaks is now in and around the visitors' car park near the house; another is located to the east of the lake and the last two are in an area called 'Thorns' on the nineteenth-century OS map and in the Park Quarry, now a wood. The 'Thorns' area is most reminiscent of ancient pasture woodland and indeed has, besides some ancient oaks, many hawthorns and now also ash and sycamore (*Acer pseudoplatanus*). A large maiden oak with a girth of 7.44 m stands outside the former park at the edge of Todd Wood, not far from the site of an old mill.

Kimberley Park in Norfolk has been mentioned as a deer park since around 1401. In the nineteenth century, it was still a large landscaped park with parkland, a wood (Great Carr), a lake and an avenue. The present park is reduced in size, especially in the western part, by arable fields near Park Farm. Most of the recorded ancient oaks are in a remnant of parkland in the eastern part of the park near Kimberley House, and the largest here are 7.00–8.50 m in circumference. A large oak stands west of the house and was measured as having a 10.70 m circumference in 2010. This measurement was found to be incorrect on my visit in April

Figure 4-21. Ancient pollard oak in Kimberley Park, Norfolk

2014; the largest oak in the park is 8.50 m in girth and is not that tree. For a detailed description see Chapter 9.

Knole Park in Kent was first mentioned as a deer park around 1360. The present Tudor mansion, enclosed by a rectangular walled garden, is situated in the large park, which has been extensively landscaped with open parkland, woods on the higher ground and grassland valleys in chalk downland. It is still a functioning deer park, but large sections of former parkland are now a golf course. The house and park are owned by the National Trust. Most native trees in Knole Park are beeches, but there are also oaks. Several

of these are recorded as ancient and veteran maidens or pollards, six have been measured as >6.00 m in girth.

On the nineteenth-century OS map, **Lullingstone Park** in Kent is a park of 240 ha with open parkland and Upper and Lower Beechen Woods inside an ancient park boundary. Farm fields have taken out a section along the northern perimeter, but the park still extended from east to west from Lullingstone Castle to Park Gate. More arable fields have since broken up the park and the core of what remained is now occupied by an 18-hole and a 9-hole golf course, but with some preserved areas of chalk grassland and woodland. There are many ancient maiden and pollard oaks, now mostly situated in the partly wooded roughs of the golf course and some also in Upper Beechen

Wood, indicating that this was originally parkland (or pasture woodland) and not a managed wood with coppice and standard oaks. Lower Beechen Wood may also have been pasture woodland, but was converted to plantations in the eighteenth century, mainly with beech and sweet chestnut (Pitt, 1984). These woods are also important for hornbeam pollards. The largest oak at Lullingstone has a circumference of 10.36 m; 16 more living oaks measure between 6.00 m and 9.00 m in girth. Lullingstone is not listed as a medieval park by Cantor (1983) who used a cut-off date of 1485, but it has been described as a medieval deer park (Pittman, 1983) and may have been imparked in the late fourteenth century; the manor house dates from 1497. There was a deer park until the 1930s; during World War II it was used by the Royal Air Force who set up a decoy airfield to divert German bombers. Both this activity and the golf courses laid out in the 1960s and 1970s destroyed some ancient oaks and much of the unimproved parkland. Forestry plantations in the 1940s and 1950s (and perhaps earlier) have also destroyed ancient oaks and this, and the growth of secondary woodland, have further altered the ancient park. The woodlands are an SSSI but the golf courses are exempt from this notification.

Moccas Park in Herefordshire is an ancient deer park, but there is no historical record of it prior to the seventeenth century (David Whitehead in Harding & Wall, eds., 2000). The nineteenth-century OS map shows the remains of a motte and bailey castle situated at the eastern boundary of the park, but this may never have had a stone building. Instead, Moccas Park may have been the park belonging to Geoffrey de Bella Fago in 1316 and listed by Cantor (1983) as being located in the Parish of Dorstone. The castle there was about 2.5 km from the park but the wooded Dorstone Ridge, on which part of the present Moccas Park is situated, may have been the only land in the area suitable for a medieval deer park. This park, now an SSSI and an NNR and still privately owned, is one of the most important sites in England for ancient oaks. This park is described in more detail in Chapter 9.

Nettlecombe Park in Somerset is an ancient deer park connected with the manor of Nettlecombe, which was granted to Hugh de Ralegh in around 1160. This park was studied by Rose & Wolseley (1984), who were interested in the rich lichen diversity associated with the ancient oaks. These authors were able to reconstruct the extent of the medieval park on the basis of the ancient trees still present but also on the presence of lichens associated with ancient trees. The outline of the medieval deer park has been disrupted by the creation of fields and woods, and only a small part of the park is presently unimproved grassland.

The nine recorded large oaks >6.00 m in girth are mainly present around Nettlecombe Court, now a field centre, with a few in the fields to the south, among which is the largest with a girth of 8.52 m.

Penshurst Place in Kent was first mentioned as a deer park in 1290. It was a large park extending from Penshurst Place, a hall dating back to 1341, to a bridge on the Enfield Road crossing the River Medway; a short section of the park ditch and bank can still be seen close to the bridge. Perhaps the most famous inhabitant and owner of Penshurst Place was Sir Philip Sidney (1554–86); an ancient oak with a 9.42 m girth, now approaching the end of its life, is named after him. It stands a short distance beyond the lake as seen from the house. The largest oak, with an 11.17 m girth, is some distance away in a corner of Rookery Wood, and may have stood on the boundary of the deer park. Augustine Henry measured this oak, as one does now under the huge low limb, as having a girth of 10.16 m in 1924. Landscaping and forestry plantations have much altered the park and only eight recorded oaks of 6.00 m or greater are now known to exist.

Sacombe Park in Hertfordshire is not mentioned by Cantor (1983) as a medieval deer park, and its locality is not marked on the map of medieval parks in Hertfordshire by Rowe (in Liddiard, ed., 2007, p. 129). However, in a text on the website www.hertfordshire-genealogy.co.uk/, it is stated that "The Park is said to be one of the first which were enclosed in England" and that "it contains some gigantic oaks, which show very high antiquity." The two largest ancient oaks here are recorded as having girths of 7.51 m and 7.15 m, both are pollards; six other ancient or veteran oaks are between 6.00 m and 7.00 m in girth. A storm in 1786 is reported to have destroyed "many very large trees" that "had been standing in perfect security for some centuries." The present park is very fragmented and mostly turned to arable farming and some woods, but its (medieval?) shape is still visible in hedgerows and field boundaries. The manor of Sacombe has been in the possession of noble families from Peter de Valognes (1045–1110) onwards, so the presence of a medieval deer park at some time before 1485 seems plausible.

Scadbury Park in Chislehurst, part of the London Borough of Bromley, is probably an ancient deer park. A moated manor house was built there in 1314 by John de Scathebury. There is no mention of a deer park in Cantor (1983) and the historical records do not seem to mention it. Owing to surrounding and encroaching farm fields, which can be seen on the nineteenth-century OS map, and recent urban development, the original shape of the park boundary is difficult to reconstruct, but there are some hints of a typical deer park shape. There are several veteran or ancient (pollard) oaks in Park Wood to the west of the moat, as well as one or two elsewhere in the park. Most of the trees that I have seen here are not truly ancient; the largest oaks here range from 5.00 m to 6.50 m in circumference.

Sheriff Hutton Park in North Yorkshire is mentioned here despite the fact that is has only one ancient oak >6.00 m, with a girth of 8.15 m, and around 20 ancient or veteran oaks of between 3.00 m and 5.90 m girth recorded by the ATI volunteers (2006–13). This park has been

well researched by the Sheriff Hutton Women's Institute Community Pale Project, and the results of this work were published in a book (Dennison, ed. 2005). From this evidence, it would appear that a few more oaks of around 6.00 m in circumference exist but are not recorded in the ATI. Ralph de Neville obtained a licence to enlarge an existing park in 1334 and the ruins of a new castle begun in 1382 still dominate the village and park. The park of *c.* 300 ha has almost all been converted to farmland, but remains of the park pale have been traced. The "4,000 decayde and decayinge okes, the most of them headed" surveyed by John Norden in 1624 had mostly disappeared towards the end of the seventeenth century. Subsequently, the park was

Figure 4-22. Pollard oak in Lullingstone Park, Kent

compartmented and turned over to separate farms. Sheriff Hutton Hall and its immediate surroundings remained as a park, in the 19th century still of substantial size, but it is now much smaller. The only truly large ancient oak that remains is near the hall.

Shobdon Park in Herefordshire is a large park with a country house, Shobdon Court, on the site of a medieval priory. On the nineteenth century OS map it was marked as a deer park and its outline, in part marked by country lanes, could indicate an ancient park. It is not listed as such in Cantor (1983), but evidence of a motte near the mansion and toponyms indicate ancient occupation and the existence of a (perhaps smaller) park. It is not, however, marked as a park on seventeenth and eighteenth-century maps of the county. Parts of the park are now farmland and there is industrial development near the mansion. There are three oaks here that have girths over 9.00 m and three others measured at 7.60 m, 7.78 m and 8.25 m, plus a number of smaller veteran oaks, so it is at least likely that pasture woodland was present here in the Middle Ages, probably in the context of a deer park.

Woburn Park in Bedfordshire is a large landscaped deer park famous for its exotic deer herds, among which are a herd of Père David's deer. It belonged originally to a Cistercian abbey founded in 1145 which had a deer park. In 1547 the abbey was given by Henry VIII to John Russell, First Earl of Bedford; its conversion to a country seat dates from after 1630. Near Star Lodge stands the Domesday Oak, with a 9.30 m girth and two ancient oaks of just over 8.00 m in circumference are in what is now a safari park. These oaks may have stood in the medieval park, but by the late sixteenth century, there was no longer a deer park at this location according to Speed's map of Bedfordshire (Nicolson & Hawkyard, 1988). Indeed, the many oaks in the present deer park, located to the south of the safari park, are not ancient and only two are recorded with girths of around 6.00 m.

Shute Deer Park, now known as **Woodend Park**, is a partly wooded park west of Shute in Devon. On the nineteenth-century OS map it is partly parkland, partly farm fields and woods. Parts of the park's original boundary can be recognised from aerial photography as wood banks or wooded lanes. The park is reputedly a medieval royal deer park, and the large *c.* 10.00 m-circumference 'King John's Oak' is presumably associated with this theory, but the park is not listed in Cantor's (1983) gazetteer of medieval parks. There is, however, a well-preserved section of the medieval park pale. There are over 40 ancient or veteran maiden and pollard oaks with girths ≥5.00 m at this site, mostly concentrated on and around the central

hill, which contains most of the remaining parkland as well as traces of ancient hedgerows. These trees can also be interpreted as indicative of an ancient deer park. This park was described by Harding (1980) as important pasture woodland for the conservation of wildlife. A more detailed description appears in Chapter 9.

Woodstock Park in Oxfordshire has been mentioned as a deer park since the early twelfth century. It was imparked by Henry I and owned by the Crown for several centuries. Its site is now occupied by the still larger landscaped park of Blenheim Palace, one of Lancelot ('Capability') Brown's grandest designs. Despite the extensive landscaping, there are still many ancient pollard oaks in the Lower Park ('Little Park' on the nineteenth-century OS map) and especially in the High Park, where the largest oak, which has a 10.37 m circumference, stands. Unlike the landscaped part of the park, High Park is still pasture woodland, although not presently grazed and partly mowed to suppress bracken. These areas probably more or less correspond with parts of the medieval deer park, which has been described as having been surrounded by a 7-mile (10.2 km) stone wall and a main gate facing the (old) town of Woodstock (Richardson in Liddiard, ed., 2007). A detailed description of High Park in Blenheim is given in Chapter 9.

Tudor deer parks

In this chapter so far, we have explored ancient and veteran oaks in medieval deer parks. After 1350 there was a sharp drop in newly created parks and this decline continued in the fifteenth century (Chapter 7, Figure 7-2). This did not mean that most parks disappeared, although many did, but that imparking by the landed gentry and nobility declined, probably mostly for economic reasons. The deer parks that remained often became larger as their owners imparked more land that had been abandoned after the Black Death by a greatly reduced peasant population, which took almost three centuries to return to a pre-1350 level. There may also have been a significant decline of parks during the Wars of the Roses (1455–87), when the old feudal powers of the nobility began to weaken as rival factions fought each other and allegiances often changed dramatically and more than once. Was there a revival of deer parks in Tudor times? The Tudor monarchs certainly had a keen interest in hunting and deer, and as before, would set the trend for their entourage of courtiers and would-be followers. Elizabeth I, in particular, preferred shooting deer in a park with a crossbow over the wild chase on horseback through the woodlands and fields of the Royal Forests. She visited many parks apparently for this purpose. I have found in this survey of parks that a significant number of ancient

Figure 4-23. Veteran oak on the lawn at Woburn Abbey, Bedfordshire

and veteran oaks are associated with parks that were not mentioned before 1485, Cantor's (1983) latest date for the establishment of a medieval park. The fact that these parks were not mentioned before 1485 does not mean that they were not in existence before this date. Other evidence, some of it historical, may be brought to bear to establish the likelihood of imparkment after 1485 but before 1603, when Elizabeth I died. A new kind of land owner, the wealthy merchant aspiring to the status of 'gentry' had begun to acquire parks in the fifteenth and sixteenth centuries, and the sixteenth century saw a substantial increase in imparking by these new land owners (Fletcher, 2011). I have called such parks Tudor deer parks.

More often than parks in the Middle Ages, Tudor parks had a substantial mansion within; the owner of the estate created a park around his house. Early small-scale landscaping features, such as ornamental ponds created by damming a stream and open vistas and lawns, began to appear in this period. However, trees were still not or rarely planted or seeded, and if so, these trees, perhaps with the exception of sweet chestnut, would have been native trees only. Ancient avenues or rows of sweet chestnuts, such

as those in Croft Castle Park and Brampton Bryan Park, probably date from the late sixteenth or early seventeenth century. Deer keeping was still a primary motive of imparking, although the parks sometimes had other uses such as cattle grazing. Late sixteenth and early seventeenth-century maps of the English counties have been helpful in locating such parks; a particularly informative source are the maps published by John Speed in 1616 (Nicolson & Hawkyard, 1988), which according to Speed's index included 749 parks in England. Parks containing ancient trees that were not listed in Cantor's (1983) gazetteer, and not revealed by close scrutiny to be medieval, are here considered to have originated in the Tudor period (1485–1603). Useful sources listing many of these post-medieval parks are also provided by Shirley (1867) and by Whitaker (1892). Some deer parks created in the Tudor period are of significance especially for veteran oaks. Below, I review a selection of the Tudor deer parks, most of which have at least five ancient oaks ≥6.00 m in circumference.

Althorp Park in Northamptonshire has, except for one corner, the perfect rounded shape of a medieval deer park, but it was created in the early sixteenth century by

Figure 4-24. Veteran oak on the lawn at Woburn Abbey; drawing by the author (2015)

Pedunculate Oak
Quercus robur
Woburn Abbey, England
Aljos Farjon
fecit Istanbul Maio 2015

Sir John Spencer, who bought the land in 1508 and built the first house here. This 180-ha park was created around the house, which is typical for the Tudor period. It was landscaped first by André le Nôtre in the 1660s and in the eighteenth century by Henry Holland, with further alterations in the nineteenth century. As a result of these alterations, most of the ancient oaks have gone and those that remain are concentrated in a small area of ancient parkland near West Lodge, believed to be the site of the medieval hamlet from which the park's name derives. This village was abandoned by the time the first country house was built and the village fields had been turned to pasture for sheep. The largest oak (all are maidens), with a girth of 8.04 m, stands in the northern part of the park. It is likely that the three largest oaks in the park (7.22–8.04 m) were already mature trees at the beginning of the sixteenth century. Althorp is one of the earliest locations of oak planting in England, with some plantings dating from the second half of the sixteenth century.

Bolton Abbey Park is in Wharfedale, North Yorkshire. The Augustinian priory of Bolton ('Bolton Abbey') was dissolved in 1540 and thereafter fell into ruin. The gate house was extended to become a residence by the Cavendish family, the Dukes of Devonshire, in the late seventeenth century. The High and Low Parks that make up Bolton Park are situated to the north across the River Wharfe, and can be reached by a wooden bridge. The ancient oaks are mostly in the Low Park between Bolton Park Farm and Laund House, with one 7.10 m oak beyond a house called High Laund; the oaks range between 6.30 m and 7.30 m in girth. It is therefore likely that whoever owned the estate between 1540 and the latter part of the seventeenth century established a deer park here, unless the monks had one.

Croft Castle in Herefordshire is not mentioned as a medieval deer park in Cantor (1983), although a castle or fortified manor house has existed here since the late fourteenth century. The actual park at present is limited in extent, with landscaping near the house and remains of seventeenth-century sweet chestnut avenues. Most of the estate is either farmland or coniferised forest plantations, the latter leased to the Forestry Commission. On the escarpment summit in the north is an Iron Age hill fort, Croft Ambrey, which has open parkland and rough grassland extending on the slope to the west. There are many ancient oaks scattered throughout what may have been a medieval common and later a Tudor deer park, whose date of imparking remains problematic (Whitehead, 2001). Much of the present park shows ridge and furrow from a medieval arable field system. A deserted medieval village may have been situated to the southeast of the little church, with its commons on higher ground beyond the fields towards and including Croft Ambrey. Many of the ancient and veteran oaks in the conifer plantations are dead or dying; a rescue operation involving felling of the conifers began in 2014. The largest oak is recorded and measured by the ATI with a 10.97 m circumference; it is known as the Croft Castle Oak or the Quarry Oak (perhaps because it stands in the quarry where stone for the first castle was taken?). Depending on locality, the ancient and veteran oaks here are assigned to common (COM) or to Tudor deer park (TDP) in the database of oaks in England with ≥6.00 m girth.

Devereux Park in Herefordshire was first recorded in 1575, and a century later, it was subsumed into the Stoke Edith estate. There is no country house in or near the park, which was connected to Stoke Edith by a carriage drive from Devereux Pool designed by Humphrey Repton around 1800. Park Farm sits in an area of open fields exploited

mostly as grassland surrounded by former coppice woods, now partly coniferised. This division has remained the same since the nineteenth century. Five ancient and veteran oaks with girths of between 6.35 m and 8.40 m stand in the open fields or their margins, and some smaller veteran oaks were recorded in some of the woods.

Dudmaston Hall in Shropshire has a landscaped park with a string of artificial lakes around a seventeenth-century mansion. Older maps indicate the presence of a moated manor house on the same site. It may have been Francis Wolryche (1563–1614) who established a park here, using wealth obtained from the wool trade. The early maps do not, however, show a deer park but only the house. Lodge Farm and Park Farm may both indicate the former existence and perimeter of an early park. There are eight veteran oaks, with girths from 6.21 m to 7.38 m, recorded in the park, which is now owned and managed by the National Trust.

Felbrigg Park in Norfolk is a large park surrounding a seventeenth-century Jacobean mansion. Earlier houses had existed on the site since the late Middle Ages; one sixteenth-century resident, Thomas Wyndham, was a councillor to Henry VIII. By his time, a small deer park existed and in 1674 this covered 60 ha. Maps prior to Faden's map of Norfolk (1797) do not appear to show any parks at all in

Figure 4-25. The base of the Quarry Oak or Croft Castle Oak at Croft Castle, Herefordshire

Norfolk, despite the fact that these existed. The nineteenth-century OS map shows a park with numerous parkland trees around the house and an extensive wood, Great Wood, in the northern half. According to Rackham (1986, p. 141), this wood was once part of Aylmerton Common. The present park covers 440 ha and was extensively landscaped in the mid-eighteenth century. Numerous veteran trees have been recorded here but many are not old (recording enthusiasm) and only six oaks are ≥6.00 m in girth; only one of these is recorded as ancient, but this may also be biased towards recording trees as 'veteran'. The park and house are owned by the National Trust.

Grimsthorpe Park in Lincolnshire is a large landscaped park located between Little Bytham and Grimsthorpe. Most of the ancient and veteran oaks in Grimsthorpe Park are concentrated in a preserved section of unimproved parkland on either side of Creeton Riding. The largest oak in the park (with 9.32 m girth) is located in a deep ditch just to the north of Steel's Riding. Grimsthorpe Park was granted by Henry VIII to the Eleventh Baron Willoughby de Eresby in 1516; at least some of the land had belonged to an abbey of which only some earthworks and ponds now remain. Two adjacent deer parks have been mentioned, which partly included pasture woodland; they are therefore considered to date from the Tudor period. The section near Creeton Riding especially is a good example of ancient pasture woodland and forms part of an SSSI, the other part being similar but younger oak parkland north and south of Steel's Riding. A detailed description is given in Chapter 9.

The large deer park at **Helmingham Hall** in Suffolk is not listed as a medieval park in Cantor (1983). The Tudor house of around 1510 is surrounded by a moat and there stood an earlier house on this site which may have had a park. There are many native oaks, some of which are ancient pollards; the largest of these is now dead (visited October 2013). It thus seems likely that these oaks once stood in an early Tudor deer park. I also noticed several ancient pollard field maples in this park. Helmingham Park could be counted as a true ancient deer park as it has not been landscaped, save for the erection of a mound with an obelisk and the creation of an avenue to the house; neither has the grassland been improved as it has in parks such as Ickworth Park, also in Suffolk. Oaks have been planted here for several centuries, especially in a central area known as Oak Grove, but the oldest trees, including the field maples, should be considered spontaneously grown. However, there are not many ancient oaks in the deer park, and the size of the park and the central position of the mansion also point to a later creation date than 1485.

Holme Lacy Park in Herefordshire is marked as a deer park to the south of Holme Lacy House on the nineteenth-century OS map. At this time, much of this park was parkland, but later conifer plantations were created that are now being felled to restore the parkland. Other sections have been transformed to arable farm fields. The Scudamore family had a house here from the fourteenth century onwards; the present large mansion (now a hotel) dates from the seventeenth century. When exactly the deer park was created is unclear, but it may have been in the sixteenth century as it is present on Speed's map of Herefordshire published in 1610. There are two large oaks, one of 9.75 m girth and another, known as The Monarch and clearly a maiden tree, with 9.25 m girth. In February 2015, Brian Jones recorded twelve further ancient and veteran living oaks in the former deer park, two with girths >8.00 m and six with girths >7.00 m. There are also several dead standing oaks or stumps.

Killerton Park in Devon is on a hill which rises 100 m above the River Culm (28–128 m a.s.l.). Killerton House is on the south side of the hill and the river on the steeper north side. These slopes are shown as woodland on the nineteenth-century OS map and are still wooded. The ancient and veteran trees do not occur on the riverside slopes in this woodland, which indicates that it was probably ancient coppice wood, although now partly coniferised. The present house replaced an Elizabethan mansion, which probably had a park with deer. Landscaping in the eighteenth and nineteenth centuries was largely limited to planting trees as the hill would not allow water features. The parkland therefore has many non-native trees, among which are conifers, but retains a number of ancient and veteran oaks, twelve of which have girths of between 6.00 m and 7.10 m. It is owned by the National Trust.

Near **Leighton** above the River Burn in North Yorkshire is a small fragment of ancient pasture woodland located just north of Leighton Reservoir. It is pictured in the book by Muir (2005: 19), who stated that the land was once owned by Fountains Abbey. On the nineteenth-century OS map, the land on which it stands is called Leighton Park and just south of it is the "Site of Leighton Hall", now partly drowned in the reservoir that also covers much of Leighton Park. This park is presumably the same as a park around Healey called New Park on Speed's early seventeenth-century map of the North and East Ridings of Yorkshire (Muir, 2005: 120). The c. 20 recorded oaks of Leighton are nearly all small veterans ranging from 3.00 m to 5.51 m in girth and growing at c. 200 m a.s.l.; they are probably only a few centuries old. There appear to be both maidens and pollards in a field that is reputedly still grazed

Figure 4-26. A dead ancient oak in Helmingham Deer Park, Suffolk

by cattle. A 6.00 m maiden oak stands near the present farm buildings in farmed grassland.

Longleat Park in Wiltshire surrounds the large Elizabethan house Longleat and was landscaped by Lancelot ('Capability') Brown. The house, owned by the Marquess of Bath, was built in the sixteenth century on the site of an Augustinian priory. The house with its park is now a major visitor attraction. The park is shown on a map of Somerset published in 1575. There are a number of ancient and veteran oaks, among which is the 'Longleat Oak', a maiden tree with a girth of 8.11 m. In the safari park, several of these trees are dead or dying, possibly as the result of damage inflicted by the exotic animals; the remaining live oaks are now better protected. It is possible that there was already a deer park at Longleat in medieval times, but it is not recorded in Cantor (1983). The priory/park was also close to Selwood Forest, a medieval hunting forest.

Oakly Park near Bromfield in Shropshire is situated in a bend of the river Teme. A moat on the other side of the river in Bromfield indicates a medieval fortified manor house, but Oakly was first mentioned as a 'new park' belonging to a separate manor in 1490. It had been part of a chase belonging to Ludlow Castle before this imparkment. In the nineteenth century, Oakly was still a large park, but much of it has now been converted to farmland, leaving only some early nineteenth-century features such as avenues and coppice woods or coverts. Only a section near the house, separately called Oakly Park on the OS map, remains as parkland, and here 19 oaks with girths ≥6.00 m (seven ancient or veteran oaks are ≥7.00 m) are recorded (two appear dead). About ten ancient or veteran oaks remain in a few corners of the former (larger) park. Conversion to agriculture in the twentieth century must have destroyed many oaks (see Chapter 11).

Packington Park in Warwickshire is a large deer park surrounding the sixteenth-century Old Hall, with a late seventeenth-century second mansion, Packington Hall, in its southwestern corner. The present park was laid out by Lancelot ('Capability') Brown with wide open spaces and a series of artificial lakes. Most of the old parkland trees probably date from this eighteenth-century landscaping, but the ATI records include 35 ancient or veteran oaks with girths of between 6.00 m and 8.77 m. These oaks could date from the sixteenth century or perhaps earlier. On the Packington Estates website (www.packingtonestate. co.uk/), it is claimed that the deer park is a remnant of the ancient Forest of Arden. This was a privately-owned forest

in late Anglo-Saxon time and had virtually disappeared by the thirteenth century (Rackham, 2003b), so none of the present oaks in the deer park would have any connection with this forest. No medieval deer park in this location is mentioned in Cantor (1983), but since most of the large oaks are concentrated near the Old Hall, a park may have been a feature of this manor house. Unfortunately, the

area with the greatest density of ancient and veteran oaks (nearly all maidens) has been developed into an 18-hole golf course. Other parts of the deer park have fewer ancient oaks but are largely unimproved grassland, and apart from a few plantations and a large 'pool', little landscaping has been carried out in this section of the park. Exceptionally, the deer have access to the golf course, too.

Parham Park in West Sussex has an Elizabethan house built in 1577 after the estate was confiscated from the Monastery of Westminster by Henry VIII in 1540. The 120-ha deer park surrounding the house may have been imparked in the sixteenth century, but the earliest mention of a herd of fallow deer is from 1628. It is not mentioned in the gazetteer of Cantor (1983) and, upon this evidence, is probably an ancient but not a medieval deer park. Parham Park has a good number of ancient oaks, the largest of which is 7.69 m in circumference; most are in the 5.00–6.00 m range. This site is an SSSI on account (in part) of the ancient oaks and associated biodiversity. It is described in more detail in Chapter 9.

Pull Court Park near Bushley Green in Worcestershire is in the northern part of the parish of Bushley. The mansion there is now Bredon School. This area was once part of Malvern Chase, in which Bushley Park was imparked in 1296 (Cantor, 1983). Pull Court was a separate manor in the Middle Ages, held by the Abbots of Tewkesbury until the dissolution of the monasteries. It then passed through a succession of owners, but there was no mention of the manor having a park before the seventeenth century. The shape and nature of the park on the nineteenth-century OS map as well as the current presence of 10 ancient and veteran oaks with girths of between 6.00 m and 7.20 m, make it likely that there was a deer park here, possibly created after the land moved from the monastery into private hands.

Ripley Park in North Yorkshire is a Tudor deer park probably created by Sir William Ingleby, who built the Tudor tower of Ripley Castle on the site of a medieval manor house in 1555. A small park is indicated at Ripley on John Speed's map of Yorkshire dated 1610. However, the park has, besides several smaller oaks, three great pollard oaks (of 8.60 m, 8.67 m and 9.08 m in girth) which could date back to the Forest of Knaresborough, which in the fourteenth century extended to Ripley from Knaresborough. The park still has deer and although small (some parts are now a farm), it retains the essential landscape features of an ancient deer park. Through the park runs a partially hollow track on the banks of which stand several ancient oaks, none of these reach 6.00 m in girth. There is also evidence of ancient field boundaries inside the deer park. The park also has a few ancient sweet chestnuts and large sycamores planted later. The grassland is mostly unimproved, with patches of nettles and rushes but almost no bracken. A small stream runs through the park and a lake has been created between the deer park and the castle. The park is used for events, but the ecological impact of this use seems slight.

Spetchley Park in Worcestershire has belonged to the Berkeley Family since 1605; prior to this, it belonged to the Sheldon and Littleton families. The Berkeleys, from a wealthy merchant background, could have started a deer park here, but none is indicated on Speed's map of the county, which is based on Christopher Saxton's map of 1579. This could mean that previous owners had no deer

Figure 4-27. Ancient maiden oak in Packington Park, Warwickshire

park here but that there was one by the early seventeenth century. The park is certainly old, and the deer park in the southern part indicated on the nineteenth-century OS map is the site of nine of ten oaks with girths between 6.00 m and 6.90 m. Three are pollards, the others are maidens. In addition, there are a larger number of smaller veteran oaks. A hedge and woodbank with a rounded shape along parts of the perimeter are also indicative of an old park. Marks of ridge and furrow in the deer park point at medieval ploughland, so the park was established partly on farmland. This agrees with the sizes and possible ages of the largest oaks, some of which (the pollards?) were probably present at imparkment and others may have established after 1600.

Stoneleigh Deer Park in Warwickshire was imparked in 1616, shortly after the close of the Tudor period, and is now mostly converted into an 18-hole golf course with the River Avon running through it. There are many ancient and veteran oaks and a wood around the site of the deer keeper's lodge. Another part of the park ('Abbey Park') is in use as a business park and has modern buildings to accommodate this activity. Several oaks exceed girths of 7.00 m; the largest oak with 8.60 m girth is in a corner of the golf course and another oak with 8.42 m girth is at an old gate house to the park. A 9.24 m-girth maiden oak near Stoneleigh Abbey was not in the deer park of 1616. This park was still a true deer park in the nineteenth century, but the nature of an ancient deer park has been destroyed save for the presence of several ancient oaks. As in Packington Park, the ancient oaks are attributed to the Forest of Arden on several historic websites, but this forest had long ceased to exist at the beginning of the seventeenth century. Perhaps some of the land south of the Avon, known as Cloud Field, was used as a wooded common by the villagers of Stareton prior to imparkment. The large oak near Stoneleigh Abbey has been assigned to a manor in the database.

Studley Park in North Yorkshire is situated next to Fountains Abbey, one of the largest (ruined) Cistercian monasteries in Europe, which was founded in 1132. In 1539, after the Dissolution, the abbey and 200 ha of land were sold by the Crown to a merchant, Sir Richard Gresham. The deer park, still with herds of red, fallow and sika deer (*Cervus nippon*), was imparked in the sixteenth century when a Tudor-style house, known as Studley Royal House, was built in the northwest corner of the park. No large house exists in the park at present. A lake and a water garden were created between the deer park and the abbey

ruins. Other landscaping and building include an avenue of limes and a nineteenth-century memorial church at the park's western end. Despite these additions, parts of the park retain the unimproved parkland setting with ancient and veteran trees: many sweet chestnuts, but also eight oaks with girths of between 6.00 m and 8.30 m and several

smaller ones. The estate was designated a World Heritage Site, mainly because of the abbey ruins and their historic landscape setting.

The Vyne in Hampshire is a Tudor mansion with gardens and a park. It was built for William Sandys, Lord Chamberlain to Henry VIII, who was entertained three times here. Originally the park was larger than at present, by the eighteenth century it had already been reduced to an area west of the Bramley-Sherborne road adjacent to Morgaston

Figure 4-28. Pollard oak of 9.08 m girth in Ripley Park, North Yorkshire

Wood. According to the 1998 edition of the *National Trust Guide to The Vyne* by Maurice Howard, a medieval park existed here before Tudor times, but no records of this exist and Cantor (1983) does not mention it. (The moated Beaurepaire House, situated 1 km north of The Vyne, had a deer park from 1369 onwards.) The earlier park may have been around Vyne Lodge Farm, where a small moat exists, but this was already farmland by the early seventeenth century. The deer from the later Tudor deer park of The Vyne are all but gone, but several ancient and veteran oaks remain, one of which is severely leaning and now stands propped up in the garden; it has a girth of 7.58 m.

Whiddon Park in Devon is a small elongated park with many oaks and some beeches and ashes; seven of the eight oaks that have a circumference of 6.00 m to 6.90 m are recorded as ancient. This was most probably established as a deer park in Tudor time but not earlier. On the nineteenth-century OS map the park is much more open than it is now, as neglect and lack of grazing have increased the tree cover. In the northeast and north, the park borders Whiddon Wood and the River Teign.

The medieval and Tudor parks with ≥5 ancient oaks described in this chapter do not comprise a complete survey of England. Although the survey is nearly comprehensive, 'new' parks with ancient oaks are being recorded as I write and will be found in the future. In October and November 2016, after I closed the database used in this book, in Suffolk two more parks with several ancient oaks were visited by me and one, Henham Park (a Tudor deer park) could, on further analysis, become a 'most important site for ancient oaks' as described in Chapter 9. Other parks may turn out to have more ancient oaks than currently recorded and thus merit inclusion in this chapter. I should be grateful if owners, as has happened in Suffolk, would

be willing to share such data on the ancient oaks as remain unrecorded in the ATI.

This survey concludes my investigations into the relationship between the ancient and veteran oaks and the ancient deer parks, medieval and Tudor, of England. It spans the period from 1066 (or a mention in Domesday Book (1086), which in a few cases may date back to late Anglo-Saxon time) to 1603, or roughly 550 years. The surveys in this chapter demonstrate the importance of this connection, an observation that is backed up by statistical data in Chapter 5. Few if any of the oaks now alive can possibly date from the beginning of this period (see Chapter 2), but many were there when it ended about 415 years ago. Throughout the surveyed period, as before and since, oak trees have germinated, grown and died in a continuing cycle; it is important here to demonstrate the likelihood that this continuum has existed on a particular site for longer than the oldest oaks now present. This is why history, and the evidence it provides relating to the landscape, have been emphasised in the accounts of parks given in this chapter. Because oaks were mostly not planted or seeded before *c.* 1600, I hope to have demonstrated: 1) that for Tudor parks this continuum extended back into the Middle Ages; and 2) that for medieval parks the continuum dates back into Anglo-Saxon times and possibly beyond to prehistory. In Chapter 10, we shall see how important this continuum is for biodiversity, and how biodiversity itself is evidence for this continuum. But before we can investigate biodiversity, we must first look at other features of historical landscapes which have affected the establishment and survival of ancient and veteran oaks. This will be done in the next chapter, which will end with some conclusions about the historical landscape context of ancient oaks derived from evidence given in both chapters.

5

Distribution of ancient and veteran oaks in England explained: Royal Forests, chases and other historical connections

The Royal Forests

Young (1979) presented a list of 71 named Royal Forests that existed in 1327–36, together with a map of their location and approximate shape and size; this was based on an inventory by Nellie Neilson published in 1940 (see reference in Young, 1979). By the fourteenth century, the number and extent of the Royal Forests were already past their peak, which occurred during the reign of Henry II towards the end of the twelfth century. Rackham (1976) gave a figure of 90 forests in which the king had exclusive right to hunt the "beasts of the forest" (mainly deer). These forests may have covered up to a quarter of all England, but much of this land did not bear trees. Unlike the situation under the Frankish kings on the continent, where *forestis* implied uncultivated land without a clear owner (Buis, 1985; Vera, 2000 and references therein), which by default belonged to the king, in England, a Royal Forest often extended onto land that had an owner. This was because Forest Law was imposed on Anglo-Saxon ownership and often included settlements with cultivated land (Rackham, 1980). The most powerful magnates, in imitation of the king and with his consent, also established

Figure 5-1. Ancient and veteran oaks in Bear's Rails, Windsor Great Park, Berkshire

forests; these were usually but inconsistently known as chases (Cantor, 1982). A comprehensive inventory of all known medieval forests and chases has been compiled by John Langton and Graham Jones at St. Johns College, University of Oxford (info.sjc.ox.ac.uk/forests/Index.html). The difference between 'legal forest' and 'physical forest', as pointed out by Rackham (1976), means that the latter was usually much smaller than the former.

A continuous process of disafforestation and 'assarting' throughout the Middle Ages and especially later has greatly reduced the areas of nearly all physical forests and destroyed many altogether. Rackham (1976) presented an example for the counties of Essex and Middlesex, where Waltham (now Epping) and Hatfield Forests are reduced but extant and Hainault, Writtle and Kingswood Forests and Enfield Chase have almost disappeared. By the thirteenth century, these Royal Forests were themselves remnants of the legal Forest of Essex, founded around 1100 and encompassing most of the county. Another example is Cornwall and Devon; both counties were declared to be Royal Forests, almost in their entirety, by King John in the late twelfth century, but widespread resentment by barons and commoners alike caused disafforestation. In 1204, the Forest was reduced back to the original eleventh-century Forests of Dartmoor and Exmoor, and there was a further disafforestation of all land throughout England not owned by the crown in 1215 (Marren, 1990; www.legendarydartmoor.co.uk/fore_dart.htm).

The distinction between common and Royal Forest could be diffuse; commoners would have had access to and use of the land for subsistence needs before it was afforested, and often continued to use it after afforestation. Nevertheless, Forest Law did impose restrictions, apart from a hunting ban, which included limited or even no access to the trees; this could, and sometimes did, lead to conflicts. Usually, a *modus operandi* was agreed that allowed commoners to make limited use of timber and wood, often from licensed coppices, and serious trespass was fined. Grazing of animals, pannage and collecting of firewood had to continue because the English forests were never far from habitations, which, together with their ploughlands, were expanding from the eleventh to the fourteenth centuries.

Over the centuries, the incompatibility of these uses by a growing population with what, in effect, amounted to nature conservation has led to forest degradation and to reduction of most of the Royal Forests. The number of squatters' 'assarts', that is, plots of forest land turned to arable use, increased while royal control and resolve wavered (Muir, 2005). In forests near densely populated

Figure 5-2. Ancient pollard oak at West Lodge (Forest of Bere), Hampshire

London especially, Forest Law was difficult to enforce. The county of Norfolk, one of the most densely populated in the Middle Ages (Whittock, 2009), had no Royal Forests at all. The period of Elizabethan rule to the end of the Civil War (1558–1651) was particularly destructive, not only because of the growing needs of the common people but also because the demands for ship building timber and wood for iron smelting grew exponentially. A few Royal Forests remain today, although with much reduced area, most notably the New Forest and the Forest of Dean, although

the latter has been much altered by activities such as mining and plantation forestry.

Royal Forests usually included wooded areas, but others such as Exmoor in Somerset and Devon, Mendip in Somerset, Dartmoor Forest in Devon, and Pickering in North Yorkshire did not. Wooded Royal Forests, as defined by Rackham, were those that had more woodland than the surrounding country, but even these forests usually had large treeless areas ('wastes') such as heaths. "To the medieval a forest was a place of deer, not a place of trees" (Rackham, 1986). Evidence of the extent of wooded forest is difficult to obtain, and the situation changed constantly. It has been estimated that between 1250 and 1334, the area

of Royal Forests shrank by a third (Cantor, 1982). Often, the enclosed coppice woods that existed within a matrix of open pasture, wooded or not, survived as ancient woods in a farmland setting, either arable or pasture. From an agricultural point of view, the coppices were still useful while the rest of the forest was not and was made useful. It is therefore difficult to determine whether ancient oaks that still exist were once part of a wooded forest landscape. Brief reviews of (remnants of) more or less wooded Royal Forests, several of which I have visited, are given below, with descriptions relevant to the presence or absence of ancient and veteran oaks. The list given in Table 5-1 is not a complete list of all Royal Forests that existed, but provides a good sample and includes those that are likely to be relevant to existing and/or 'historical' ancient oaks.

Survey of English Royal Forests

The **Forest of Aconbury** near King's Thorn in the south of Herefordshire was an ancient hunting forest that had Aconbury Hill with its Iron Age hill fort at its centre. Within this forest, Athelstans Wood appears to refer to the Anglo-Saxon king Aethelstan (924–939). King's Pitts Wood seems to be another reference to the forest's royal connections. Wallbrook Wood, another ancient wood, is on the hill slopes to the east of Aconbury Hill. The woods are now mostly plantations, some with conifers, and there are no ancient or veteran oaks.

The **Royal Forest of Alice Holt** south of Farnham in Hampshire was never large and now comprises 851 ha, mainly planted with Corsican pine (*Pinus nigra* ssp. *laricio*). Around 140 ha of pedunculate oak, dating from 1815, are a remainder of planted oak forest felled during the two World Wars. There are some other broadleaved trees but no ancient or veteran oaks apart from a now dead trunk near Alice Holt Lodge. In 1826 the 'Holt Forest Oak' with >12.00 m girth was recorded here (Strutt, 1826); and in 1837, Loudon mentioned the 'Grindstone Oak' with 10.00 m girth. There is now much emphasis on family recreation in this forest, and the Forestry Commission operates a research centre in Alice Holt Lodge. Just south of Alice Holt is Woolmer Forest; this was already a mostly treeless heath in the Middle Ages because it is situated on poor sands of the Folkestone Beds. Gilbert White in his book *The Natural History and Antiquities of Selborne*, published in 1788, described Woolmer Forest as "without having one standing tree in the whole extent." (To White, the word 'forest' implied trees, as it does today, but in the Middle Ages it did not.) The heath was broken up by enclosures and planted, in the nineteenth century with oaks and later with conifers; most of these trees have been felled and the

Figure 5-3. Nath's Oak near Minety, Wiltshire is a lone relict of the Royal Forest of Braydon

Figure 5-4. The Oakley or Mottisfont Oak in Hampshire may be associated with Buckholt Forest

land restored to heath, but this is now used for military exercises and consequently much damaged despite being an SSSI and (partly) a Special Area of Conservation (SAC).

Ashdown Forest was a medieval hunting forest situated on the highest sandy ridge of the Weald in East Sussex. Created soon after the Norman conquest of England, by 1283 it had been imparked with a pale enclosing some 5,200 ha, but 34 gates and hatches in the pale allowed commoners and their livestock inside to exercise their rights of use. For a time, Ashdown was effectively a very large royal deer park used by the English kings up to Henry VIII, but with common access — presumably the cattle and other domestic animals were moved out when the king wished to come and hunt deer. Although Rackham (2003b) listed it as 'wood pasture', Ashdown must have degraded to heath fairly soon after its imparking considering its geology and topography. On a map of Ashdown Forest by John Kelton dated 1747 (Map Room, Royal Geographical Society, London), 'commons' and 'copses' or 'woods' are shown but no symbols of trees for pasture woodland are evident. There are a few large beech pollards in the woods that are reminiscent of those in pasture woodland, but most of the present woods are dominated by planted Scots pines and there are no ancient or veteran oaks.

The **Forest of Bere** (by Porchester) was situated near the coast east of Southampton in Hampshire, stretching from the River Test in the west to the Sussex border. It was divided into a West Walk and an East Walk, separated by Waltham Chase and Hambledon Chase, which belonged to the Bishop of Winchester. Agriculture and later urbanisation greatly reduced and fragmented this forest. The rural landscape at present is a mosaic of farmland, downland, heath and woods, the latter mostly planted with conifers or a mix of broadleaved trees and conifers. A few ancient coppice woods survive, most notably those with small-leaved lime near Wickham. Several of the woods are within the South Downs National Park and one of these, in the Meon Valley near Wickham now known as West Walk, has two ancient pollard oaks. This wood, now administered by the Forestry Commission, is the largest fragment of the physical medieval forest that remains. Much of it has been coniferised but it retains some native broadleaved trees, especially in the North Boarhunt section which has an early nineteenth-century plantation of native oaks that has now become a mature oak-beech wood. The two pollard oaks, one with 9.55 m circumference, are on private property at West Lodge, once the location of the Verderer's Court (Preston & Wallis, 2006). They are the only two ancient oaks with girths ≥6.00 m that remain in the large area once occupied by the Forest of Bere (by Porchester).

The **Royal Forest of Bernwood** was situated northeast of Oxford around the castle and village of Brill in Oxfordshire and further southeast into Buckinghamshire, from at least Norman times to the early thirteenth century. This area had been used by royals for hunting well before the Conquest, but the Forest was gradually reduced after 1300 (the castle was abandoned in 1327) and ceased to exist in the early seventeenth century. The wood that is presently known as Bernwood Forest (*c.* 100 ha) is partly ancient coppice wood but mostly planted with conifers by the Forestry Commission, under whose management it now resides. A few areas are more open terrain, resulting from the felling of conifers, with invading trees such as birch. There are no ancient or veteran oak trees that have reached 6.00 m in girth, despite an assertion that "some fine old oaks survive around Brill" (Whitlock, 1979). The nearest veteran oak, with a girth of 5.75 m, was recorded at Wootton Underwood. A few other woods in the area are of similar nature and none of these resemble medieval pasture woodland or unmanaged woodland, even though several are now managed as nature reserves by the Berks, Bucks & Oxon Wildlife Trust. The present area known as Bernwood Forest is an SSSI.

The **Royal Forest of Braydon** (or Braden) in Wiltshire yielded mainly fallow deer to the king (Rackham, 2003b), which makes it sound almost like a deer park. This forest once extended across northern Wiltshire from near Malmesbury almost to Berkshire, and a survey of 1222 ranked it second in all England for its number of timber trees, among which must have been many oaks. Little of it remains except for forest plantations (partly conifers)

in several woods and names on the map, such as Braydon Wood, Braydon Side, Braydon Green Farm and Gospel Oak Farm, which hints at the existence of an ancient oak. The woods are all surrounded by farmland and there is one ancient oak of 9.20 m girth, Nath's Oak, standing solitarily in a field at Gibb's Farm in the parish of Minety. This parish was legally excluded from Braydon Forest as late as 1573, so this oak may be associated with the Royal Forest.

Buckholt Forest (Bocolt in Domesday Book (1086)) was situated between the River Test and the Hampshire–Wiltshire border and between Broughton in the north and the River Dun in the south. On Joan Blaeu's map of 1645, it is reduced to an area west of Bossington and the southern parts have disappeared. The area is now a mosaic of fields and woods, several labelled as 'copses' indicating coppice woods, with Spearywell Wood the largest complex of plantations. Nearby is Mottisfont Abbey, which was an Augustinian priory from 1201 until the dissolution of the monasteries. Just to the north, along the smaller west arm of the River Test, at Oakley Farm stands one of the great ancient oaks of England, the Oakley or Mottisfont Oak, which has a girth of 10.74 m. The course of this river arm seems to have been diverted to flow past the priory and the oak could have stood on the eastern edge of the forest. Two smaller ancient or veteran oaks are nearby.

The **Royal Forest of Chute** was situated just south of the larger Savernake Forest in Wiltshire and Hampshire. Like Savernake, it was in an area of heavy clay-with-flint, which in medieval times was difficult to plough, and consequently remained mostly wooded. Harewood Forest near Andover, which is shown on some ancient maps, was adjacent to or

THE "NEWLAND" OAK, 41 Feet in Girth.
(One of the largest in the Kingdom.)   W. PORTER, PHOTO.

Figure 5-5. The Newland Oak in the Forest of Dean, pictured here in a postcard from ca 1900, no longer exists

a part of Chute. The remaining woods there are 'copses', coppice woods that were excluded from the forest and enclosed, and are now partly coniferised. Little if anything remains of the Forest of Chute. It lives on in the name of a parish and village, which has some ancient woods; and near Conholt Park are hanger woods that may date from medieval times. Coldridge Wood is part of a larger wooded area managed by the Forestry Commission; this wood is not coniferised and has an 'oak pollard' toponym in one corner on the Ordnance Survey Map. The adjacent Collingbourne Wood and other parts of the Forestry Commission wood are copses. There are no ancient or veteran oak trees recorded in the area.

The **Royal Forest of Clarendon** in Wiltshire included Melchet Wood (also called Forest or Park), which had been imparked prior to a perambulation taken in 1278–79 but was then part of this Royal Forest (www.british-history. ac.uk) and Buckholt. A smaller deer park was created later; in the nineteenth century, there was still a park but much of this is now farm fields, although the park boundary is still evident. The Forest of Clarendon extended to the north of the New Forest as far as the wooded areas east of Old Sarum and later Salisbury. Of these, Bentley Wood is the largest remaining, but it consists mostly of conifer plantations. Another area that once belonged to the Forest of Clarendon is Groveley Wood near Great Wishford (see below under Groveley Forest). This is now mainly beech forest with inserted conifer plantations on a hill between the valleys of the rivers Wylye and Nadder. The building of Salisbury and its cathedral in the thirteenth century apparently resulted in the felling of most of the oak timber trees left in the Royal Forest (Whitlock, 1979). In the seventeenth century, the felling of oaks for the Navy further depleted the resource. Hatton Gardner visited Clarendon Park in 1976 and recorded a 'hollow' oak with 8.30 m girth, the (presumably ancient) 'King John Oak' of 6.30 m girth, and two other oaks of 6.15 m and 5.55 m girth, with no comments on their status (pers. comm. Owen Johnson, by email 7th August 2015).

The **Forest of Dean** in Gloucestershire is one of the best known Royal Forests in England. As a wooded forest, it probably dates from Anglo-Saxon times and was perhaps the closest to 'wildwood' remaining in England at the time of the Norman Conquest (Rackham, 1976). It may have been the last refugium for native wild boar (*Sus scrofa*), the other 'beast of the forest' besides deer, until exterminated by order of Henry III around 1260; this animal has been (illegally) reintroduced into The Forest of Dean recently. Until recently, many sheep were grazed and the reintroduction of cattle and ponies is being considered.

Extensive mining and since the late seventeenth century plantation forestry, first with oaks for the Navy but later with conifers, have greatly altered the character of this forest and have mostly destroyed the native vegetation (Rackham, 1976). Plantations include native oaks of both species, but mostly pedunculate oak, sweet chestnut and beech; the predominant conifers are Douglas fir (*Pseudotsuga menziesii*), Norway spruce (*Picea abies*) and Japanese larch (*Larix kaempferi*). This has left so few ancient or veteran oaks that the only oak recorded as an ancient tree in Dean is the Verderer's Oak, a 7.20 m girth maiden oak near Speech House, mapped as tree No. 25810 on the Interactive Map of the ATI. That other ancient oaks existed is attested by images and descriptions of the great Newland Oak (Figure 5-5), a massive pollard which was still alive in the 1950s and that stood on a farm which was once part of the Forest of Dean (Miles, 2013).

Delamere Forest (with adjacent Mondrem Forest) was situated in the Forest of Mara on an elevated sandy area of the Cheshire Plain; it must have been turned, by cutting and grazing, into heath and scrub at an early stage of its history. This is or was the forest of meres, shallow watery remains of melted ice blocks from the retreating glaciers of the last Ice Age. The deciduous and mixed woods here are all secondary, and the wooded areas are dominated by conifer plantations. In the central part of Delamere Forest is Blackmere Moss, which was (re)flooded after clear-felling in 1998, having been drained for (unsuccessful) forestry in around 1815. North of it is an area known as Kingswood on the nineteenth-century OS map; it was mostly turned to farmland. The great oak of Marton village (Figure 2-5) may once have been at the edge of this forest. The timber-framed church of Marton dates from 1343, so a settlement existed here at that time, but the oak, one of the greatest in the land, is likely to be substantially older than the church.

The Forest of **Duffield Frith** in Derbyshire was confiscated by Henry III in 1266 and was subsequently used as a hunting forest by the Plantagenet kings (Wiltshire *et al.*, 2005). At this time, Duffield Frith was supposedly a wooded forest (Muir, 2005) with large numbers of oaks, but a survey conducted in 1560 found them much depleted, although still counting 59,412 'large oaks' and 19,736 'dottard oaks', the latter "only fit for fuel" (Dowsett, 1942; Whitlock, 1979). These were presumably ancient oaks. In 1587, another survey revealed only 2,764 large oaks and 3,032 small ones; figures that throw doubt on the veracity of those reported 27 years earlier. Charles I tried to maintain his royal hunting rights over Duffield Frith Forest, but in 1643 during the Civil War, the commoners took possession of what remained of the ancient forest and turned it into a common.

Parliamentary enclosures put an end to this in 1786, and the entire forest was subsequently converted to farmland and some small coppice woods, now often planted with conifers. In the southwest corner of the area is the medieval deer park of Kedleston, created around 1400 (see Chapter 4). This is the only area that still has ancient and veteran oaks with girths ≥6.00 m; whatever many ancient oaks there were in Duffield Frith Forest have now all gone.

In the Middle Ages, **Epping Forest** was part of the much larger legal area of Waltham Forest. It has maintained the

Figure 5-6. Wyndham's Oak near Silton, Dorset once stood in the Forest of Gillingham

largely elongated shape and size seen on a survey map of 1772–74, which was reproduced in Rackham (1976). It was situated entirely in Essex but is now in Greater London and Essex, with an area of 2,476 ha. Since 1878, the forest has been administered by the City of London Corporation as Conservators of Epping Forest. In the Middle Ages,

Figure 5-7. A young oak has been planted at the site of the Doodle Oak in Hatfield Forest, Essex

it was a landscape of pasture woodland (*silva pastilis*), glades and heath; the trees were mostly pollards, with beech on gravelly hill tops, hornbeam on lower slopes, and oak in between (Rackham, 1976). Commoners' rights to wood and pasture maintained this medieval landscape until the early nineteenth century and the City of London Corporation has since prevented its total destruction. Nevertheless, enclosures made before the twentieth century and management for nearly two centuries have altered the woodland. Tall beech trees (often lapsed pollards) now take precedence over oak and hornbeam pollards almost everywhere, and heath has become overgrown with birch and pedunculate oak. Only recently has a better understanding led to improved management and attempts to restore the ancient forest, but the full scale at which this is necessary defies the means of the Conservators. The uses that created the forest are no longer of economic importance. The result, almost everywhere, is tall beech forest in which the oak particularly is in decline. Nonetheless, with assistance from the Heritage Lottery Fund, the Conservators have carried out crown restoration work on 1,200 lapsed pollards in the past decade, of which

over 600 are ancient and veteran oaks. In addition, with financial support from Natural England's Stewardship Scheme, the Conservators are restoring 390 hectares of pasture woodland, clearing young holly and birch from around the ancient oaks, pollarding veteran hornbeams and creating thousands of new pollards. The oaks in Epping Forest, including maidens and pollards, are not large and none reach 6.00 m in circumference; the largest of the 7,500 oak pollards mapped is 5.30 m in girth (Dagley & Froud, unpubl.). Many are already senescent, mainly due to competition from beeches and hornbeam. Some large pollard oaks existed in the past, such as the Fairmead Oak pictured by Miles (2013). Some of the current largest lapsed oak pollards, with girths of between 4.00 m and 5.40 m (which I measured on 26th February 2016), exist in the area of Warren Hill, which on the nineteenth-century OS map consisted of hedged fields with scattered trees. These were early illegal enclosures of forest land, which the oaks pre-dated, and which the City of London Corporation was able to buy back and restore to the forest in the 1870s. At least one positive aspect is that Epping Forest has not been touched by conifer plantations. It is heavily visited, as its surrounding area is now largely urbanised. Grazing with longhorn cattle has been introduced in several areas of Epping Forest, alongside the tree management described above, and so restoration of the key elements of the forest's pasture woodland landscape is well underway.

Feckenham Forest in Worcestershire, also known in medieval times as the Forest of Worcester, is still a distinct hunting forest on William Kip's map (after Saxton) of the county dated 1607; it was disafforested in 1629. In the late eighteenth century, Feckenham Forest was no longer indicated on maps. On the nineteenth-century OS map, the country around Feckenham is all farmland with one exception: Berrow Hill, which has an existing coppice wood on its south side. A 'Great Oak' once stood at the convergence of several lanes in the village. A great pollard oak near Stoke Court with 10.40 m girth is very likely a remainder of this forest. A pollard oak with 7.30 m girth stands in a field in Bradley Green; in 1628 there were apparently still "110 acres of … waste of Bradley in Feckenham Forest" (British History Online: Fladbury Parish) and this oak could have been growing on this uncultivated land. The medieval episcopal park of Hanbury had once been part of the Royal Forest, and the ancient oaks there are assigned to the park (see Chapter 4).

Fulwood Forest in Lancashire was a remnant of the southeast part of the Forest of Amounderness and consisted of undeveloped lands drained by Savick Brook and its tributaries. It was partly moorland and partly wooded; on the nineteenth-century OS map, there are still many hedged fields with hedgerow trees, presumably mostly oaks. The area is now heavily urbanised and open spaces are occupied by a golf course and other recreational grounds. There are no ancient or veteran oaks.

To the north of the ancient city of York was the Forest of Galtres, which was located west of the River Derwent and south of Yearsley Moor in the Vale of York. According to Muir (2005), this was one of the more wooded forests of the north of England, but now it is almost all farmland. Depletion of the forest and illicit enclosures accelerated in the later Middle Ages, until some time after 1660, Galtres was finally disafforested. Strensall Common remains among smaller fragments of heath and woodland. Along Cram Beck near Welburn stands an ancient pollard oak with a girth of 7.20 m that may be associated with the forest. Nearer Castle Howard are several other oaks larger than 6.00 m in girth, but these are possibly associated with a medieval park that may have been there (see Chapter 4).

The Royal Forest of Gillingham was originally one of the divisions of Selwood Forest in Somerset; it was situated in the northernmost part of Dorset. In the sixteenth century it was said to be "four miles long and one broad", but in 1625 it was disafforested and the deer were removed. The countryside around present-day Gillingham has almost no woodland except for Kings Court Wood near the site of the ancient royal palace, but there are a few solitary ancient and veteran oaks that may once have been part of the remaining Royal Forest. The largest and best known tree is Wyndham's Oak at Silton (Figure 5-6), a squat pollard measured as 9.86 m in girth in the Tree Register of the British Isles (TROBI). It is situated in a field 150–200 m east of St. Nicholas Church and has been named after Sir Hugh Wyndham (1602–84), who bought Silton Manor in 1641. Tradition has it that Sir Hugh, a judge of the 'Court of Common Pleas' used to sit on a bench under this tree, looking out over what then still physically remained of the Royal Forest (Pollard & Brawn, 2009). Other ancient or veteran oaks are at Feltham Farm and Eastview Farm to the northwest and at Woodwater Farm near Gillingham; all three trees measure over 7.00 m in circumference.

Grovely Forest in Wiltshire, now known as Grovely Wood, is situated between the Rivers Wylye and Nadder on a chalk plateau, and was once part of the Forest of Clarendon. It was crossed by a Roman road and there are various ancient earth works, such as Grim's Ditch and Grovely Castle, as well as field systems, that indicate its prehistoric settlement and cultivation. The surrounding chalk slopes, as well as some large fields inside Grovely Wood, are mostly arable fields. In the eastern half of Grovely Wood some large beech pollards remain but no ancient or veteran oaks have been recorded. The boundary outline, with concave borders typical of ancient wood commons, has not changed since the nineteenth century or probably earlier, but sections of the wood have been planted with conifers in the twentieth century.

The Royal Forest of Hainault was situated a little distance to the southeast of Epping Forest in Essex, which is much larger. Together with Epping and Havering Forest to the east, it formed part of the greater Waltham Forest, which was in turn a remnant of the earlier Forest of Essex. In the mid-nineteenth century, Hainault Forest was disafforested by an Act of Parliament and much of the old growth woodland was cut down. The contractions have continued since that time, and the 136 ha that remain is now known as Hainault Forest Country Park, which is administered by the London Borough of Redbridge and mostly given over to recreation. A broad strip of

woodland (113 ha) on the north side, mostly in Essex, is owned and managed by the Woodland Trust and an SSSI. The only woods present in Hainault Forest Country Park are secondary. The Woodland Trust part is the only part still in a more or less primitive state, but with much secondary tree growth in what would be more open pasture woodland if still in a medieval state. In its northern half, it has numerous hornbeam pollards but few veteran pollard oaks, and none of these are large. However, until 1820 when its remains finally blew down, this forest had at least one great oak, known as the Fairlop Oak, which by the eighteenth century had become a major tourist attraction (Miles, 2013). Apparently, it was killed by fire (vandalism?) in 1805, as witnessed by an illustration in Elizabeth Ogbourne's *History of Essex*, published in 1814.

Hatfield Forest in Essex is a small Royal Forest (now 425 ha) which remains in its medieval state as pasture woodland, with enclosures of (mostly) hazel coppice with oak standards, grassland glades or plains, small fens and a lake. By the Middle Ages, this was already an almost completely 'compartmented' forest with coppice woods separated from central plains for grazing, and consequently with very few large trees retained. Hatfield Forest was scarcely a hunting forest because of these uses; the Crown had given up hunting deer here in 1446. It was studied and described in detail by Rackham (1976) in his book *The Last Forest*, and it has been owned by the National Trust since 1924. There are very few beeches here but oaks, ash,

hornbeam, field maple, hazel and thorns are abundant. During a late phase of private ownership around a century ago, several horse chestnut trees and a few conifers were planted. Most oak trees are standard oaks in coppice enclosures, a few are pollards but almost none of these are large or ancient; around 20 recorded oaks are between 5.00 m and 6.00 m, and eight are between 6.01 m and 7.27 m in circumference. Instead, there are several very large oak coppice stools, some with large stems and one with a 9.60 m circumference. These occur in Lodge Coppice, in some places mixed with old coppiced ash trees, a few coppiced hornbeams and many coppiced hazel trees, all uncut for many years. This site is unusual as the coppicing of oak was uncommon in lowland England. At the southern fringe of the forest is a tall oak, possibly a high pollard, of 7.02 m in circumference. A very large oak known as the Doodle Oak once stood in the northernmost part of Hatfield Forest; it was dead by 1858. Its estimated circumference is marked around an interpretative panel (Figure 5-7) and it is said that 850 annual rings were counted in the remaining stump in 1949. This seems implausible as the tree would have been hollow.

The **Haye of Hereford** in Herefordshire was a small forest just to the south of Hereford that has nearly all been converted to farmland; with the farms called Haywood and Haywood Lodge and a house called Forest Gate perhaps the only reminders. On Joan Blaeu's 1664 map of Herefordshire, it still exists as Hawood

Figure 5-8 (far left).
The Fingringhoe Oak
stands in the village
of that name, Essex
Figure 5-9 (left). The
Great Oak in Spye
Park, Wiltshire once
stood in the Forest of
Melksham

Forest; on the nineteenth-century OS map, a section of Wellington Coppice is called Haywood Forest. Some of the woods that are present here as coppice woods on the nineteenth-century OS map still exist as woods with native trees, but others have been converted to farmland. There are two veteran pollard oaks with girths >6.50 m near Haywood Lodge.

Inglewood Forest in Cumbria extended southeast from Carlisle to Penrith; its eastern boundary was the River Eden, close to where it joins with the River Eamont, and in the west this Royal Forest bordered on the fells of what is now the Lake District. This was therefore mostly a lowland forest, although much of it was moorland (heath) from before its creation. Woodland may have been confined to the sides of streams, such as the River Caldew near Hawksdale and (in the southeast corner) the confluence of the Rivers Eden and Eamont. In these parts, some ancient and veteran oaks with girths of between 6.00 m and 7.50 m exist today which are not associated with ancient parks and may be connected with the Royal Forest. Near Hawksdale, the fields have (remains of) ancient hedges with hedgerow trees of ash and oak, perhaps indicating a wooded landscape in the past. Most of the former forest has been converted to farmland, and conifer plantations have more recently been established on several remaining fragments of moorland. Ancient native woodland is almost absent apart from a small section along the River Eden known as Edenbank Wood.

From 1154, **Kingswood Forest** in Essex was a small Royal Forest surrounding the ancient town of Colchester. According to Rackham (2003b, p. 181), it yielded large quantities of oak timber to the kings, which may have been one reason why it was disafforested in around 1535. The town now occupies almost the entire area of the former forest, with only Donyland Woods and Heath above the Roman River and High Woods in the north of Colchester possibly remaining, although developed as coppice woods rather than pasture woodland. Wivenhoe Park was created in 1427 by the Earl of Oxford in the remains of the forest. Until 1760 the Great Oak, an ancient pollard, stood near Mile End Hall in Colchester. Just beyond the Roman River, which formed part of the boundary of Kingswood, in the village of Fingringhoe, stands a large ancient oak but this site was not part of the former forest.

Kinver Forest in south Staffordshire and Worcestershire was a small forest west of the River Stour near Kinver. South of this ancient village a sandstone ridge, from which the village's originally Welsh name may derive, extends towards Kingsford in Worcestershire. Much of this ridge is still wooded, now with tall beech-oak forest and conifer plantations. On the nineteenth-century OS map much of this woodland is marked as open common, that is, heath or grassland with shrubs; it merges with Blakeshall Common, which has the typical convex edges of an ancient common. The remains of a castle on Castle Hill and woods to the west of this hill may also be associated with the Royal Forest. Some ancient and veteran oaks have been recorded but none reach 6.00 m in girth.

In the Middle Ages, **Leighfield Forest** was known as the Royal Forest of Rutland, created by Henry I soon after 1100. It was largely disafforested towards the end of the thirteenth century, and what remained in the county of Rutland became known as the Forest of Leighfield. Final and complete disafforestation followed around 1630 under Charles I and almost all of the land was converted to agriculture. A small number of ancient woods remain which were once within the forest; these have been managed as coppice woods for a long time. In Owston Woods, one of these former coppices, is a huge small-leaved lime coppice stool, now with large stems and a circumference at its base of 20 m. Lime stools of this size are possibly early medieval in age and certainly date back to the time of Henry I (Rackham, 2006). There are no recorded ancient or veteran oaks in the area of this former Royal Forest.

The **Forest of Melksham** in Wiltshire was a small Royal Forest mostly situated to the east of Melksham. In the early thirteenth century, when King John is said to have hunted there, this forest was at its largest and extended perhaps

Figure 5-10. The King Charles Oak in Salcey Forest, Northamptonshire

as far as Calne. Most of this forest was disafforested in the early seventeenth century. The area is now mostly farmland with scattered woods, the most extensive being the Great Wood on the Bowood estate and the wood in Spye Park. Both of these woods are partly coniferised, but in Spye Park there remain a few large ancient pollard oaks that probably date from the time of the Royal Forest. The largest living tree at Spye Park has been measured at 10.67 m in circumference (Figure 5-9). This tree now stands in the margin of a conifer plantation. Nearby, a very squat pollard (of 9.80 m girth) and a now rotten stump (formerly of similar size) locally dubbed The King's Oak were visited by the author on 29th May 2013. In a field close to a school in Bowerhill, just south of Melksham, grows another ancient oak measured at 8.70 m in circumference. As at Bere, Feckenham and Gillingham Forests, these ancient trees may be the only extant remains of the physical medieval Royal Forest of Melksham.

The small **Myerscough Forest** in Lancashire was a remainder of the Forest of Amounderness and has long since been converted to farmland. In 1297 the forest earned 20 shillings a year for the Earl of Lancaster, presumably from grazing. From 1314 part of the forest was enclosed as a deer park. A short distance west of Bilsborrow stood Myerscough Hall, now part of a college, but this is an eighteenth-century house. An older lodge had existed nearby since at least the seventeenth century, when James I visited to hunt in 1617; this is now Myerscough Lodge Farm. In the sixteenth century, the park was described "with a pale on the moor side" and the only woods are now along the Old River Brock (canals have been dug through the moorland including the new River Brock). Presumably the deer park would have been partly wooded but no ancient or veteran oaks remain in the area. The sprawling buildings of the college have all but destroyed what was left of the park, but a few fragments of the woods remain.

The small **Forest of Neroche**, in the Blackdown Hills in the south of Somerset, had the castle of Neroche at its western end. It was still largely intact in the early sixteenth century according to a map produced by Mick Aston (www.nerochescheme.org/archaeologyPlaces.php) but was disafforested in the seventeenth century. All that is left of it is Curland Common, which is now wooded and was woodland around 1500. Nothing remains on the reconstructed late medieval map of the extensive open land or pasture woodland, or of the other woods and (deer)

parks in and around the Royal Forest. Only a few veteran oaks are recorded in its former area, the largest with a girth of 6.12 m. Forestry Commission conifer plantations and farmland dominate the rural landscape.

The **New Forest** is the largest of the medieval Royal Forests still in existence. It is situated in Hampshire and Wiltshire, to the west of Southampton, and has an area of 37,100 ha. It forms the greater part of the New Forest National Park and is administered by the New Forest National Park Authority. The New Forest is mostly unenclosed pasture woodland and heath, grassland and valley bogs. In the wooded areas, however, there are enclosures for plantation forestry, mostly with conifers, and there are several villages and hamlets with enclosed fields around the houses. Common rights, especially grazing and pannage, are still exercised and are regulated by the Verderers of the Forest. Open pasture woodland covers 14,600 ha, heath and grassland 11,800 ha, wet heath and bogs 3,300 ha, and enclosures with tree plantations 8,400 ha, of which 8,000 ha have been planted with conifers since 1920 by the Forestry Commission on Crown land. During the Middle Ages, extensive areas were encoppiced. The woodland is now dominated by beech trees as the oaks were felled extensively over several centuries, mainly to supply the Navy (Tubbs, 1986); beech is also more shade tolerant and outcompetes oak if left uncut. As a result, there are relatively few large pollard oaks, most around 7.00 m in circumference; the largest measured oak in The New Forest is tree No. 88534 near Woodgreen, with a girth of 8.78 m. There are 161 ancient and veteran oaks with girths ≥5.00 m and some oak woods, both open and enclosed, with much younger trees. There are many old beech pollards which have not been cut for 200 years or longer. No trees are now removed from the unenclosed woodland and the old beeches are now often senescent or dead. Tree regeneration is poor because of excessive grazing pressures. A more detailed account is given in Chapter 9.

The small **Pamber Forest** in Hampshire, on the border with Berkshire, was a forest separate from Windsor Forest in the reign of Edward I (1272–1307). On Saxton's map of Hampshire (1575), this forest is depicted as pasture woodland with solitary trees, as well as what looks like closed canopy woods and open terrain. It then occupied an area between Wasing in Berkshire, Silchester and Pamber Green. It was disafforested under James I in the early seventeenth century. Assarts had fragmented the forest, and by the nineteenth century most of it had been converted to arable fields and coppice woods, with no pasture woodland left. Torbary (Tadley), Pamber and Silchester Commons are heaths with scattered trees on the nineteenth-century OS map, with Pamber Common being enclosed and subdivided into farm fields. Silchester and much of the western half of Torbary became urbanised in the twentieth century. The copses of Pamber Forest are now a mix of deciduous woods and conifer plantations. There are no ancient or veteran oaks.

Parkhurst Forest on the Isle of Wight was a hunting forest preserved for the lords of the island from the time of the Norman Conquest or earlier until 1293, when it became a Royal Forest. Historically, part of this small forest was pasture woodland and another part was heath. Parkhurst Lodge was a hunting lodge which was still extant in the nineteenth century; there is now a small wooded area in the middle of a meadow on its site. On the nineteenth-century OS map there is still open terrain with heath and shrubs, but in the eastern part this is all now converted to planted deciduous woods, plantations with conifers, or farmland. There is just one veteran oak with a girth of 5.00 m on the edge of the forest.

The **Royal Forest of Rockingham** extended legally across 20,000 ha between Kettering, Corby and Peterborough in Northamptonshire and Cambridgeshire. The area was already well populated in the Middle Ages and royal grants from the forest rarely refer to timber trees (Whitlock, 1979). Rockingham was disafforested at the end of the eighteenth century and was largely converted to agriculture or disturbed by ironstone mining (Marren, 1990). Almost nothing of the physical forest remains, but some of the isolated woods that stand amidst farmland were once part of it. Those that are managed by the Forestry Commission, which took over the remaining public woodland in 1923, were turned to plantation forestry, mostly with conifers, and lost their medieval characteristics. Privately owned woods often have a mix of broadleaved trees and conifers. Exceptions are Bedford Purlieus west of Peterborough, now a National Nature Reserve, and the small Old Sulehay Forest near Wansford, which is probably also a remnant of the medieval forest with a few oak pollards and some very large oak coppice stools. Elsewhere, a few scattered oak pollards remain but the largest specimens recorded (of 7.00+ m in circumference) are now dead stumps.

Salcey Forest was a small Royal Forest in Northamptonshire just south of Northampton; its remnant is administered by the Forestry Commission. It consists of woodland, mostly planted, with sections of broadleaved trees and other sections planted with conifers. In its centre lies Salcey Lawn, a large meadow with a few scattered mature to ancient oak trees and with a private house. Some parts are more traditional woods with coppice, mostly neglected, but recently this use has been resumed in a few

sections. There is much recreation at Salcey with a visitor centre and a treetop walk. There used to be a number of large ancient pollard oaks, including the Salcey Great Oak mentioned by Loudon (1838), but most of these have died; on a visit in April 2011, all that remained of those still indicated on a tourist leaflet map were rotten stumps. There are a few old oaks left along the inner edge of Salcey Lawn, one of which is an ancient pollard in a paddock; this is now the largest (10.62 m in girth at 1 m) and presumably the oldest living oak pollard in Salcey Forest (Figure 5-10). These trees are on private land and there are no living ancient oaks left in the public areas.

Savernake Forest in Wiltshire was established as a Royal Forest in the twelfth century; it is mentioned in the 1130 Pipe Roll as 'the Forest of Marlborough'. It once legally encompassed land to the extent of 39,000 ha, extending as far south as Harewood Forest near Andover in Hampshire. The present Savernake Forest is only 1,800 ha and is privately owned but on a 999-year lease to the Forestry Commission. It is situated on a chalk plateau with a clay-with-flints soil in most places. In the eighteenth century, when long beech avenues were planted through the forest, Savernake Forest was still nearly 10 times its present size. Clear felling during the World Wars was often replanted with conifers, but thanks to restraints imposed by the private owners, the Earls of Cardigan, coniferisation has been limited and much of the core forest is dominated by beech or oak. More recently, the Forestry Commission has acknowledged the values of ancient trees and woodland and has changed its management to preserve what is left. A small part of the forest is being restored to pasture woodland with cattle. There are still good numbers of ancient and veteran trees, mostly beech and fewer oak, at Savernake. The best-known oak is the Big Belly Oak (of 11.18 m in girth). Only one other presently living oak exceeds 10.00 m in circumference; most are <6.00 m in girth, indicating younger age and perhaps establishment after the sixteenth or seventeenth centuries. There are remains of an old park pale, indicating the imparkment of a section of the Forest as a deer park. However, there are also assart fields and more conifer plantations in this part of the forest than elsewhere; these land uses have probably resulted in the relatively few oak pollards there. Oak pollards are also rare in the most southern part of Savernake, which is mostly tall beech forest with scattered conifers and secondary woodland with birch and oak. This forest is described in more detail in Chapter 9.

Selwood Forest on the Wiltshire–Somerset border was a large forest in Anglo-Saxon time. As a Norman Royal

Forest, it extended at least from the present Picket & Clanger Wood (an SSSI) south of Trowbridge southwards to Bourton. Longleat Park was imparked partly from this forest, probably in the Tudor period, by which time the forest was reduced to the hills forming a ridge running between Longleat and Stourton. The Longleat Woods (an SSSI) have been the largest area of woodland here since the eighteenth century; from there, a more or less continuous belt of woods extends south to Stourton. Much of this belt is planted with conifers, but the Longleat Woods consist mostly of tall forest broadleaved native trees, predominantly oak and ash and frequently with hazel underwood. Few ancient or veteran oaks remain in the area once covered by Selwood Forest, as indicated for example on a 1575 Somerset map, on which the forest was already reduced. Three ancient oaks have been recorded ranging from 6.00 m to 6.66 m in girth.

Sherwood Forest was once one of the largest of the Royal Forests, first mentioned in 1154. Situated on the Triassic sandstone areas in Nottinghamshire, it consisted in the Middle Ages of heath and scattered pasture woodland; in 1215, it had "15 woods and groves" with mostly birch and oak. Today, Sherwood Forest National Nature Reserve encompasses *c.* 182 ha of woodland with some scattered open grassy areas and heath in the northern part. Other

remnants of the Royal Forest are managed by the Forestry Commission, which established plantations of Scots pine after the Second World War, often on heath. The only substantial woodland area in which ancient and veteran pollard oaks are still present is known as Birklands and Bilhaugh, near Edwinstowe; this area covers 797 ha and includes the National Nature Reserve. Parts of this area are notified as an SSSI and an SAC. Conifers (pines) have been planted here since the 1920s, but mostly not in large blocks of monoculture, and current management aims to replace them gradually with native trees. There are also some oak plantations and there are small areas with heath. Some other parts of Bilhaugh especially are more thoroughly coniferised or in use as military training grounds. The famous Major Oak (10.66 m in girth) is in the reserve; it is much visited by tourists, with a Robin Hood oriented visitor attraction area nearby. Away from the crowds, parts of the woodland are medieval in character with many ancient and veteran oaks; some sections of this wood are being grazed again. A survey carried out during the period 1996–99 identified 1,643 standing veteran oaks, of which only 991 were still alive (Clifton, 2000). Sherwood Forest probably has the greatest proportion (40%) of standing dead veteran oaks of any large remaining pasture woodland in England, mainly due to past intensive forest management

Figure 5-11. A veteran oak in Birklands, Sherwood Forest, Nottinghamshire

Figure 5-12. An opened area with ancient oaks in South Forest, Windsor Forest, Berkshire

but possibly also related to the lowering of the ground water table in the area. The Birklands and Bilhaugh area is described in more detail in Chapter 9.

Shirlett or **Shirlot Forest** in Staffordshire extended from Broseley and Much Wenlock to Monckhopton; it included Shirlot Common and the later Shirlot High Park, as well as Spoonhill Wood, and smaller woods near Acton Round. By the thirteenth and fourteenth centuries, assarting and the establishment of villages by Wenlock Priory had already begun to eat away the forest; several settlements, such as Monk Hall in Monkhopton, were then abandoned during the Black Death. Other parts, such as Willey Park and

Shirlot High Park, were imparked and added to private estates. Much of the former forest is now farmland, and the heaths that remained until recently have been planted with conifers. Hardly any ancient oaks remain, but two are notable: a 7.69 m girth maiden oak on Muckley Cross Farm and the 10.04 m girth Acton Round Oak, a split-bolling pollard standing along an ancient track from Acton Round to the Monk Hall settlements.

In the thirteenth century, **Shotover Forest** in Oxfordshire included the parish of Headington as well as modern Shotover and Stowood. Headington is now part of the Oxford District and urbanised. Shotover is now a public Country Park, with a hill of that name, a house and a private park that reaches up to the A40. The Royal Forest extended north from there to Woodeaton and Noke; woods such as Noke Wood (Boscus de Oke), Woodeaton Wood

(Boscus de Wode Etone), Stanton Great Wood (Boscus de Hornle et Sydele) and Brasenose Wood (Boscus de Chalfe) were part of Shotover Forest during the reign of Henry II, but most had been disafforested by 1298 (Steel, 1984). These woods are now embedded in farmland or are at the edge of urban sprawl, and they have been managed as coppice or more recently partly coniferised. In 1642, little more was left of the Royal Forest than what is now Shotover Country Park and Shotover Park and estate. Brasenose Wood is still a traditionally managed wood (Marren, 1990). Shotover Park has a veteran pollard oak (of 6.54 m in girth). There is also an ancient pollard oak measured to be 9.14 m in circumference at Holton, which was a medieval manor house and hamlet near the southeast corner of the forest and had a deer park surrounded by the forests. Otherwise, no ancient oaks dating to the Middle Ages exist in the area at present. On Shotover Hill, the last but overgrown remnant of pasture woodland, are a number of veteran oaks but all are much smaller than 6.00 m in girth.

'West Derbyshire Forest' may have been a group of rather small areas or parks: Simonswood, Croxteth Park, Toxteth Park and Burtonwood, in what is now Merseyside. The medieval deer park Knowsley Park is first mentioned in 1302; Croxteth and Toxteth were imparked in the thirteenth century (Cantor, 1983). A small fragment remnant of Simonswood is called Woodwards Plantation on the nineteenth-century OS map, this was apparently not coniferised and consisted of open secondary deciduous woodland. Knowsley Park is still a large park with a similarly vegetated area, the deer park, in its centre. Croxteth Country Park is much smaller and also has an area of open parkland but is entirely surrounded by the housing estates of West Derby. The other parks and woods in this heavily urbanised region have gone completely. There are no ancient or veteran oaks and the small woodland areas are probably secondary.

Whittlewood Forest extended from the south of Northamptonshire to the border area with Buckinghamshire, mainly between the villages of Syresham, Silverstone and Potterspury. Hazelborough Wood and Bucknell Wood, managed by the Forestry Commission, are the largest woods at present; there are also numerous smaller woods and copses in the area; all are managed (ancient) woodland embedded in an agricultural landscape. The larger woods have been coniferised — converted to conifer plantations from coppice woods — or sometimes converted to a mixture of conifers and broadleaved trees. The profusion of bluebells (Hyacinthoides non-scripta) indicates ancient coppiced woods with standard oaks, and this management has been reinstated in some of the woods.

There are two recorded ancient oaks in Whittlebury Park. This park, now with a hotel and golf course, was part of the hunting forest of Whittlewood until the early nineteenth century.

The medieval **Royal Forest of Windsor** included a large area near Windsor in Berkshire, Surrey and into Middlesex. A large section of it, directly south of Windsor Castle, was imparked in the Middle Ages and eventually became the Home Park (near the castle) and Windsor Great Park. The latter was one of the largest medieval deer parks in England and is currently c. 2,000 ha, but only part of this is still a deer park. On seventeenth-century maps of the county of Berkshire, the Royal Forest of Windsor extended from White Waltham in the north to Winkfield in the south. Windsor Forest, like the Great Park, is administered by the Crown Estate; it is now a much-reduced woodland and planted forest area to the southwest of the Great Park between Oakley Green west of Windsor and Cheapside. What remained of the medieval open Royal Forest was 4,735 ha enclosed in 1817–18 (Menzies, 1904}. Swinley Park between Bracknell and Bagshot was also part of it; Bagshot Forest was the heath to the south of this pasture woodland with oaks. The woodland is principally managed as timber-producing commercial forest and is consequently largely made up of plantations. Early plantations were mostly native oak, but upon their felling, the oaks were often replaced with conifers; Scots pine was planted on the heath near Swinley and Bagshot. Swinley Park was a post-medieval deer park and it has a number of ancient oaks, possibly dating from the seventeenth century or a few from an earlier time. Most of these trees are 4.00–6.00 m in circumference but the largest oak is 8.40 m. This park was also planted with Scots pines and with beech. Ancient and veteran oaks are few in much of the present Windsor Forest. In the southern part of Windsor Forest, known as South Forest (Figure 5-12), stand a 7.65 m living oak pollard and a 10.24 m dead one nearby, both are at least 120 m outside the medieval park pale of Windsor Great Park. Remains of the park pale indicated on the Explorer 160 map 1:25,000 (Ordnance Survey, 1998) north of Forest Lodge curve around Ranger's Lodge where they divide, one following the A332 but another turning west between the woodland of former Cranbourne Chase and the arable fields of Flemish Farm, demonstrating that Cranbourne Chase was once part of Windsor Forest, not of the Great Park. This woodland, free of plantations and now (again) grazed by longhorn cattle, has remained close to the medieval pasture woodland and is the richest area in all of Windsor for ancient and veteran oaks. (See Chapter 9 for a more detailed description of Cranbourne Chase.)

Forest	Pasture woodland	Ancient woods	Modern woods	Ancient oaks
Aconbury	No	Yes	Yes	None
Alice Holt	No	No	Yes	None
Ashdown	No	Yes	Yes	None
Bernwood	No	Yes	Yes	None
Bere	No	No	Yes	Two
Braydon	No	Yes	Yes	One
Buckholt	No	No	Yes	Two
Chute	No	Yes	Yes	None
Clarendon	No	Yes	Yes	None
Dean	No	No	Yes	One
Delamere	No	No	Yes	None
Duffield Frith	No	No	Yes	None
Epping	Yes*	Yes	No	Yes
Feckenham	No	Yes	Yes	Two
Fulwood	No	No	Yes	None
Galtres	No	Yes	Yes	One
Gillingham	No	Yes	Few	Few
Grovely	No	Yes	Yes	None
Hainault	No	Yes	Few	None
Hatfield	Yes	Yes	No	Few
Haye of Hereford	No	Yes	Yes	Few
Inglewood	No	Yes	Yes	Few

Table 5-1. The presence today of pasture woodland, ancient woods, modern woods and ancient or veteran oaks in the 44 Royal Forests reviewed in this chapter. Wood pasturing may have lapsed in the past but if reinstated (*) it scores positive. Ancient woods are here woods that date from medieval time and have not been altered by plantation forestry. Modern woods can be ancient woods so altered, or established after 1600 by the planting of trees. The presence of ancient or veteran oaks is based on recordings on the Ancient Tree Inventory's Interactive Map and the author's field observations since *c.* 1998.

The **Wrekin** is a prominent hill between Shrewsbury and Telford in Shropshire — an outcrop of Precambrian lava flows and ash deposits rising from the Severn Plain, with an Iron Age hill fort on its summit. This hill is still mostly wooded, in part now coniferised. The Wrekin formed the western part of a Royal Forest, known by the Normans as Mount Gilbert Forest, which extended westwards to Shrewsbury and included Houghmond or King's Wood (with Houghmond Hill) as well as the Forest of Wrockwardine. Attingham Park, created in the eighteenth century from the medieval manor of Atcham (Attingham), is a part of Wrockwardine and its ancient and veteran oaks are probably associated with that section of the forest. Some veteran oaks are also present on Houghmond Hill and on The Wrekin, although large yews and beech are more common on the latter. Elsewhere, the transformation to agriculture was already complete in the nineteenth century and veteran oaks are scarce, with just one exceeding a girth of 6.00 m.

The **Forest of Writtle** in Essex was a Royal Forest situated west of Chelmsford between Blackmore, Edney Common and Fryering, although its legal extent was once much greater (Rackham, 1976). In its centre were coppice woods and these were surrounded by open heath and pasture woodland (plains), different from Hatfield just to the north where this was the reverse. The heath and pasture woodland have nearly all been converted to agriculture, while the woods remain and have been expanded by further enclosures. These are mostly sweet chestnut coppice woods, with some hornbeam and oak, and have been in private ownership for centuries. Some of the larger woods have partly been coniferised, but much retains its ancient woodland character, with wood banks, hornbeam and some oak pollards. Millgreen Common is a remnant of the open areas of the forest, now mostly overgrown with birch and oak. Maple Tree Lane has ancient wood banks with large coppice stools of hornbeam and some large oak trees, but there are few ancient or veteran oaks recorded in this area.

Forest	Pasture woodland	Ancient woods	Modern woods	Ancient oaks
Kingswood	No	Yes	Yes	None
Kinver	No	Yes	Yes	Few
Leighfield	No	Yes	No	None
Melksham	No	Yes	Yes	Few
Myerscough	No	No	Yes	None
Neroche	No	No	Yes	Few
New Forest	Yes	Yes	Yes	Yes
Pamber	No	No	Yes	None
Parkhurst	No	No	Yes	One
Rockingham	No	Yes	Yes	Few
Salcey	No	Yes	Yes	Few
Savernake	Yes*	Yes	Yes	Yes
Selwood	No	Yes	Yes	None
Sherwood	Yes*	Yes	Yes	Yes
Shirlett	No	Yes	Yes	Few
Shotover	No	Yes	Yes	One
West Derbyshire	No	No	Yes	None
Whittlewood	No	Yes	Yes	One
Windsor	No	Yes	Yes	Yes
Wrekin	No	Yes	Yes	Few
Writtle	No	Yes	Yes	Few
Wychwood	No	Yes	Yes	None

The **Royal Forest of Wychwood** in Oxfordshire has been mentioned since Anglo-Saxon times, when it formed a boundary region between two kingdoms (Hooke, 2010). It became a Royal Forest shortly after the Norman Conquest and is mentioned as a demesne forest of the king in Domesday Book of 1086 (Hinde, 1995). A perambulation in 1300 ran from Woodstock in the east to beyond Burford in the west and from Chadlington in the north to beyond Witney in the south. Far from being a 'trackless wooded wilderness' resembling a 'wildwood', Schumer (1999) has demonstrated that it was highly compartmented at least from Domesday Book onwards, with enclosed coppices interspersed with open 'launds', areas of pasture woodland and even heaths. It was crossed by numerous paths and tracks, including Akeman Street, a Roman road still in use in medieval time. This forest has almost entirely been turned to farmland. A map of 'Whichwood Forest' drawn in 1815 for the then Lord Warden Ranger, Francis A. Spencer, shows that the Forest was still a large area made up of many coppice woods interspersed with parkland (for pasture), extending from Cornbury Park to Fulbrook, so much of the destruction dates from the nineteenth century onwards. At present, all that remains is a wooded area of 870 ha, still called Wychwood Forest, that is part of the Cornbury Park Estate adjacent to Cornbury Park, a house

and ancient deer park near Charlbury. These woods are mostly planted and managed tall forest; the main native trees are ash, oak and some beech, and there is also Turkey oak (*Quercus cerris*), sycamore and a few small conifer plantations. The wood has only a few veteran oaks. The element 'copse' in several toponyms in this wood indicates coppice, but there is now little evidence of former coppice woods. Coppice management probably dates back to the thirteenth century or earlier, but in the ancient woods that remained, coppicing has been replaced by plantations since the Enclosure Act of 1867, which finally put an end to the last remains of the Royal Forest. The many rides attest to hunting, which is now mostly pheasant shooting. In effect, the medieval Wychwood Forest no longer exists. Cornbury Park still has several ancient and veteran oaks and was once part of the forest, as was Woodstock Park (now Blenheim Park, see Chapter 4).

Most of the 44 Royal Forests reviewed here are/were south of the River Trent, the ancient boundary between the two Forest Eyres administered by the two justices of the forest (Young, 1979). This is partly because many Royal Forests in northern England were in non-wooded uplands, but also because more land was afforested in the south. It is apparent from Table 5-1 that nearly all ancient pasture woodland of the Royal Forests has disappeared, first

overexploited to become scrub or heath and then converted to agriculture or forestry plantations, or sometimes lost to urbanisation. The exceptions are Epping, Hatfield, New Forest, Savernake and Sherwood, and of these, only the small Hatfield Forest and the large New Forest retain a proportionally large area of this medieval land use. In three of these forests, management for woodland pasture has only recently been reinstated. It is traditionally associated with pollard trees (Rackham, 1976, 1986), but only Epping, the New Forest, Sherwood and Savernake still have substantial numbers of ancient and veteran oaks, many of them not pollards but maidens; Hatfield has eight and Epping has none ≥6.00 m in girth. The latter forest now has mostly beech and hornbeam pollards, which are characteristic of ancient pasture woodland remnants north of London, instead of oaks.

There is causation between the (long) absence of wood pasturing and the scarcity or absence of ancient and veteran oaks. Beech will grow large if its pollarding is discontinued, and seeds itself into oak woodland. As a much more shade-tolerant tree, it outcompetes oak. Detailed studies in the north German unmanaged lowland oak-beech forest reserves of Hasbruch and Neuenburg confirm this succession from oak dominance to beech dominance (Koop, 1981). It even occurs in lightly managed forest (Farjon & Farjon, 1991; see also Guy, 2000 and Vera, 2000), although mature oaks can hold their own for a long time, as observed for example, in the North Boarhunt section of

the Forest of Bere. This competition and its bearing on ancient oaks in England are discussed in more detail in Chapter 8.

Most Royal Forest remnants still have ancient woods (32 of 44), but there is no clear correlation between the presence of these remnants and the presence or absence of ancient and veteran oaks. In most cases, these ancient woods were intensively managed as coppice with standard oaks (Rackham, 2006) and the oaks were felled at a youngish age. If there are both ancient woods and ancient oaks in the remnants of a Royal Forest, they are usually in different places. Nevertheless, ancient and occasionally large coppice stools of oak, such as those in Hatfield and Rockingham, can sometimes occur in lowland England but are more common in upland areas such as the Lake District and Scotland. There is a clear correlation between modern woods and the absence or scarcity of ancient and veteran oaks. This is caused by the conversion of pasture woodland and ancient woods to plantations, often of conifers. In some instances, as in Savernake, Sherwood and Windsor Forests, straggling or dying and dead ancient or veteran oak pollards within the plantations are evidence of converted pasture woodland. Elsewhere, as in Alice Holt and the Forest of Dean, none, or just one or two ancient oaks are left. Ancient oaks can be present in Royal Forest remnants which have no pasture woodland at present, as they are in Bere by Porchester, Buckholt and Gillingham. In Bere, two ancient pollard oaks stand in the private grounds of West Lodge, and in Buckholt and Gillingham, the ancient oak pollards are now on farmland. There are many ancient oak pollards standing solitarily on English farmland, and it is often difficult to reconstruct the ancient context in which they grew up. In a few cases, the pasture woodland of the Royal Forest seems to be most likely, but elsewhere medieval deer parks could have been the historic landscape (see Chapter 4).

In general, the present distribution of ancient and veteran oaks seems congruent with the distribution of medieval Royal Forests, as noted in Chapter 3, but when investigated in more detail, the correlation is not so clear. This is most

County	Tree number	Tree form	Circumference	Locality
Cornwall	12217	Maiden	11.60 m	Darley Farm
Devon	113993	Pollard form (natural)	8.00 m	Great Oak Cross
Devon	111211	Pollard	8.10 m	Teigngrace
Devon	2348	Maiden	8.16 m	Escot Park
Devon	114243	Pollard	9.00 m	Herner
Devon	3574	Maiden	9.09 m	Rockbeare Park
Essex	9209	Pollard	8.12 m	Herd's Farm

Table 5-2. Ancient oaks ≥8.00 m in girth which are possibly associated with the twelfth-century Royal Forests of Cornwall, Devon and Essex by default (see text). They are not assigned to a Royal Forest (RF) in the database from which the statistics at the end of this chapter are derived. Tree numbers from the ATI database (www.ancient-tree-hunt.org.uk/).

likely due to losses. If wooded Royal Forests all had pasture woodland in the Middle Ages (Cantor, 1982; Rackham, 1976, 2003a) and if this included the most common native oaks, all wooded Royal Forests would have had oaks, either pollards or maidens. If these are now absent or scarce, they were lost at some time in the history of the Royal Forest. The causes of this loss are several.

Disafforestation meant the repeal of Forest Law and, more concrete, the privatisation of parts of the forest, usually to generate income for the royal purse (Young, 1979; Grant, 1991; Clifton, 2000). Disafforestation usually led to (or followed) the creation of enclosures, 'wastes', assarts or dwellings and caused the eventual disappearance of the physical forest. This process accelerated in the seventeenth and eighteenth centuries when whole Royal Forests disappeared upon their conversion to farmland or, if unsuitable, to intensively grazed commons. From the late eighteenth century onwards, the remains of the forests were often converted to plantations for timber production. These forestry efforts, sometimes spurred on by political necessities, converted the remaining Royal Forests to timber forests, at first by encourageing or planting native tree species, among which were oaks for ship building, later by planting conifers, which grow faster and straighter, especially on poor soils. The Forestry Commission and its predecessors, the Crown Estate and private owners, all saw 'forest' almost entirely in terms of timber, so until very recently, they planted conifers in ancient pasture woodland, heath and ancient woods. Even today, there is still a need to convince some forest managers of the importance of ancient trees (Green, 2001; 2010b). Slowly, the tide is turning, but often it comes too late.

Discontinuation of the practices of wood pasturing and/ or of pollarding is another cause of the demise of ancient oaks in remnants of Royal Forests. Grazing animals suppress re-growth of trees, especially from seedlings. If these animals are removed, the forest develops to a dense stand of trees, and in many woods, competition for light favours the more shade-tolerant beech and hornbeam at the expense of light-demanding oaks. The cessation of pollarding, often as long as 150–200 years ago (Edlin, 1971), has given especially beech pollards huge crowns. This may be the main reason why Epping Forest now has relatively few large ancient oaks; they were suppressed and eventually killed by beeches. Intensive exploitation of oaks, including 'dodders' (that is ancient pollards), for ship building in the seventeenth and eighteenth centuries promoted beech in the New Forest (Tubbs, 1986), which then acted to suppress the remaining oaks, as in Epping. Similar exploitation, as well as the planting of exotics, has depleted the ancient oaks of the Forest of Dean. Apart from Scots pine and larch, the planted conifers are more shade tolerant than native oaks, to the detriment of the oaks. All this has left only three Royal Forest remnants of major significance for ancient oaks today, The New Forest, Savernake and Sherwood, with Epping and Windsor in a class below these. In all five cases, it seems that the change in forest management policy that took notice of the importance of ancient trees for biodiversity conservation almost came too late (e.g. Clifton, 2000; see also Chapter 11), as it had done in other Royal Forests.

It is appropriate here to explain another reason why I have not been able to assign many ancient oaks to Royal Forests. This is early disafforestation, starting shortly after or even while the Royal Forests had reached their greatest extent near the end of the twelfth century. The most important examples are Cornwall, Devon and Essex, already mentioned in the introduction of this chapter. These counties were under Forest Law for less than a century, in the case of Cornwall and Devon for a shorter time than the reign of King John (1196–1216). Few ancient oaks now alive can be confidently dated back to that period of the greatest extent of the Royal Forests. Instead extant ancient trees must be associated where possible with land use after

Figure 5-14. The mysterious Darley Oak in Upton, Cornwall

disafforestation, unless they are situated in locations in which the Royal Forest existed longer. Foremost among these are medieval deer parks, many of the older of which were imparked from Royal Forests (Chapter 4). Chases were often similarly granted from Royal Forests and when Forest Law no longer applied, other parts, though still owned by the nobility, may have been administered and used as commons. Still later, several Tudor deer parks were created in what was once part of a Royal Forest. In such cases, I assigned the ancient oaks to these later land uses, not to the forest. What could we do if an ancient oak, which may be 800+ years old in Cornwall, Devon and Essex, cannot be assigned to any of these land uses? Could they be associated with the legal (not necessarily physical) Royal Forest of the twelfth century? I find this doubtful given the short time that these greater forests legally existed, but it is not impossible. The examples I have found of such oaks are listed in Table 5-2. If smaller than 8.00 m in girth, these oaks are almost certainly too young (see Chapter 2).

Chases

Chases, the mostly uncultivated lands belonging to the Crown and given to magnates or high-ranking churchmen in the king's favour for the hunt of the 'beasts of the forests', were quite numerous in the Middle Ages. Most often, they were carved out of Royal Forests — the king allowed private hunting by a favourite and his entourage in a designated part of a forest. He could withdraw this privilege and would have done so if the noble beneficiary fell out of favour. Chases could therefore be short-lived, or be conferred upon others. Many chases have been traced and mapped by John Langton and Graham Jones at St. Johns College, University of Oxford (info.sjc.ox.ac.uk/forests/Index.html). These lands were utilized much like Royal Forests, but Common Law rather than Forest Law was applied to chases, and was administered by the lords to which the land was given for the hunt. Commoners had the use of resources that were not considered to be of importance to the deer, as in the forests. These rights are explained in more detail below for commons. Here, I give examples of chases which have (or may have had) ancient and veteran oaks. As for the Royal Forests, this survey excludes most upland chases which were largely treeless in the Middle Ages.

Survey of chases

Askerton North Moor Chase in Cumbria is an unlikely chase to have had many oaks as it was almost entirely moorland until conifer plantations, known as Spadeadam Forest, covered large parts of its hitherto uncultivated eastern half. From the River Irthing, which was the eastern and southern limit of the chase, the moor rises north from 158 m at Birdoswald Roman Fort on Hadrian's Wall to 313 m close to the traces of a Roman road and signal station. However, and intriguingly, a large ancient pollard oak with 9.80 m girth stands in the edge of a strip of woodland on the right bank — the chase side — of the River Irthing at Gilsland. In local folklore or history, it is known as the Trysting Tree — meaning meeting tree, but also from the Middle English tryster "an appointed station in hunting". According to Almond (2009: 72), a large tree was considered to be an ideal hiding place for the noble archer towards whom the quarry was driven.

    Bringewood Chase in Herefordshire was centred on a ridge south of the confluence of the Rivers Clun and Teme and was owned by the de Mortimer family, the Earls of March. The ridge crest rising to 363 m was still open pasture in the nineteenth century, but its north-facing slopes were wooded; these woods and the ridge itself are now mostly coniferised. In the west is Hunstay Hill, still open pasture and parkland with many ancient and veteran trees,

Figure 5-15. Ancient maiden oak of 9.10 m girth at Downton Walks, Herefordshire

among which are five oaks that have girths from 6.03 m to 9.10 m. The largest oak (Figure 5-15) is in the edge of a secondary wood called Downton Walks, near a recently ploughed field in which also stand two ancient large-leaved limes (*Tilia platyphyllos*). Hunstay Hill is now part of the Downton Castle Park and Estate, which has a partly landscaped park created in the eighteenth century along the gorge of the River Teme.

    Byrkley Park in Needwood Forest in Staffordshire is all that remains of what was still a sizeable park in the nineteenth century, now being destroyed by industrial development. Two small areas of parkland remain in which there are several ancient oaks with girths of between 5.50 m and 7.35 m. The shape of the former park is reminiscent of an ancient deer park, but there is no record of it in the

Brockton Coppice near Stafford, part of **Cannock Chase** Country Park in Staffordshire, is not coppice woodland but former pasture woodland and was used as a common. The chase had once been part of a Royal Forest which became fragmented by assarting from numerous settlements, leaving Cannock Chase as the largest uncultivated area in 1300 (Cantor, 1982). It was given to the Bishop of Coventry and Lichfield and remained with the Church until the Dissolution. There are a number of maiden oaks that grew in open canopy (Figure 5-16) but none are ancient or large and much of the woodland has degraded to heath, now again returning to open or closed wood with mostly birch and oak regeneration. One oak measured as 6.00 m in girth at 30 cm above the ground is in fact a lapsed coppice stool. In between the older oaks there are recent plantations of native oaks. If there were ancient oaks here, they were probably cut in connection with coal mining during the Industrial Revolution, with the smaller oaks or sapling oaks that were left becoming the oldest oaks now present. The remainder of Cannock Chase became heath, now mostly planted with Scots pines and other forestry conifers. There is a herd of fallow deer. At the northern edge of the former chase, where the high sandy ground of Oakedge Park slopes down to the River Trent, stands one of the very large ancient oaks of England with a girth of 11.00 m (Figure 5-17). This tree has an oval diameter and two large stems; there may have been a third limb. If it was a pollard, it would have been a very low-cut one.

Corse Chase in Gloucestershire belonged to the Earls of Gloucester in the twelfth century and was presumably part of the great manor of Tewkesbury, with which it descended until the sixteenth century. It was connected with Malvern

archives of the Staffordshire Historical Collection (Cantor, 1983). Needwood Forest was a private forest or chase owned by the Berkeley family of Berkeley Castle, who gave their name to Byrkley Park and, within it, to Byrkley Lodge, a thirteenth-century hunting lodge. The park later reverted to the Crown, together with the estate and forest. A survey in 1684 counted 38,218 "good timber trees" in Needwood Forest outside the parks (Whitlock, 1979), most of which must have gone to the Navy not long after. The park seems to be of a post-medieval date, and was perhaps associated with an eighteenth-century rebuild of the lodge and subsequent disafforestation.

Figure 5-18. Pollard oaks in pasture woodland near Cranbourne Tower, Berkshire

Chase and extended between the Rivers Leadon and Severn. On Saxton's map of Worcestershire, published by John Speede in 1610, the chase in the vicinity of Forthampton in Gloucestershire is called Corswoode. Bushley Park, a medieval deer park, lay to the northeast in Worcestershire. A large field of grassland near Home Farm belonging to Forthampton Court has many ancient and veteran oaks; on a visit 25th July 2015 with Brian Jones, we observed 17 oaks alive (some just) and 23 dead. There appears to be no protection of these trees from farm animals and damage to the root bases was evident on most oaks. Near Alcock's Farm stands a healthy ancient oak with a girth of 8.65 m in a field intensively farmed for grass silage.

Cranborne Chase in Hampshire, Wiltshire and Dorset covered a large region west of the New Forest, but its core area was on the chalk hills south of the River Nadder; to

the west it extended to the River Stour. Cranborne Chase passed from ownership by the Earl of Gloucester to John I upon the king's marriage to Gloucester's daughter, Isabel, in 1189. It remained in royal possession until James I granted it to Robert Cecil, First Earl of Salisbury, in 1605. In 1296, there were just 1,433 acres (573 ha) of woodland. The boundaries of Cranborne Chase are difficult to establish as they changed often; an area to the east of Yeovil known as Blackmoor Vale was called Blackmore Forest (Whitlock, 1979). There are still woodlands in the chase area; one is known as Chase Wood. Most of these woods have long been managed as coppice, especially with hazel (Rackham, 2003b), and coniferisation has occurred in most of them in recent times. On the chalk hills and their surrounding vales, more ancient yews than oaks have been recorded, as can be expected. Presumably the hills were mostly covered in chalk grassland with abundant shrubs, and large whitebeams (*Sorbus aria*) were also commonly recorded. If high forest developed here, beech not oak would have been the dominant tree. Despite this, ten ancient oaks in the database with girths of between 6.00 m and 8.75 m can be associated with this chase. One of these oaks stands near Old Wardour Castle near the northwestern boundary at the edge of the chalk plateau. During its varied and sometimes confiscated ownership, this hexagonal castle may at times have served as a hunting lodge.

Cranbourne Chase in Berkshire was a part of Windsor Forest to the west of Windsor Great Park. The area has been variously marked on maps as Cranbourne Wood, Cranbourne Chase or Cranbourne Park, and much of it has been considered part of Windsor Great Park for at least a century. It seems that the area was only partly wooded and merged to more open terrain with heath in what is now an area of plantation forestry of the Crown Estate known as Windsor Forest. An area demarcated by Forest Road off the A332, by an old park and field boundary towards Flemish Farm, by the medieval park pale back to the A332 and by a strip of Windsor Great Park from Ranger's Lodge to Forest Lodge across the road, historically belonged to the chase. This area is still rich in ancient and veteran oaks. There are farm fields, both grass and arable, and parkland in this chase, but one section in particular has many ancient oaks. This part of the ancient chase is being managed as pasture woodland. The largest oaks in Cranbourne Chase are 8.00–9.20 m in girth, while a still larger oak known as King Offa's Oak stands along Forest Road just outside the area in Windsor Forest. A more detailed description of this site is given in Chapter 9.

Figure 5-19. The Sun Oak is a large lapsed pollard with a girth of 9.20 m associated with St. Leonard's Forest, a chase

Enfield Chase in Middlesex, now within Greater London, was situated to the west of Epping Forest. It was first recorded in 1325, when it was owned by the king (Henry IV), but it was given as a hunting ground —perhaps with a park, Enfield Old Park, now Grange Park and mostly built over— to the Mandeville family, Earls of Essex. In 1777, it was divided among parishes and private estates and ceased to be an open hunting ground. By the nineteenth century, it was nearly all farmland with a few parks and isolated woods. Trent Park was then the principal park, leased by the Crown to Sir Richard Jebb, then sold to others and until 2012 housing a campus of the University of Middlesex. This park has some woods, with a few veteran oaks and hornbeams remaining as possible relicts of the situation before 1777. A maiden oak with a girth of 6.15 m stands in front of the mansion. Much of the farmland has been built over in the twentieth century and urban expansion seems set to continue into the twenty-first century.

Hampton Court Chase in Surrey was an area south of the River Thames across the river from Hampton Court Palace. It extended south to Cobham and included a small park there. To the west, it joined with Windsor Forest and must once have been a part of this. Until late in the nineteenth century this land was still largely open, with fields as well as commons and woods. Then villages became commuter towns and, together with golf courses, roads and other developments, have transformed the countryside into a suburban landscape. Yet several ancient and veteran oaks remain, from 6.33 m to 8.15 m in girth, most likely associated with the chase and its undeveloped landscape of grassland, heath and pasture woodland.

Nidderdale Chase was one of a series of chases in the Yorkshire Dales and was situated in the River Nidd drainage, from Middlesmoor through Pateley Bridge to Kettlesing. In the upper dales chases — Langstrothdale, Bisshopdale and Coverdale — oaks are absent or rare, and ash and sycamore are most common among veteran trees. In Nidderdale, there are some ancient and veteran oaks, although ash and sycamore, joined by beech, remain the more common trees in upper Nidderdale and in higher side valleys. Up on the moors, trees are absent or scarce. Sheep grassland and hedgerows with trees dominate in the valley and ancient woods are left on steeper slopes, many of them now coniferised. About ten large ancient and veteran oaks have been recorded in Nidderdale, with girths between 6.00 m and 7.00 m. Muir (2000; 2005) has associated these and other pollard oaks with (late) medieval hedgerows. Ancient hedgerow oaks were reported and mapped near the village of Hampsthwaite and in fields around Clint, but had not been included in the ATI by

April 2016. The large and ancient oaks reported in 1871 near Elton Spring in Birstwith have almost gone; one oak is now 6.10 m in girth and two near Clapham Green are 6.00 and 7.00 m. It would appear from Muir's surveys that the medieval chase had been enclosed by the sixteenth century, with most oaks limited to hedgerows.

Saint Leonard's Forest near Horsham in West Sussex is part of the forested Weald of early medieval time. At its western end, Sedgwick Castle existed by 1086; during the Middle Ages, the various lords of this castle and Bramber Castle held the hunting and timber rights in the forest. Sele Priory held rights to the underwood (coppice). Later, the forest was imparked but it remained a chase as it also contained settlements with farmland and commoners had access through several gates. In the sixteenth century, much of the forest was cleared, mostly for charcoal used in ironworks; deer had disappeared by 1553. On John Speed's maps of Surrey and Sussex (surveyed by John Norden in about 1595; see Nicolson & Hawkyard, 1988), it is surrounded by a pale and depicted as a very large deer park. Most of the ancient and veteran trees today are beech pollards. Yews are also common and there are some large small-leaved limes, but few ancient oaks have been recorded. The largest oak is known as the Sun Oak, a pollard with a girth of 9.20 m, which is marked on the nineteenth-century OS map in St Leonard's Park near the end of Sunoak Road. On a visit on 22nd December 2015 with Tim Pearce, we re-measured this great oak and counted eleven live limbs and four stumps of limbs.

Stourfield Chase in Hampshire is now completely engulfed by the city of Bournemouth, which on the nineteenth-century OS map is still a small town on the coast, surrounded by commons and fields. Yet there are still a number of mostly veteran oaks, the largest of 6.54 m in girth, which all have the shape of parkland trees. These trees are most likely associated with the chase and commons, and stood along lanes through the fields or in hedgerows. Other and perhaps larger oaks may have existed but if so fell victim to the urban sprawl of the twentieth century.

Tydenham Chase in Gloucestershire was situated to the south of the Forest of Dean between the estuary of the Severn and the River Wye; in the north, it bordered on the Royal Forest. The land slopes up gradually from the broad and shallow Severn estuary to the gorge of the Wye, but in the southern end of the chase beyond the village of Tydenham, the two rivers meet on a low plain. Offa's Dyke runs from the shore of the Severn to the Wye and follows the escarpment upstream. In a 330° bend of

Figure 5-20. Pasture woodland with young oaks on Ashtead Common, Surrey

the Wye, forming a narrow-necked promontory known as Lancaut, partly open and partly wooded, are several ancient and veteran oaks, the largest with a girth of 7.50 m. This peninsula was part of the chase in the Middle Ages; the deserted village has existed since Anglo-Saxon time and was at first Welsh; at present, there is only one working farm and the ruins of a twelfth-century church in what remains of this village. Lancaut is an SSSI and a nature reserve. Elsewhere, the chase is mostly converted to agriculture and managed woods, with a few areas left uncultivated but without ancient or veteran trees.

Yardley Chase in Northamptonshire was a Norman hunting chase, but it has been changed substantially and now has woods, pasture and arable land; only a small part of the northern section is still parkland with ancient and veteran trees. A 5.5 km avenue was laid out from Castle Ashby in a southsouthwest direction to the chase. Military use during and after the Second World War has added rail tracks (now dismantled) and numerous small bunkers, concrete huts and other buildings, many surrounded by wet ditches and earthworks. There are tracks used by vehicles and rides through the woodland, which is mostly hazel coppice with oak/ash standards and a few Scot's pine plantations. The site is owned partly by the Ministry of Defence (MOD) and partly by the owners of the Compton Estate (Castle Ashby). Unimproved grassland, with c. 250 mature planted oak trees, lies on either side of the avenue, covering c. 30 ha; some of this extends into the MOD site where more bramble bushes and thorn shrubs are allowed. An avenue of pedunculate oak and horse chestnut stands across the more open section of the park; this is an extension of a star-shaped pattern of rides laid out in a wood in what is now the MOD part of the chase. In the more open parkland section owned by MOD, the SSSI notification (1990) mentions "large over-mature pollards and maidens of ash and pedunculate oak"; on a visit in March 2016, two staff of Natural England and myself estimated that 30–35 veteran oaks were present, nearly all maidens. Of these, only three exceeded 6.00 m in girth, with the largest at 7.24 m. Two large oaks with girths of 8.33 m and 9.11 m were recorded here in 1943 but are now dead and decayed.

Commons

There are other possible associations of ancient or veteran native oaks with historical components of the rural landscape. Probably the most obvious are commons. Commons were traditionally lands on which the commoners, i.e. the rural population not being land owners, had rights of usage. This usage could vary, but most often included rights to graze livestock, to collect surplus vegetative materials such as dead wood, brush wood and litter, to cut gorse or furze (*Ulex europaeus*) and heather, and sometimes to cut or dig peat if present. At least in England, the trees, i.e. the timber, were reserved for the land owner, usually the lord of the manor; communal ownership was not rare in some parts of the continent, an example being the 'marken' in Gelderland, The Netherlands (see Buis, 1985, Vol. 1).

What interests us here is a particular type of common, what Oliver Rackham has called "wood pasture common". These were only a subset of the commons and were wooded but grazed, resulting in more or less open parkland not unlike a medieval deer park or, indeed, a wooded Royal Forest. Royal Forests were often declared in areas that included pasture woodland commons (Muir, 2008); this implies that the commons prevailed but restrictions were imposed concerning the 'venison and vert'. If we go back far enough in history, almost all woodland must have been 'common' before sections of it were afforested and/ or imparked. As far as the utilization of trees on pasture woodland commons was concerned, the normal practice was pollarding, which allowed simultaneous grazing of animals and production of wood, although not much large timber for building. Commoners could obtain permission to use such timber ('housebote'), but this required either felling the pollards at a fairly young age or taking oak trees grown as standards in enclosed coppices, but the latter were rare on commons (Rackham, 1986). Requirements for timber may also have been satisfied from hedgerow trees. Commoners also used pasture woodland commons for pannage, the feeding of pigs on the fallen acorns in the autumn, but this would only be of significance in a mast year. These rights were not exclusive to commons and could also apply to (parts of) Royal Forests and even to deer parks (Rotherham, 2007a), so we cannot be sure that areas where ancient oaks occur were once pasture woodland commons, even if we know that pannage occurred there; villagers using deer parks for pannage often paid a rental fee, or provided services in lieu of payment, to the lord of the manor (Bond, 1981).

The identification of former pasture woodland commons is also difficult because while commons are often indicated on the OS maps, these maps do not distinguish pasture woodland commons from the more prevalent commons. These used to be more open and totally unimproved and unenclosed grazing lands, typically chalk grassland with shrubs, or heaths and moors on more acid soils. Trees are indicated on the nineteenth-century OS maps, but only if they were large and stood solitarily or in hedgerows. Using toponyms such as those with the suffix 'lea' or 'ley', which may have meant open woodland, to locate former pasture woodland commons is fraught with uncertainties (Muir, 2008; see also Hooke, 2010).

Finally, the "tragedy of the commons" (Hardin, 1968) made it likely that many pasture woodland commons would gradually degrade to grazing lands indistinguishable from the more abundant and widespread type of common, which would only have been more wooded in a distant past. According to Rackham (1986), by 1300, pasture woodland commons that may have dated back to Anglo-Saxon times and earlier had greatly diminished or if not overgrazed to the extent that they became heath or grassland, had been imparked. Enclosure in later centuries often eliminated what remained of the commons, converting them to agriculture or plantations. Historical records rarely mention whether the common was wooded or not (Muir, 2005), which makes evaluation in the present context difficult. The presence of ancient or veteran pollard trees on commons provides some evidence for pasture woodland dating back at least to the approximate age of the pollards. Large, and thus presumably ancient, pollard oaks are very rare on existing commons; in some locations, there is just a single large oak which may have had some local significance that spared it from the axe. Beech pollards are more often encountered on commons; this is certainly the case on chalk substrates such as those in the Chiltern Hills and on the North Downs. Examples of commons which have (or may have had) ancient or veteran pollard oaks are given and discussed below.

Survey of commons

Ashtead Common in Ashtead in Surrey is an ancient pasture woodland common. It is a National Nature Reserve and SSSI of 200 ha owned and managed by the City of London Corporation. The common has many oak pollards, but most of these trees are not very large in terms of trunk circumference and perhaps only 300–400 years old. There are some areas without pollard oaks which would have been more open grassland or perhaps heath with clumps of gorse and thorn, but as with all neglected commons, there is much young tree growth (here mostly birch) turning the common into a forest, with the pollard trees a reminder of the pasture woodland. Hazel coppice is present within some of the areas with oak pollards, suggesting (temporary) enclosure; coppicing and other traditional management, including grazing, has been resumed here and there, but the only browsing or grazing animals in the areas with most of the oak pollards are wild deer. Bracken and brambles dominate the ground cover; the bracken is now mowed in places to restore open glades and reduce the fire risk. This site is described in more detail in Chapter 9.

Berkhamstead Common near Berkhamstead, Hertfordshire is part of a larger complex of pasture woodland which had the irregularly concave outlines typical of commons. This complex extended from Ivinghoe to Frithsden and included Frithsden Beeches and Ashridge Park, a medieval deer park now mostly transformed into a golf course and a small landscaped park around Ashridge College. The proximity of the eleventh-century Berkhamstead Castle, which became a favourite haunt of the Norman and Angevin kings, could mean that royal hunts were held at Berkhamstead, but the common did not become a Royal Forest. Beeches and sweet chestnuts are the most common large trees and there are several ancient or veteran oaks, but only one on the actual common reaches 6.10 m in circumference. Two further oaks >6.00 m in girth stand in the former Ashridge Park. In the wooded parts, there has been much secondary growth of trees, especially birch, since grazing and cutting ceased. In some former copses plantations of conifers have been established, but most of the formerly more open commons retained their native trees. There is currently no grazing of domestic animals in the woodland; the National Trust only grazes the downlands of the larger Ashridge Estate. The woods are variously managed for timber, with access for recreation, or as nature reserves, or for a mixture of these aims. A herd of fallow deer roams freely on the Ashridge Estate, which includes the common, the former park and farmland; a modern deer fence excludes these animals from the grounds of Ashridge College.

Binswood in Hampshire, *c.* 6 km southeast of Alton, is a nature reserve of 62 ha owned and managed by the Woodland Trust. It is apparently ancient pasture woodland, surrounded by medieval wood banks and hedges. These boundaries provide evidence that the northern third of Binswood has been converted to fields, a situation already apparent on the nineteenth-century OS map. It was once part of Woolmer Forest and, according to its shape, was more likely a deer park than a common. A farm at the northern end is called Park Farm, but there are no remains of a manor house, a moat, or a park pale. Commoners' grazing rights date back to the reign of Henry IV, that is, the early fifteenth century (Lovett Jones & Mabey, 1993), so it is possible that Binswood was formed by the disparkment of a royal park, hence the absence of a manor house. The Woodland Trust states that there are "ancient oaks and beech trees", but no oaks larger than 5.20 m in girth were recorded by the ATI. The site is managed as a pasture woodland common, but there are large areas of hazel coppice.

Brookhill Plantation in Norfolk is situated between Thursford and Little Snoring, and is presently a wood bisected by the road (A148) connecting these villages. Towards Little Snoring, across a small stream (the River Stiffkey), are plantations on what was a common — called

Figure 5-21 (far left). The Druid's Oak in Burnham Beeches, Buckinghamshire
Figure 5-22 (left). Oak coppice in Burnham Beeches, Buckinghamshire

for example, 'Old Gorse'. On nineteenth century maps, the common was adjacent to Thursford Park, the farmhouse shown close to the boundary is now a group of modern farm buildings that separate the common and the park completely. Thursford Wood, between Brookhill Plantation and Thursford Park, was known as Lawn Plantation on the nineteenth-century OS map and lay within the sixteenth-century park; neither this wood nor the park has any ancient or veteran oaks. Brookhill Plantation and part of Thursford Wood are a nature reserve owned and managed by the Norfolk Wildlife Trust. There are other plantations on the OS maps, and on the nineteenth-century OS map some heath is still not planted with trees. There are six ancient or veteran oaks with girths of between 6.00 m and 7.70 m and a number of smaller veteran oaks. The fact that they are all pollards points to an ancient pasture woodland common.

Bucknell Hill, at Upper Lye in Shropshire, was a common in the parish of Bucknell, situated upstream from the village and its fields above the River Redlake. The ancient and veteran pollard oaks stand in the valley or on lower slopes in grassland already compartmented by hedges in the nineteenth century. The higher slopes are now planted with trees, mostly conifers. The two largest oaks have girths of 7.20 m and 8.10 m and could date back to when this land was commonly grazed.

Burnham Beeches in Buckinghamshire, just north of Slough, is a medieval pasture woodland common (Rackham, 2003b). It is a National Nature Reserve and SSSI of 220 ha owned and managed by the City of London Corporation. It is now mostly densely forested, but in earlier centuries, it would have been much more open woodland and even heath with scattered trees. Some small more or less open areas with heather or peat moss (*Sphagnum*) bog have been restored. Grazing by cattle has resumed in designated and fenced parts, and more recently in a larger area with 'virtual' fences. The ancient, as opposed to the veteran and younger, trees are all pollards. The majority of the *c.* 500 living pollards are beeches, but about 15% are oaks, although only five of these are between 5.00 m and 6.00 m in circumference, with just one exceptionally large at 8.89 m in girth (Figure 5-21). In one area, there are remains of old oak coppices (Figure 5-22), which were enclosed in the seventeenth century. Pollarding and coppicing are again carried out here and there; but the old lapsed pollards will only be cut back higher up in the crown, for 'crown restoration'. Unlike on Ashtead Common, most of the mature 'maiden' trees are beech and oak, often with a secondary canopy of holly. As a result, many ancient and veteran oak pollards were in danger of being killed by over-shadowing, which would turn the reserve into a beech-dominated high forest. Efforts to 'halo' these oaks to set them free from overshadowing are now being made. The oaks are included in a regular work programme for checking, halo clearing and restoration cutting.

Croft Ambrey and the higher parts of Croft Castle Park in Herefordshire were not a park but a pasture woodland common in the Middle Ages. The Iron Age hill fort and its steeper slopes to the north and west, as well as the north slope of Yatton Hill, are still more or less open parkland and a few ancient oaks remain on the site of the hill fort, one with 8.07 m girth. The gentler slopes to the south of the hill fort have been planted with conifers, to form Ladyacre Plantation; the conifers have killed most of the ancient and veteran oaks that grew there. Parts of this area were still pasture woodland in the nineteenth century, but the OS map of that time also shows compartmentisation (enclosures) and farm fields where today not even stumps of oaks remain in the conifer plantation. This is an example of the enclosure and conversion to farmland, now forestry, of ancient pasture woodland commons. In 2014, the felling of conifers began in an attempt to save what oaks survive; the National Trust (landowner) and the Forestry Commission (leaseholder) are working in partnership in this effort.

Ebernoe Common in Sussex, a few kilometres to the north of Petworth, is considered to be an ancient pasture woodland common (Rose, 1993). Grazing with cattle has resumed now that the common is a nature reserve managed by the Sussex Wildlife Trust, but this commoner's right had apparently not been exercised for a long time previously. Much of the common is now high beech-oak forest (Figure 5-23) with a dense understorey of holly; there are only a few small

Figure 5-23. Oak-beech forest in Ebernoe Common, West Sussex

glades, perhaps of recent origin, partly kept clear by the cutting of trees. Although there are some large oaks with spreading crowns, as well as large rotten stumps of oaks that indicate the existence of more open woodland in the past, there are no ancient or veteran oak pollards. In some areas, there is hazel coppice with standard oaks, indicating managed woods in association with a brick works and an earlier iron furnace. There is also an area in which oaks were planted for timber. The common has probably been compartmented since the sixteenth century (iron works) or later, separating the grazing animals from the woods.

Fritton Common in Norfolk is a small common that now looks more like a municipal park, with a road through the

length of it and planted trees. Rackham (2003b, p. 174) mentions "a handful of pollard oaks still to be seen" here, but the nineteenth-century OS map shows no trees and none have been recorded by the ATI. It is possible that this common was pasture woodland in medieval time but has been robbed of its old trees. A small wood on the common near Manor Farm did not exist in the nineteenth century.

Furze Hill near Mistley in Essex is currently owned and managed by Mistley Parish Council. From the early eighteenth century, it was in the private ownership of the Earl of Oxford and then the Rigby family. At this time, Furze Hill was imparked with a ha-ha and hedges, but earlier it may have been a pasture woodland common. There is no evidence of a medieval deer park. The name Furze Hill indicates the abundant presence of gorse which, in Britain at least, is an indicator of (over)grazing which can linger in woodland long after the grazing has ceased.

Figure 5-24. The ancient pollard oak known as Old Knobbley on Furze Hill near Mistley, Essex

The Mistley Tree Survey conducted in March 2003 (www.oldknobbley.com/old_knobbley/other_trees/) recorded 17 oak pollards with a girth >5.00 m, among which is Old Knobbley with 9.60 m girth; these are certainly or very probably older than 350 years. The planting and pollarding of oaks continued in the eighteenth century (Miles, 2007) and several of these younger oaks remain here too. There are very few planted non-native trees. Like other abandoned commons, Furze Hill is now overgrown forest where it has not been turned into hedged farm fields or sports grounds.

Hampstead Heath in Middlesex (now part of Greater London) is now a non-build space that is hemmed in on all sides by urban development and dissected by roads. Its western part is the remnant of an ancient common that belonged to the manor of Hampstead (Farmer, 1984). Up to the late nineteenth century Hampstead Heath was much more rural, with Hampstead and Highgate still separate from London. Parliament Hill was a series of fields with hedges not too dissimilar from the present situation, but the common proper was much more open heath and grassland than it is now. The ancient and veteran oaks, three of these with girths of between 6.00 m and 6.40 m, are typical maiden trees that grew in open fields, hedgerows or on the heath, not in the nearly

Figure 5-25 (left). Pollard oaks at Oak Pin near Risby, Suffolk

Figure 5-26 (above). A pollard oak in autumn leaves, Richmond Park, Surrey

closed canopy forest now present in many places. On the nineteenth-century OS map, these trees are situated in those open field areas or in heath. Some of the old trees (not only oaks) stand in a line, a feature quite common on former commons elsewhere too, indicating that they were once part of former hedges or boundaries now completely gone in the closed-canopy forest. Parts of Hampstead Heath including Parliament Hill are now a landscaped park with a string of artificial lakes, Highgate Ponds, that were created for Kenwood House. Kenwood is an ancient wood but has no ancient oaks as it was managed as coppice with standards. None of these areas belonged to the common of Hampstead and there are no ancient or veteran oaks there.

Oak Pin near Risby in Suffolk is at first sight an odd locality for ancient or veteran oaks. It is a more or less triangular field of grassland of *c*. 9 ha, grazed by sheep, surrounded by a hedge and with a farm and two houses in one corner. The surrounding land is farmland, mostly arable or seeded grassland, but this is a fairly recent reclamation from heath in the sandy border region of The Brecks, which lies north of the River Lark. According to Rackham (2003b) other small fields with pollard oaks can be found here, and this is indeed the case. The nineteenth-century OS map also shows remaining fragments of heath in the area. Brakey Pin just north of Oak Pin, where I have seen a few oak pollards, is another of these fields with veteran oaks. Oak Pin has about 110 oak pollards; most should be classified as veteran with perhaps 2–4 ancient; the largest I measured in April 2014 had a girth of 5.44 m. The village of Risby is mentioned in Domesday Book (1086) and was possibly settled by Norse invaders; it remained small and on the nineteenth-century OS map the surrounding fields are much smaller than the large fields on the former heath in which Oak Pin lies. Hengrave Hall is about 3 km to the east, a large Tudor Manor House with a park in which stands a 'Queen Elizabeth Oak' marked on both the nineteenth and twenty-first century OS maps. The hall was

Figure 5-27. The Meavy Oak on the village green in Meavy, Devon

built by a London merchant and completed in 1538 as a prodigy house, visited by Elizabeth I in August 1578. The surrounding land was not farmed and remained in use as a common until the nineteenth century or even later.

Reffley Wood near King's Lynn in Norfolk was an ancient pasture woodland common (Rackham, 2003b) in which the trees belonged to a third party, not the lord of the land as was usual. In the nineteenth century it was surrounded by farm fields, but urbanisation (the town of South Wootton) has now encroached upon it, taking 'a bite off'. The wood has been coniferised and only a few oak pollards, none of them reaching 6.00 m girth, remain on the northeastern boundary, which has a wood bank on or near which they stand.

Richmond Park in Surrey (Greater London) was created by Charles I in 1637 from lands that had been used as commons by six surrounding villages or hamlets. There were also a few scattered farms and enclosures with coppice woods. Along Beverley Brook were hay meadows. By the time of imparking, much of the commons were open heath and grassland with gorse and thorns, but some parts were still pasture woodland with pollard oaks. Some ancient oaks stand in lines marking old boundary hedges. When Richmond Park became a deer park, large herds of deer were introduced but commoners retained access through gates in the brick park wall. The planting of oaks and sweet chestnuts in the park, the creation of lakes and enclosures, and the building of lodges have altered the park, but it retains much of the earlier medieval landscape of the commons. A detailed description is given in Chapter 9.

Sellack Common in Herefordshire is situated on the flood plain of the River Wye between the river and the hamlet of Sellack. There is a parish church, the church of "Baysham" first mentioned in 1291, and a few houses on higher ground; nearby is Caradoc Court, a mansion on the site of a medieval manor house. On the nineteenth-century map the flood plain is not indicated as a common, but on the current OS map, the area where the oaks are is called "common". The land is unenclosed grassland for grazing and a public right of way leads to a footbridge across the river. The ancient and veteran oaks stand mostly in lines along water ditches that drain the land towards the river. The recorded oaks have girths from 4.60 m to 7.60 m and the largest trees, measuring ≥6.00 m, are ancient pollards.

The Mens in Sussex is hailed by Rackham (2003b, p. 175–6) as "one of the best surviving wooded commons". The boundary shape is indeed consistent with that of a common, but the tree cover and flora provide little evidence of this land use. Only the scattered, declining woodland hawthorns (*Crataegus laevigata*) under tall beeches and oaks are indeed indicative of more open pasture woodland vegetation in the past; the ubiquitous holly trees only seem to be increasing in number, as on nearby Ebernoe Common. In November 2013, I walked two-thirds of this common, mostly off the paths, in search of ancient or veteran oaks or pollards but I could not find any; indeed, just one pollard oak with a girth of 6.35 m has been recorded at the northern end of the common near Idehurst Farm by the ATI volunteers. This oak is pictured in Tittensor (1978). Neither could I see any beech or hornbeam pollards, although there have been reports of such. In summary, the pasture woodland common of The Mens is history, of which few traces remain on the ground. Increased enclosures in the late nineteenth century were used to transform much of The Mens to a forest of oak and beech that were grown as timber trees (Tittensor, 1978); these are now well over a century old and tall trees. Current management by the Sussex Wildlife Trust does not involve grazing, which ceased more than a century ago; the existing strategy of 'laisser-passer', i.e. natural succession, will turn The Mens into a beech forest rather soon as many of the oaks appear

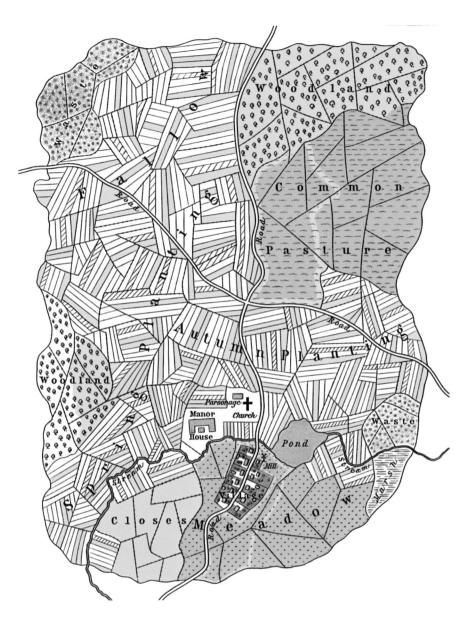

Figure 5-28. A medieval manor with its various uses of the land. Source: Economics of Agriculture in the Middle Ages (Wikipedia The Free Encyclopedia online)

to be in decline. The non-intervention policy results in increasing amounts of standing and fallen dead wood. In a few centuries, we might learn here what a 'natural' forest in southern England will be like and what organisms occur in it.

West Wickham Common in Greater London is a small common mostly on the slope of a hill adjacent to the larger Hayes Common, which is on a plateau. Most of the ancient and veteran oaks occur in a narrow strip of woodland on the slope, with a few isolated oaks on the plateau. Two pollards exceed 7.00 m in girth; the others are smaller than 6.00 m. On the nineteenth-century OS map a house named Coney Hall, built in the 1650s by the Lord of the Manor of West Wickham, appears isolated on the road at the foot of the hill; this is now surrounded by suburban housing. Perhaps because of the proximity of this urban development, the slopes are dissected by an excess of foot trails and signs of vandalism are apparent on several of the oaks. This site is owned and managed by the City of London Corporation, who have tried to stop the abuse by fencing off and blocking gaps in some of the ancient oaks. Management action to 'halo' some of the ancient oaks and to reduce the weight of limbs is also undertaken.

These examples demonstrate the difficulty in defining the status of a piece of land as common pasture woodland since the Middle Ages. Commoners' rights may have existed even if the land had been imparked, as is probably the case with Binswood. Common land may have been appropriated by an individual and passed on through one or more families before it was returned to public use or ownership, as has been the case for Furze Hill and Ashridge Park. Randomly scanning the OS map of England online reveals that large parts of the country have no commons at all so called. This is at least in part due to the Enclosure Acts of the eighteenth

century, which privatised many of the commons. Commons are now mostly found in a wide arch around London, in the North and South Downs, in the Chiltern Hills, in Surrey, Sussex and Kent on the Weald, in Essex and in The Brecklands of Norfolk where they are called warrens. Many of the lowland heaths, now almost always planted with Scots pine, may once have been commons but also parts of Royal Forests, they have not been called commons on the OS maps since the late nineteenth century or even earlier. Few former pasture woodland commons now appear to have ancient oaks; on chalk soils, there are more often ancient beech pollards and yews.

The village green is a variant of sorts to the common where oak pollards may have been preserved. Greens are now only used for recreation, but in the past their potential for grazing livestock could not be neglected, and consequently the few trees present were likely to be pollards. Village greens that have ancient pollards may also be remnants of larger pasture woodland commons (Rackham, 1976). Examples of ancient oaks that stand on village greens are the Meavy Oak in the village of Meavy, Devon (Wilks, 1970) and the Tilford Oak in the village of Tilford, Surrey. The gigantic Cowthorpe Oak, already dead and gone in the late nineteenth century but depicted by J. G. Strutt in his *Sylva Britannica* (1826) and by other artists, stood on a village green. This tree, however, must have been older than the village and its church (built 1456–58) and was perhaps a relic of the Royal Forest of Pickering in North Yorkshire.

Manors

During the Middle Ages, agriculture was primarily organised in manors under a feudal system of land tenure. The land was given out by the Crown to the nobility of the realm, or to the Church and leading monasteries, and was farmed under the manorial system. These tenants-in-chief (around 180 of them at the time of Domesday Book) would in turn have subtenants, each holding a manor. The centre of the manor was the manor house or monastery, from which farming and all other land use was organised and regulated. A portion of the arable land was farmed for the exclusive use of the lord of the manor as demesne land; its produce was mainly for trade. The remainder of the manorial land was divided first into various uses, depending on the needs of the manor and its population and on the physical properties of the land. A manor always had arable land, and usually pasture, woods, meadows (for hay making), 'waste', one or more fish ponds and, depending on the topography, a permanent stream or at least a natural source of water. The arable land was usually cultivated as open fields, in which peasants had narrow strips aggregated into furlongs for ploughing and seeding with the same crop spread over several fields. About one third of this was for spring planting, another third for autumn planting, and a third was laid fallow in rotation. The peasants would have to work on the lord's land for several days; the remainder of the six-day working week was dedicated to their own fields, the produce of which was taxed in kind (or later in cash) by both manor and Church. Woods and pasture were exploited communally, as they had been before the manorial system, but in the woods, the right to the timber was reserved for the lord of the manor.

The peasants usually lived in a village, as part of the manor. The lord and his family lived in the manor house, which in the Middle Ages was usually fortified to some extent, with a wet moat or a curtain wall for example. The church was shared among the lord and his tenants, as was the hall of the manor house. The lord of the manor had jurisdiction over his tenants. Under this more or less self-contained system, a manor was a microcosm of medieval existence, with the nearest market town the only place outside the manor that a peasant was likely to visit

Figure 5-29. The Eaton Manor Violet Oak, Shropshire belonged to a medieval manor

(unless he was conscripted for war). There were of course variants to this pattern; where the economy was centred on pasturing rather than on arable farming, as in the north and west, a manor could be quite different. Yet in large parts of lowland England, the manor with a nucleated village and open arable fields was the predominant kind (Whittock, 2009). A 'typical' manor is illustrated in Figure 5-28, showing the divisions mentioned. The disposition of the various land-uses was based on frequency of access. The more labour-intensive land was closest to the village, whereas land with uses that needed less attention, such as the woods and pastures, were more distant from the village, usually along the parish boundaries. With more or less primitive ploughs and limited pulling power, the natural properties of soils, as well as their fertility, also determined the location of the various land uses. Pasture

and woods tended to be situated on the soils less suitable for arable farming.

I have looked for evidence of a manor house or its subsidiary (e.g. a grange to a monastery) during the medieval and Tudor periods (i.e. before 1600) in the vicinity of existing oaks whose girths are recorded as ≥6.00 m (≥400 years old on average) in the database. Taking into account parish boundaries, I have assigned to manors any of these oaks that were not already assigned to deer parks, Royal Forests, chases or commons of the period. A total of 647 oaks in the database are thus associated with manors. These trees, like everything else on his land, would have been the property of the lord of the manor. The manor category has perhaps a lesser relevance to the presence of ancient and veteran oaks than the others used here. The sources do not usually tell us if there was pasture woodland on the manor, and even if pasture was mentioned, it need not have had trees. With the information I have consulted, of which British History Online information about parishes and manors therein is a major source, only the approximate extent of manors is indicated at best. I often had to estimate from maps, especially the nineteenth-century OS maps, whether the oaks stood on land belonging to the manor.

Early manors were often very large, and often later divided, or then again sometimes united. In the Middle Ages after the Norman conquest, all agriculture in England was manorial; in lowland England, oaks, if not in a forest or a chase, would have been present on a manor. It could be argued that I should assign all ≥6.00 m girth oaks not in forests or chases to manors, but I have assigned to this category only those oaks growing on identifiable manors that had not already been assigned to another category. The rationale for including manors as a category is mainly that they have often remained in the private ownership of generations of the same family; thus, like parks, their land tenure has remained relatively unchanged and their oaks are more likely to be preserved. An indication of this land ownership stability could be that a majority of oaks on manors (349 or 53.9%) are maiden trees; only 225 or 34.8% were recorded as pollards (see Table 8-1 in Chapter 8). Pollarding was mostly done by villagers where they had rights of such use of the trees. Many manors became country houses for the squirearchy and developed a park after 1600, a landscape change that may well have contributed to the relative paucity of large oak pollards today. At these locations, I have assigned the ≥6.00 m oaks present to the manor, not to the park, because they are likely to be older than the park. Only two sites which I have assigned to the category 'manor' have more than a few ancient oaks and are briefly described below.

Figure 5-30. Pollard oaks in spring leaves, Richmond Park, Surrey

Survey of manors

Horner Wood (Horner Combe) in the northern part of Exmoor National Park in Somerset has been mentioned in the context of pasture woodlands (Rose, 1993) and the National Trust has recorded many ancient or veteran oaks there; these have been uploaded onto the ATI database. On inspection of this site on 30 May 2013, I found that for the most part, this wood is not ancient pasture woodland nor are there many large oaks. Deer come down from the moor in winter into the woods, but that of itself does not qualify the area as pasture woodland. There is mostly oak coppice on the slope up to the moor and (hazel) coppice wood with oak standards in the valley bottom. A few locations appear to have been more open in the past, but are now invaded by birch. These woods formed part of the large Holnicote

Frostenden Grove in Suffolk is a meadow surrounded by woodland strips or wooded banks and ditches; there is a pond on the north side. To the west stands a house, 'The Grove', that is surrounded by a small park. In the woodland strips are *c.* 20 ancient and veteran oak pollards, the largest of which has a girth of 7.62 m and three others are >6.00 m in girth. On the nineteenth-century OS map solitary trees are indicated in the meadow, but only three oaks and a stump stand there now. There is evidence of an ancient hornbeam hedge along the east side and to the south along the road, where there is a large hornbeam pollard. Apart from the large oak trees, the woodland strips are young and were mostly planted in the second half of the twentieth century. Frostenden was mentioned in Domesday Book (1086) as having wood for 40 pigs; and in the sixteenth century, it was recorded as a "wood pasture region". By this time the manor was probably divided in two, one part held by the Dean and Chapter of Westminster; later, the land was divided still further. Frostenden Grove may have been the last remnant of manorial pasture woodland when it was enclosed with a hornbeam hedge.

Other historical landscape features

Hedges are still a very common feature in the English rural landscape. In 'Ancient Countryside', fields are smaller than in 'Planned Countryside'; as a result, there are at least twice as many hedges in the former (Rackham, 1976). Ancient hedges can be several metres wide and can resemble narrow strips of woodland; they are further characterised by irregular shapes and old stools of native species formed by hedge laying (Muir, 2008). Hedgerow trees, especially oak and ash (and elm still in some places), are common and have been so at least since Anglo-Saxon times, although explicit mention of trees in hedges is rare in the charters (Hooke, 2010). Clear evidence of the presence of trees in hedges appears on detailed maps of the English countryside from the late sixteenth century onwards. In an enclosed agricultural landscape, where the exploitation of woods was often restricted, hedges were a place where the villagers could have access to trees for lop (hedgebote) or even timber (housebote). These trees could be maiden trees but were often pollards, or more rarely shredded trees.

Only since the second half of the sixteenth century were these trees commonly planted, usually when new hedges were created (planted and laid) in open fields of communal ploughland (Muir, 2005). Sir Anthony Fitzherbert, in his *Book of Husbandry* (ed. of 1534) wrote: "And set thy oak-settes and the asshe .x. or .xii fote a-sonder, and cut them as thou dost thy other settes, and couer them ouer

Estate, a manor first mentioned in Domesday Book when an area of "2.5 virgates", which referred only to the ploughland, was held by two nuns. We do not know when the manor first had woods belonging to it, but we do know that it came into lay hands and greatly expanded in later centuries. The woods must have been intensively exploited for their timber and underwood. Horner Wood is now owned by the National Trust. Just seven oaks have girths of between 6.00 m and 7.00 m and most are much smaller; some low coppice stools could also be ancient. The large oaks are all 'multi-stem', resulting from high coppicing, so their girths were measured at between 0.5 and 1.3 m above the (sloping) ground. They were recorded as 'pollards' but they are not properly (lapsed) oak pollards at this height of past cutting.

Tree number	Tree form	Girth	County	Locality	Oak name
12217	Maiden	11.60 m	Cornwall	Darley	Darley Oak
105209	Pollard	12.80 m	Herefordshire	Newbury Farm	Gospel Oak
49847	Maiden	9.10 m	Norfolk	Tasburgh	No name
67865	Pollard	9.40 m	Norfolk	Woodgate	No name
142090	Maiden	9.50 m	North Yorkshire	High Mains Farm	No name
88446	Pollard	11.11 m	Shropshire	Crow Leasowes Farm	Crowleasowes Oak
22	Pollard	12.75 m	Worcestershire	Yell's Farm	Great Witley Oak

Table 5-3. Ancient oaks >9.00 m in circumference standing solitarily. These trees have no other ancient or veteran oaks present within a radius of 1 km and no known historical connection to a medieval deer park, a Tudor deer park, a Royal Forest, a chase, a common or a manor. Tree numbers from the ATI database.

with thornes a lyttell, that shepe and cattle eat them not." In earlier times, trees that had been dispersed as seeds naturally and germinated in hedges were simply spared to grow, often to become large trees (Rackham, 1976). This practice was probably continued in later times in the new hedges; where oaks and other trees occur in the landscape, their seeds will often germinate. The most common hedge tree (in most parts of England) is the native oak (*Quercus robur* and *Q. petraea*); this was also the case in the Middle Ages, as oaks were the most valuable trees to keep for timber. Of course, timber trees were not maintained until their ancient stage but would have been cut before they reached that age. Pollard oaks in hedges could survive longer and, unlike timber trees, could be freely used by the peasants. They are relatively rare in hedges at the present time, partly because pollarding has not been practiced for 150–200 years (Edlin, 1971), but mainly because they were often destroyed and replaced by timber trees after around 1800 (Muir, 2005). The locality of ancient or veteran pollards on the now-fenced border of a field may indicate that they once stood in a hedge. In other places, their alignment may reveal their origin in a hedge.

The nineteenth-century editions of the OS map are much superior in detail that is useful for our purpose when compared to modern editions; they carefully depict prominent hedgerow trees as well as solitary trees in fields. Estate maps drawn in the sixteenth and seventeenth centuries also show hedgerows with individual trees (examples in Muir, 2000; 2005). These maps sometimes make it possible to confirm ancient or veteran oaks as hedgerow oaks by plotting them on the current map using a GPS location, and then looking at this situation on the earlier maps. Performing this task for all of the ≥6.00 m girth oaks in the database that have not been allocated to any of the land use categories used here would be very laborious, with I suspect only limited positive returns, in part because many localities given in the ATI are not accurate within tens of metres. In fact, hedgerows may even have existed here and there within the other categories, for example, along roads or cattle drives through chases, Royal Forests or commons. When parks incorporated farmed fields or ancient tracks, as was especially common in Tudor parks or in parks that were enlarged after 1350 when the rural population had crashed, hedgerows and their trees may also have been incorporated. Examples are Ripley Park in North Yorkshire (Muir, 2005: 122) and Richmond Park in Surrey. In such cases I have allocated the oaks to a park or a common, not to a hedgerow.

Boundary oaks were trees that stood in an open landscape and marked the boundary of a manor, parish, Royal Forest, chase or a common. They were already mentioned in Anglo-Saxon charters (Rackham, 1994; Hooke, 2010), following Roman tradition. Only around 11% of the trees mentioned in these charters were oaks; the most commonly mentioned tree is apparently hawthorn, indicating that boundary trees often did not stand in hedges, as a hawthorn in a hedge would not have been particularly conspicuous. Oaks and other trees were used as boundary markers in perambulations, which could be written down but were often passed on orally. The young men of a parish or village would walk the boundary, often led by the local priest, to learn the markers (among which were trees) and thereby the limits of the parish. These trees would have been already there; it would have made little sense to plant one and wait for many years until it became large enough to be a landmark. Also, if trees had been planted as future boundary markers, we would see an increase of trees used as markers over time. If a marker tree was blown down and died, however, it may sometimes have been replaced. Many other kinds of landmarks

Figure 5-31. The Gospel Oak at Grendon Bishop, Herefordshire

were also used to define boundaries and trees were never predominant. Nevertheless, boundary trees would be preserved, and pollarding achieved this while still allowing use of the wood.

A possible variant of trees that were used as landmarks are trees that marked places of assembly, in Anglo-Saxon time especially 'hundred' meeting places (Hooke, 2010). A 'hundred' ('wapentake' in the north) was a basic unit of local governance and would have been delimited with a perambulation. Sometimes, an oak tree was given a name that could indicate its use as a boundary or assembly tree; an example is the Gospel Oak in Herefordshire (Figure 5-31), now an ancient pollard oak of *c.* 12.80 m circumference and thus certainly dating back to the Middle Ages. The perambulating priest may have recited from the Gospels when his young charges reached the tree, but this notion is of course speculative. 'Shire Oak' was another name that indicated a boundary-marking oak (Muir, 2008). Of the four oaks so named in the List of Historic and Named Oak Trees in Britain (Harris *et al.*, 2003), none exist today.

The problem with historical research on such 'marker' or 'meeting' oaks is that there is much storytelling and many anecdotes that relate to them; many quite implausible tales are cited in the literature on oaks. Folk history is usually not backed by evidence. The evidence would have to come from charters, legal proceedings regarding boundaries and so on, and collecting it would be a laborious procedure with limited results. For this reason, I have not used boundaries or markers as categories in the database to describe the historical context of ancient oaks.

Despite this uncertainty, a comparative history of the landscape may provide some evidence of the presence of

County	Ancient oaks	Circumference	(Former) medieval deer park	Other land use or locality
Bedfordshire	Doomsday Oak	9.30 m	Woburn Park	
Berkshire	King Offa's Oak	11.18 m		Cranbourne Chase
	Signing Oak	9.72 m	Windsor Great Park	
	6296	9.34 m		Cranbourne Chase
	28036	9.20 m		Cranbourne Chase
	35308	9.05 m	Easthampstead Park	
Cheshire	Marton Oak	14.02 m		Delamere Royal Forest
Cornwall	Darley Oak	11.60 m		Unknown
Cumbria	Trysting Tree	9.80 m		Askerton Chase
Derbyshire	Old Man of Calke	10.01 m	Medieval park?	Calke Park
	24093	9.41 m	Medieval park?	Calke Park
Devon	King John's Oak	10.04 m	Woodend Park	
	3574	9.09 m		Manor
Dorset	Billy Wilkins Oak	11.92 m	Melbury Park	
	Wyndham's Oak	9.86 m		Gillingham Royal Forest
Essex	Great Oak	10.75 m	Medieval park?	Barrington Hall
	Burry Maiden Oak	9.86 m	Easton Park	
	8470	9.68 m	Medieval park?	Porter's Hall
	Old Knobbley	9.60 m		Common
Gloucestershire	10973	11.28 m	Eastwood Park	
	10297	10.55 m		Manor
	134563	10.50 m		Manor
	Bourton Oak	10.21 m		Common
	10744	9.78 m	Painswick Park	
	3351	9.60 m	Haresfield Park	
	35381	9.40 m	Sudeley Park	
	113589	9.10 m	Brockworth Park	
Greater London	Kenley House Oak	9.56 m		Common?
Hampshire	Oakley Oak	10.74 m		Buckholt Royal Forest
	Bere Oak	9.55 m		Bere Royal Forest
	Neatham Manor Oak	9.08 m		Manor
Herefordshire	Gospel Oak	12.80 m		Unknown
	Jack of Kent's Oak	11.37 m	Kentchurch Park	
	Quarry Oak	10.97 m		Croft Park
	Old Oak (12140)	10.90 m		Manor
	Wagoner's Oak	10.25 m	Medieval park?	Shobdon Park
	2565	9.94 m		Holme Lacy
	7528	9.90 m	Medieval park?	Shobdon Park
	6050	9.50 m		Manor
	Hypebaeus Oak	9.29 m	Moccas Park	
	11112	9.28 m	Netherwood Park	
	Great Oak	9.25 m	Eardisley Park	Lower Welson
	The Monarch	9.25 m		Holme Lacy
	88597	9.21 m		Manor
	42623	9.20 m	Brampton Bryan Park	
	84867	9.19 m	Sarnesfield Park	
	134803	9.10 m		Bringewood Chase
	141651	9.10 m	Medieval park?	Shobdon Park

Table 5-4. The ancient oaks of England >9.00 m in circumference, listed by county. As regards the locations of these trees in the historic landscape, medieval deer parks, or putative medieval deer parks, are named in the fourth column; other land uses, including Royal Forests, chases, later deer parks, manors, commons and 'unknown' uses are named in the fifth column. Tree numbers from the ATI database.

County	Ancient oaks	Circumference	(Former) medieval deer park	Other land use or locality
Hertfordshire	Great Oak	10.70 m	Kingswalden Park	
	Old Stumpy	10.50 m	Hatfield Park	
	6873	9.54 m	Medieval park?	Brocket Park
	61086	9.50 m	Ware Park	
	61009	9.44 m	Hertingfordbury Park	(Panshanger)
Kent	Fredville Oak	12.18 m		Manor
	6359	11.17 m	Penshurst Park	
	48304	10.55 m		Manor
	6663	10.36 m	Lullingstone Park	
	6693	9.74 m	Medieval park?	Brenchley
	The Sidney's Oak	9.42 m	Penshurst Park	
Lancashire	31493	10.00 m		Manor
Lincolnshire	Bowthorpe Oak	12.80 m		Manor
	35685	9.32 m		Grimsthorpe Park
Norfolk	Carbrooke Oak	9.42 m		Manor
	67865	9.40 m		Unknown
	49847	9.10 m		Unknown
Northamptonshire	King Charles Oak	10.62 m		Salcey Royal Forest
North Yorkshire	38805	10.75 m	Ingleby Greenhow	
	142090	9.50 m		Unknown
	54569	9.08 m		Ripley Park
Nottinghamshire	Major Oak	10.66 m		Sherwood Royal Forest
Oxfordshire	136563	10.37 m	Woodstock Park	(Blenheim)
	144530	10.34 m	Woodstock Park	(Blenheim)
	?	10.05 m	Cornbury Park	
	Chastleton Oak	9.77 m		Tudor park?
	144526	9.47 m	Woodstock Park	(Blenheim)
	12181	9.40 m	Woodstock Park	(Blenheim)
	Radley College Oak	9.40 m	Radley Park	
	Breame Oak	9.14 m	Holton Park	
Shropshire	Lydham Manor Oak	13.01 m	Medieval park?	Lydham Manor
	Crowleaseowes Oak	11.11 m		Unknown
	Powis Oak	10.54 m	Medieval park?	Underhill Hall
	Acton Round Oak	10.04 m		Shirlot Royal Forest
	Eaton Manor Violet Oak	10.03 m		Manor
	101064	9.65 m	Medieval park?	Lydham Manor
	Holt Preen Oak	9.13 m	Medieval park?	Holt Farm
	10551	9.11 m	Medieval park?	Moor Park
	Dryton Oak	9.04 m		Manor
Somerset	King's Oak	10.74 m	Queen Camel	(Hazlegrove)
	4484	10.26 m		Manor
	25686	9.34 m	Alfoxton Park	
	135235	9.24 m	Queen Camel	(Hazlegrove)
	Domesday Oak	9.19 m	Ashton Court Park	
Staffordshire	133703	11.00 m		Cannock Chase
	123045	9.55 m	Yoxall Park	
Suffolk	145660	10.45 m		Manor
	Tea Party Oak	9.79 m	Ickworth Park	
	Haughley Oak II	9.67 m	Haughley Park	
	43246	9.50 m	Melford Park	
	Sibton Abbey Oak	9.36 m	Medieval park?	Sibton Abbey

County	Ancient oaks	Circumference	(Former) medieval deer park	Other land use or locality
Sussex	Elizabeth I Oak	12.67 m	Cowdray Park	
	8705	10.50 m	Danny Park	
	4426	10.32 m	Cowdray Park	
	Whiligh Oak	9.27 m	Medieval park?	Whiligh Manor
	Sun Oak	9.20 m		St Leonard's Chase
	14888	9.13 m	Petworth Park	
Warwickshire	3382	9.24 m		Manor
Wiltshire	Big Belly Oak	11.18 m		Savernake Royal Forest
	Great Oak	10.67 m	(Spye Park)	Melksham Royal Forest
	King of Limbs	10.25 m		Savernake Royal Forest
	Cathedral Oak	10.15 m		Savernake Royal Forest
	Pot Belly Oak	9.80 m	(Spye Park)	Melksham Royal Forest
	Nath's Oak	9.20 m		Braydon Royal Forest
Worcestershire	Great Witley Oak	12.75 m		Unknown
	Temple Oak	11.10 m		Manor
	34322	10.32 m		Feckenham Royal Forest
Totals	**115**		**60 (16 inferred)**	**55 (20 Royal Forest + chase)**

boundary or meeting oaks in some locations. An interesting case of this is the Medieval Oak in Hampton Court Park, Greater London (now dead). This park, also known as the Home Park, together with Bushy Park, has been a deer park since the sixteenth century, when it was imparked by Cardinal Thomas Wolsey and avidly used by Henry VIII for chasing and shooting deer. These parks, now divided by Hampton Court Road, were and still are surrounded by a brick wall. In stark contrast to nearby Richmond Park, which has around 130 ancient and veteran oaks, Hampton Court's Home Park has just this one (dead) ancient pollard oak, although at 12.25 m in circumference, it was larger than any oak in Richmond Park. These parks have a very different medieval history. Whereas Richmond is a deer park created from pasture woodland and open commons on hilly, gravelly sand, the Hampton Court parks were established on level and mostly arable land, for the most part after 1500. There are numerous archaeological traces of furlongs and field margins in Bushy Park especially (Longstaffe-Cowan, 2005). The Medieval Oak therefore once stood in arable fields and was most likely an assembly oak as it is not on any ancient boundary. A group of ancient oaks near the Hampton Court Gate of Bushy Park possibly represent a small area of pasture woodland.

A variant of the boundary oak, although not usually designated as a landmark for perambulation, are oak pollards on wood banks. These and other tree pollards would be larger in girth than the standard oaks inside a coppice wood and were presumably pollarded because cattle could reach them from fields alongside, especially if the ditch was silted up or the tree stood on its edge. Such pollards still exist around some ancient woods (Rackham, 2003b, p. 232) but are not common and as far as I know, very few are ancient large oaks.

There are many isolated ancient oaks in the English countryside, often standing in the middle of farm fields or on a farm that did not exist in the Middle Ages. These trees often have no obvious link to a hedge or a wood bank past or present, and no demonstrable connection to a destroyed deer park or to a former Royal Forest or chase, a pasture woodland common or a vanished manor. It is of course possible that detailed archaeological and historical local research may provide some explanation as to why they are still there and what their medieval or Tudor landscape context was. But for most, this information will probably be impossible to obtain because no ancient records refer to the trees and because later agricultural or urban development has destroyed the traces of past land use related to these trees. In Table 5-3, the 'unexplained' solitary ancient oaks with a girth >9.00 m are listed by county. I have taken solitary to mean that no other ancient or veteran oak ≥6.00 m is present within a radius of 1 km from the large ancient oak. These are also the 'unknown' oaks listed in Table 5-4.

Locations that have the largest ancient oaks
A survey of the largest ancient pollard or maiden oaks in England still alive in early 2016 may be another approach to investigate the type of landscape and land use in which these ancient oaks originated and were most common. A somewhat arbitrary minimum circumference of >9.00 m has been chosen to delimit these trees. Although there

Figure 5-32 (left). The ancient oak in Quaker Lane, Tasburgh, Norfolk

Figure 5-33 (below). The Crowleasowes Oak at Crow Leasowes Farm, Middleton, Shropshire

is a relationship between circumference and age, this relationship is non-linear for individual ancient oaks (see Chapter 2). Nevertheless, it is reasonable to assume that these largest oaks are generally also the oldest. Trees that exceed this minimum circumference are most probably over 600 years old and therefore originate from the Middle Ages. Coppice ('multi-stem') trees are excluded here because they are in a different landscape category associated with coppice woods.

For this survey, data are taken from the PDF of Great Oaks of Britain and Ireland (www.treeregister.org/membership/oak.shtml), from the Ancient Tree Inventory (ATI), from the List of Historic and Named Oak Trees in Britain in Harris *et al.* (2003), from Hight (2011), from

Miles (2013) and from field work by the author. Unverified records in the ATI database have been ignored unless there was other evidence available. Given the uncertainties about girth measurements explained in Chapter 2, some trees may be smaller or larger than the figures given here. Some oaks may thus fall below the threshold, but others recorded at just below or at 9.00 m may have been larger and should then have been listed. It is assumed that these uncertainties will not substantially influence the total number of oaks with a girth >9.00 m. The oaks are listed by county and identified by name or by their ATI number.

This inventory of the largest ancient oaks in England shows a moderate predominance of medieval deer parks in relation to their distribution. In 16 cases, a medieval park

County	MDP	%	TDP	%	RF	%	CHA	%	COM	%	manor	%	other	%	Total
Beds	5	55									4	45			9
Berks	96	42	7	3	6	2.6	75	33	6	2.6	22	9.7	14	6	226
Bristol	4	57									2	29	1	14	7
Bucks	16	64					1	4	2	8	3	12	3	12	25
Cambs			5	62					1	12	2	25			8
Ches	2	11			1	5.5					7	39	8	44	18
Corn	8	57									4	29	2	14	14
Cumb	16	23	6	8.7	6	8.7	6	8.7	8	12	14	20	13	19	69
Derbs	82	93	2	2							2	2	2	2	88
Devon	35	33	34	32					1	1	24	23	11	10	105
Dorset	9	14	3	5	5	8	9	14	9	14	26	41	2	3	63
Co Dur											7	78	2	22	9
E Yorks									1						1
E Sussex	11	23	10	21					1	2	11	23	14	30	47
Essex	28	35	9	11	8	10			4	5	21	26	10	13	80
Glos	44	33	5	3.7	3	2	15	11	4	3	39	29	24	18	134
G London	14	17	6	7			1	1	42	52	10	12	8	10	81
G Man	1	33									1	33	1	33	3
Hants	14	18	10	13	37	46			3	4	8	10	8	10	80
Hre	103	28	53	14.5	2	0.5	5	1.4	47	13	96	26	60	16	366
Herts	80	79	1	1					2	2	12	12	7	7	102
I of W			6	86									1	14	7
Kent	41	45	9	10					1	1	23	25	17	19	91
Lancs											4	57	3	43	7
Leics	3	60	1	20									1	20	5
Lincs	2	6	19	54					1	3	7	20	6	17	35
Norfolk	37	19	9	5			5	2.6	24	12	75	39	42	22	192
N Yorks	41	40	22	22			10	10	3	3	9	9	17	17	102
Northants			9	41	3	14	3	14			1	5	6	27	22
Northd			1	14							1	14	5	71	7
Notts	6	10			52	90									58
Oxon	142	82	3	2	1	0.6			7	4	10	6	10	6	173
Rutland											1				1
Salop	35	18	27	14	21	10	4	2	8	4	42	21	62	31	199
Somerset	37	45			3	3.6			4	5	18	22	20	24	82
S Yorks			1												1
Staffs	14	54					7	27			2	8	3	11.5	26
Suffolk	89	42	14	7					10	5	62	29	35	17	210
Surrey	23	40	7	12	2	3.5	2	3.5	9	16	8	14	6	10	57
T & W											1				1
Warks	27	18	82	54					2	1.3	32	21	9	6	152
W Mids	2	15									9	70	2	15	13
W Sussex	38	42	30	33			6	7	5	5.5	3	3	8	9	90
W Yorks			1	14							3	43	2	29	7
Wilts	2	2	17	14	90	75	1	1			6	5	4	3	120
Worcs	27	21	31	24	2	2	8	6	11	9	14	11	34	27	127
Totals	**1136**	**34.2**	**438**	**13.2**	**242**	**7.3**	**158**	**4.8**	**217**	**6.5**	**647**	**19.5**	**482**	**14.5**	**3320**

is inferred from traces of a park boundary still visible in the landscape from the air, and/or from traces of a moat nearby, indicating a medieval fortified house which often had a park. Twenty large ancient oaks are in (areas of) Royal Forests or chases and are most likely associated with these historical land uses. Two or three are likely to have stood in ancient pasture woodland commons. Five large oaks are in Tudor deer parks, i.e. Croft Castle Park, Grimsthorpe Park, Holme Lacy Park, Ripley Park and possibly a park, now with no remains, in which stood the Chastleton Oak. Several oaks are associated with manors (without deer parks), leaving just seven oaks >9.00 m as 'unknown' (see also Table 5-3).

Cranbourne Chase, once part of Windsor Forest and now in Windsor Great Park, is medieval pasture woodland and has some fields with many ancient oaks, most are not obvious pollards. A similar situation was found in High Park, part of Blenheim Park, which is the least-altered section of medieval Woodstock Park. Some of these 'maidens' may also have been high pollards or shredded oaks, but evidence for either of these treatments is now extremely difficult to obtain. Of the 115 largest ancient oaks in England, 66 are pollards, 45 are maidens and four are neither, with one undetermined. The figures for the rest of Europe are 28 pollards, 51 maidens and 17 undetermined.

When shown on a map of England (Chapter 3, Figure 3-8), some patterns emerge in the distribution of large ancient oaks within three landscape categories — deer parks, Royal Forest and chases, and other land use. There are only seven large ancient oaks north of a line from north Shropshire to south Lincolnshire, against 108 to the south. Cornwall and Devon have only three large oaks, so this most westerly part of England is also poor in great oaks. Most of these large oaks are well spread across the remainder of England, but there are some concentrations, with Herefordshire having the highest number (17), followed by Shropshire (9), Gloucestershire (8) and Oxfordshire (8). Those in deer parks are, with two exceptions in North Yorkshire, all in the southern half of England. Relatively high frequencies of great oaks in the

'other land use' category are found in the mid-western counties and in East Anglia, but this category includes several oaks with uncertain associations. Large ancient oaks that are associated with Royal Forests and chases are found from Dorset and Hampshire north to Nottinghamshire and Cumbria. This distribution is only partly coincident with (remains of) Royal Forests and chases and appears less complete than that of trees within parks.

Conclusions

On 1 April 2016, after my editing, the Ancient Tree Inventory (ATI) database contained a total of 3,320 records of living oaks with a girth ≥6.00 m for all of England. An analysis of where they occur gave the following statistics: Royal Forests 242 (7.3%), chases 158 (4.8%), medieval deer parks 1,136 (34.2%), Tudor deer parks 438 (13.2%), commons 217 (6.5%), manors 647 (19.5%) and other or unknown past land use 482 (14.5%). These results are summarised by county in Table 5-5. Recording is not complete, and with more detailed local historical research, the percentage of unknown past land use would probably come down, so these figures remain indicative.

To assign a locality with one or more of these oaks to one of the land use categories, I first looked at the locality on the OS maps, both nineteenth century and present, and determined (where possible) a property to which it belonged or still belongs, with emphasis on the historical evidence relevant to the tree. Particular attention was given to the landscape categories as they still appeared on the nineteenth-century OS map, or could be inferred. A variety of historical records published in books, as printed maps or online were consulted to determine whether an oak tree could be associated with one of the landscape categories. In all cases of inference, a question mark is added to the category in the database. As an example of my method, Windsor Great Park as a medieval deer park did not have the same boundary as it does today, therefore oaks in that park may instead be assigned to Cranbourne Chase, for example, depending on their locality; and what I term 'deer park' there is not the same area as that in which deer are currently kept. By taking 6.00 m girth as a minimum size, I believe that (in most cases) I have eliminated oaks younger than 400 years. This, in turn, means that the oaks in the database were not planted but are the spontaneous descendants of original 'wildwood' oaks.

The 'wildwood' became pasture woodland or managed (coppiced) woods or treeless land and in Roman times this was already the prevailing situation throughout England (Rackham, 1986). In Anglo-Saxon times, woodland gradually expanded from a low during the height of Roman

Figure 5-34. Ancient oak of 10.34 m girth in High Park, Blenheim Park, Oxfordshire, photographed on 28 April 2015 (**top**) and on 23 May 2015 (**bottom**)

exploitation, but by the time of the compilation of Domesday Book (1086), it still only covered 15% of all the land recorded. By then, all unmanaged woodland would have been pasture woodland. The Norman kings included almost all pasture woodland, and certainly all the larger areas, in the Royal Forests. Subsequently, chases and deer parks were created; in most cases incorporating pasture woodland that belonged, or once had belonged, to the forests. Less frequently, commons included areas of pasture woodland and much of this was turned to heath or chalk grassland by overexploiting the trees. Land was given out to nobility and churchmen who created manors, which often incorporated pasture woodland that could be imparked or not. If this sequence of events is correct, we might assign most ancient oaks to the creation of Royal Forests, or to land uses that

existed even further back in time. I do not believe, however, that this will give us an informative historical background that explains the presence and distribution of ancient oaks today. I have tried to indicate the historical situation that was relevant to the oldest extant oaks, but we must remember that other, younger ancient or veteran oaks also occur in the same localities. When these oaks are considered too, we find that medieval deer parks are by far the most important historical features in the landscape associated with ancient and veteran oaks. With similar purpose and use but of later creation are Tudor deer parks and together these parks, all established before 1600, account for 47.5% of all ancient and veteran oaks with ≥6.00 m girth in England. It is to the ancient English deer parks that we owe the preservation of so many of these oaks.

6

Ancient oaks in Europe

Both species of oak considered in this book, sessile oak and pedunculate oak, are widely distributed in Europe and beyond; their ranges mostly overlap (Le Hardy de Beaulieu & Lamant, 2006; Tyler, 2008). Sessile oak extends further north in Scandinavia and further east towards the Caspian Sea than does pedunculate oak. In northwest Europe, pedunculate oak is more successful in warmer and drier conditions than sessile oak, so that we see a gradual shift in frequency between the two species with sessile oak becoming more common towards the cooler and wetter North Atlantic coastal regions, including Ireland, Scotland and Norway. Although we rarely distinguish between the two species in this book, I do mention the species (where recorded) in this chapter because there appears to be a distinction among ancient oaks that is perhaps more pronounced than it is in England. Ancient oaks of these species, at least those larger than 7.50 m in circumference, do not seem to occur throughout their ranges but are concentrated in north-western and central Europe. They are not recorded from central and southern Spain, Portugal, southern France, Italy, the Balkan Peninsula, the North African coast, regions around the Black Sea except Turkey, most of Ukraine, Belarus and Finland. These regions are more or less marginal to the environmental requirements of one or both species, and other oak species predominate at least in the more southern and southeastern parts of

Figure 6-1. The Great Oak of Mercheasa near Rupea in Transylvania, Romania

Europe and into western Asia. Furthermore, and less obvious perhaps, almost all large ancient oaks (>7.50 m in girth) recorded on the continent appear to be pedunculate oak. This is also the more common species for large ancient oaks in England, but there, several belong to sessile oak.

It has been noted by several authors (Rackham, 1976, 1986; Rose, 1993; Pakenham, 2003; Tyler, 2008; Hight, 2011) that there are more ancient (oak) trees in Britain than in the rest of Europe (except Greece according to Rackham, but his assessment includes tree species other than oaks). Usually, this statement is in an introduction accompanied by little or no documented evidence. Here, an attempt is made to provide such evidence for the oaks. The 'hobby' of measuring and recording 'champion trees' of any species, native or introduced, by country, province or county may have originated in England with Alan Mitchell and Tony Schilling (I'm not sure), but it has now spread to almost all European countries and beyond. Many people spend their leisure time scouring the countryside, including gardens and parks, for the 'biggest' trees; 'big' is usually, but not exclusively, expressed terms of the circumference of the trunk. More recently, this information has been made available online via international, national or locally focused websites; sometimes the (earlier) records are published in print. The two oak species we are interested in here feature prominently in these inventories because they can grow to impressive sizes.

List of oaks >9.00 m in recorded circumference in Europe (England excluded)

After consulting a variety of web-based and other sources — Belgium, France, Germany, Ireland and Italy also have books published on their largest and/or oldest trees — it is possible to present an informed, albeit not entirely complete, inventory of ancient oaks with a circumference greater than 9.00 m in Europe outside England. A list for England is presented in Chapter 5. Only oaks that were still alive when the list was compiled in 2015 are included, and all of the trees in the list are maidens or pollards, not coppice stools.

Austria has a large pedunculate oak at Bad Blumau in the district of Fürstenfeld in Styria, but it measured just under 9.00 m at 8.75 m circumference in 2011. Switzerland has a large pedunculate oak known as Chêne des Bosses near the village of Châtillon in the Jura, but it measured 8.70 m circumference in 2007. Another pedunculate oak near Lausanne has been measured with 8.80 m girth at 1.00 m above the ground after breakage; it may have been over 9.00 m. Italy's largest pedunculate oak near Bertolio in the province of Udine has a girth of 8.00 m. Being smaller than 9.00m in girth, these oaks are not included in the list

below, but being largely intact, they may enter the list in future. A large pedunculate oak (of 12.05 m in girth) in Turkey that I visited in May 2015 is included although not strictly speaking in Europe. The tree record numbers given here refer to those in the list at the website www.monumentaltrees.com compiled by its members.

Belgium
- Liernu in Namur, Place de Liernu, maiden oak tree 1649 measured at 10.10 m in 2013

Czech Republic
- Náměstí Nad Oslavou in Vysočina, road 399, pollard oak tree 10489 measured at 10.20 m in 2009
- Petrohrad in Louny District, along the road to Stebno, maiden oak tree 12237 measured at 9.18 m in 2009
- Podhradí in Třemošnice, near Lichnice Castle, maiden oak tree 12313 measured at 9.10 m in 2010

Denmark
- Domain Corselitze in Karleby, oak tree 1810 measured at 9.54 m in 2006
- Jægersborg Deer Park near Copenhagen, maiden oak tree 15928 measured at 10.45 m in 2006*
- Jægersborg Deer Park near Copenhagen, maiden oak tree 14237 measured at 10.68 m in 2006*
- Nordskov near Jægerspris Castle in Frederiksund, pollard oak tree 1698 estimated at >9.00 m in 1965

*On a visit to these oaks 18 August 2015, I concluded that these trees are not proper pollards as recorded.

France
- Alouville-Bellefosse in Seine-Maritime, maiden oak tree 1704 measured at 9.95 m in 1998
- Belmesnil in Seine-Maritime, ferme Socquentot, maiden oak tree 13586 measured at 9.75 m in 2013
- Bulat-Pestivien in Côtes-d'Armor, Tronjoli, oak tree 1706 measured at 12.60 m in 1998
- Forêt de Paimpont (Brocéliande) in Brittany, pollard oak tree 15464 measured at 9.51 m in 2000
- La Chapelle-Montlinard in Cher, pollard oak tree 13913 measured at 10.52 m in 2013
- Montravail, Pessines near Saintes in Charente-Maritime, oak tree estimated at >9.00 m in 2002
- Plassay in Poitou-Charentes, field west of village, maiden oak tree 17039 measured at 9.10 m in 2011
- Saint-Jean-Brévelay in Morbihan, Brittany, Chêne du Pouldu oak tree measured at 9.30 m in 2014
- Saint-Vincent-de-Paul in Gironde, near church, pollard oak tree 4675 measured at 10.40 m in 2002

Germany

- Altenhausen-Ivenrode in Sachsen-Anhalt, Bischofswald, oak tree measured at 9.23 m in 2001
- Amt-Neuhaus in Lower Saxony, not specified
- Beberbeck in Hessen, Dicke Margarete oak tree measured at 9.56 m in 2002
- Belau in Schleswig-Holstein, Gut Perdöl, maiden oak tree 12376 measured at 11.97 m in 2010
- Berteroda (Eisenach) in Thüringen, An der Eiche, maiden oak tree 4406 measured at 10.06 m in 2013
- Borlinghausen in Nordrhein-Westfalen, Tränkeweg, pollard oak tree 5053 measured at 10.59 m in 2011
- Dansenau in Rheinland-Pfalz, on the bank of the River Lahn, maiden oak measured at 9.15 m in 1996
- Ebersbach in Saxony, Storcheneiche oak tree measured at 9.50 m in 1998
- Eisolzried in Bavaria, in the former castle park, maiden oak tree 11135 measured at 9.55 m in 2013
- Forsthaus Grünejäger in Lower Saxony, oak tree measured at 9.95 m in 2006
- Gollingkreut in Bavaria, along the road to Halsbach, maiden oak tree 11001 measured at 9.25 m in 2013
- Heiligengrabe in Brandenburg, pollard oak tree 12442 estimated at 9.10 m in 2013
- Herwigsdorf in Saxony, along road southeast of village, pollard oak tree 12577 measured at 9.01 m in 2008
- Hornoldendorf in Nordrhein-Westfalen, Mauereiche oak tree measured at 9.02 m in 2003
- Ivenack in Mecklenburg-Vorpommern, maiden oak tree 1758 measured at 11.59 m in 2013*
- Krügersdorf in Brandenburg, Dicke Eiche maiden oak tree 10600 measured at 10.32 m in 2013
- Lehsen in Mecklenburg-Vorpommern, oak tree measured at 9.20 m in 2001
- Minzow in Mecklenburg-Vorpommern, Kroneiche maiden oak tree 10014 measured at 9.87 m in 2013
- Nagel in Bavaria, Schloss Alte Kemenade, maiden oak tree 9526 measured at 9.30 m in 2011
- Neuenstein in Baden-Würtenberg, Emmertshof, maiden oak tree 14238 measured at 11.10 m in 2013
- Nöbdenitz in Thüringen, Dorfstrasse, Grabeiche maiden oak tree 1764 measured at 10.40 m in 1999
- Raesfeld in Nordrhein-Westfalen, Erle, pollard oak tree 1788 estimated at 12.25 m in 2011
- Thalmassing in Bavaria, Neueglofsheim, maiden oak tree 4319 measured at 9.70 m in 2013
- Volkenroda in Thüringen, near the Pfingstrasse, maiden oak tree 5058 measured at 9.65 m in 2013

*On a visit 20 April 2015, a recently installed information board was seen to give this oak a circumference of 10.96 m.

This is probably due to measurement at 1.30 m instead of 1.00 m (see under Germany below).

Hungary

- Zsennye in Vas County, along Jókay Mór Street, maiden oak tree 7692 measured at 9.80 m in 2004

Ireland

- Mountshannon in County Clare, Oak House, maiden oak tree 13013 measured at 9.36 m in 2013

Latvia

- Kaive in Tukums District, oak named Kaives Ozols, maiden oak tree 4006 measured at 10.20 m in 2010
- Kanepes in Jerceni, oak named Kanepu, maiden oak tree 15948 measured at 9.60 m in 2009
- Rigzemes in Dundaga, maiden oak tree 15951 measured at 9.10 m in 2011
- Sēja Estate in Vidzeme District, maiden oak tree measured at 9.12 m in 2011

Lithuania

- Stelmužė Manor Park, Valmiera District, maiden oak tree measured at 9.70 m in 2014

Norway

- Birkenes near Grimstad in Aust-Agder, Mollestad Eik pollard oak tree measured at 10.50 m in 2013
- Horten in Vestfold, Tordenskjold Eik pollard oak tree measured at 10.50 m in 2001
- Ullensvang on Hardangerfjord in Hordaland, Brureik pollard oak tree measured at 9.84 m in 2007

Poland

- Bakowo in Gdańsk County, along Road 391, maiden oak tree 11978 measured at 10.28 m in 2013
- Januszkowice in Krapowice County, on street, maiden oak tree 4798 measured at 10.23 m in 2013
- Kadyny in Elblag County, close to the palace park, maiden oak tree 14031 measured at 9.90 m in 2001
- Nogat in Gdańsk County, Dab Chroby, maiden oak tree 11973 measured at 9.18 m in 2013
- Rogalin in Poznań County, garden of Rogalin Palace, maiden oak tree 1778 measured at 9.30 m in 2011
- Szprotawa, Piotrowice in Żagań County, maiden oak tree 1786 measured at 10.10 m in 2009
- Weglowka in Krosno County, maiden oak tree 11967 measured at 9.01 m in 2013
- Zagnańsk in Kielce County, maiden oak tree 1789 measured at 9.45 m in 1998

Figure 6-2. The Ulvedals Oak in Jaegersborg Deer Park near Copenhagen, Denmark

- Młok in Ciechanów County, maiden oak tree 18851 measured at 9.19 m in 2014
- Szydłowiec Śląski in Opole County, maiden oak tree measured at 9.07 m in 2013
- Twardogóra Forest in Oleśnica County, maiden oak tree measured at 9.04 m in 2012

Romania
- Cajvana near Solca in northern Romania, in village, maiden oak tree measured at 11.01 m in 2012.
- Mercheasa in Transylvania, pasture woodland, maiden oak tree in field measured at 9.24 m in 2015

Spain
- Cartelos in Galicia, Pazo de Cartelos, maiden oak tree 12541 estimated at 11.00 m in 2013
- Jauntsarats in Navarra, near the village, pollard oak tree 14032 estimated at 11.00 m in 2013
- Los Tojos in Cantabria, Monte Saja, pollard oak tree No. 16 estimated at >9.00 m (no year given)
- Ruente in Cantabria, Monte Aá, pollard oak tree No. 10, estimated at >10.00 m (no year given)
- Ruente in Cantabria, Monte Rio los Vados, pollard oak tree 13378 measured at 12.20 m in 2012
- Sobrado in Galicia, Carbayón de Valentín, pollard oak tree 4290 measured at 9.87 m in 2002

Sweden
- Blombacka near Lidköping in Västra Götaland County, oak tree measured at 9.29 m in 2011
- Boxholm in Östergötland County, oak tree measured at 9.24 m in 1998
- Ekerö in Stockholm County, Ekebyhov, maiden oak tree 1699, measured at 10.35 m in 2007
- Gällstaö in Stockholm County, along Parkslingan (road), pollard oak tree measured at 10.25 m in 2007
- Höör in Skåne County, maiden oak tree measured at 9.32 m in 2008
- Kalmar in Kalmar County, near Stufvenäs Gastgiveri maiden oak tree measured at 9.69 m in 2006
- Lidköping in Västra Götaland County, oak tree measured at 9.27 m in 2005
- Mönsterås in Kalmar County, maiden oak tree measured at 10.10 m in 2007
- Nora Kvill, near Vimmerby in Kalmar County, maiden oak tree 1701 measured at 14.75 m in 2006*

- Norrtäljegatan in Stockholm County, oak tree measured at 9.87 m in 2008
- Norrtäljegatan in Stockholm County, oak tree measured at 9.75 m in 2007
- Prince Eugen Oak, Djurgården in Stockholm, pollard oak tree measured at 9.17 m in 2010
- Södertälje, Hörningsholm in Stockholm County, oak tree measured at 9.53 m in 2010
- Vellanda in Jönköping County, near Myresjö maiden oak tree measured at 9.60 m in 2010

*On a visit to this famous oak 21 August 2015, I concluded that it is not a proper pollard as recorded.

Turkey
- Karacaözü village, Kastamonu, Anatolia, Kaba Meşe pollard oak measured at 12.05 m in 2015

Ukraine
- Stuzhytsia in Velykobereznians'kyi Oblast, in forest, maiden oak tree 6383 measured at 9.60 m in 2009

United Kingdom outside England
- Scotland, Moray, Darnaway, Meads of St. John, maiden oak tree measured at 9.84 m in 2010
- Scotland, Scottish Borders, Jedburgh, the Capon Tree, a maiden oak tree measured at 9.74 m in 2002
- Wales, Monmouthshire, Crossenny, Penrhos Farm, pollard oak tree measured at 10.35 m in 2007
- Wales, Monmouthshire, Mitchel Troy, St. Dial's Farm, pollard oak tree measured at 10.72 m in 2007
- Wales, Powys, Coed Mawr, Craigfryn, pollard oak tree measured at 9.48 m in 2011
- Wales, Powys, Garthmyl, Garthmyl Hall, pollard oak tree measured at 10.15 m in 2009
- Wales, Powys, Llanfrynach, Maesderwen, pollard oak tree measured at 9.70 m in 2009
- Wales, Powys, Welshpool, north of Buttington in field, pollard oak tree measured at 11.03 m in 2009
- Wales, Wrexham, Bangor-on-Dee, Althrey Lodge, pollard oak tree measured at 9.43 m in 2011
- Wales, Wrexham, Chirk Castle, near Castle Mill, pollard oak tree measured at 9.63 m in 1999

Distribution of oaks >9.00 m in recorded circumference in Europe (England excluded)

On the website www.monumentaltrees.com, the ancient oaks in Europe have been mapped with a red symbol (Google map pointer) indicating that their size is at least 9.00 m in circumference. The oaks in the list above were sourced from this website, and are marked with a red symbol and their numbers on the map. On the map, an orange symbol indicates oaks of between 6.00 m and 9.00 m in girth, but among these are several maiden oaks on good to excellent sites that cannot be considered ancient in the sense used in this book. An example is oak tree No. 1921, which has a girth of 6.81 m and grows in the grounds of the estate Hilverbeek in s'Graveland, The Netherlands. No trees on this estate can date from before 1634, when reclamation of the Polder of s'Graveland was completed and tree planting could commence. Several maiden oaks in Białowieźa National Park have girths of between 7.00 and 7.50 m, but all are vigorously growing tall maiden trees. On the other hand, there will be ancient oaks among those mapped with an orange symbol. There seem to be no recorded living sessile oaks larger than 9.00 m circumference in continental Europe.

Germany leads the list with 24 ancient oaks >9.00 m in circumference recorded. Sweden has 14, France has nine, Poland has eight, Spain has six and all other countries have fewer largest ancient oaks recorded. Sweden has the highest number of oaks between 6.00 m and 9.00 m girth, but in this category, Germany also has many oaks.

Belgium
Belgium has just one oak with a girth above 9.00 m, the Gros Chêne de Hiernu, which was measured at 10.10 m in circumference at 1.50 m above ground in 2013. This tree now stands on a village square. There are three native oaks (all pedunculate oaks) with a girth greater than 7.50 m (Baudouin et al., 1992) in this country.

Czech Republic
The website www.monumentalrees.com records three oaks with a circumference >9.00 m in the Czech Republic. The tree along road 399 had a girth of 10.20 m in 2009; the tree near Petrohad was measured at 9.18 m in that year; and the oak at Lichnice Castle was 9.10 m in girth in 2010. None of these oaks are obvious pollards; they may have been cut only once or in part. If they are maiden trees, the ages given for them on this website are likely to be overestimated.

Denmark
One of the ancient oaks in the Jægersborg Deer Park north of Copenhagen is said to have a girth of 10.68 m (Figure 6-2). Its short trunk is fragmented, so the measurement is more an estimate than an accurate figure. This tree is the largest of a number of ancient oaks in a former Royal Forest, which was partly imparked in the seventeenth century. The ancient oak in the pasture woodland of

Figure 6-3. A maiden oak with a girth of 10.96 m in the deer park of Ivenack near Stavenhagen, Germany

Figure 6-4. The largest oak (10.96 m girth) in the deer park of Ivenack

Nordskov (North Forest) near Jægerspris Castle, named Kongeegen (King's Oak), is very fragmented and can be neither accurately measured nor dated. It was undoubtedly once a pollard.

Germany

The Deutsche Baumarchiv (www.deutschesbaumarchiv. de), in its 'Liste der dicksten Eichen in Deutschland' (e-published in Wikipedia), has recorded 23 ancient oaks with a circumference over 9.00 m at 1.00 m above ground. The level of measurement at 1.00 m does not quite compare with the d.b.h. (diameter at breast height = 1.30 m) more commonly used for other oaks in Europe, and could disqualify a few for the >9.00 m threshold used here. One

Figure 6-5. The largest oak in Sweden is the Kvill Oak at Nora Kvill in the province of Kalmar

oak in the German list was disqualified because of excessive root base growth at this height, and two of the >9.00 m oaks in Ivenack were measured at less than 9.00 m at 1.30 m more recently (www.monumentaltrees.com). On the other hand, two oaks in the German list are only a few centimetres short of 9.00 m circumference; laying the tape a second time may include them. All are pedunculate oaks; indeed with one exception, the 86 listed German oaks with a girth greater than 7.50 m belong to *Q. robur*. Four oaks with >9.00 m girth are found in Bavaria, more than in any other of the Länder. After England, with 24 Germany has the largest number of >9.00 m girth ancient oaks in Europe; other countries have far fewer. The Ivenacker Tiergarten (deer park) in Mecklenburg-Vorpommern is the most important site in Germany for large ancient oaks (Voß & Rüchel, 2014) because it has several ancient oaks in close proximity. Yet there are only about seven oaks with girths >6.00 m (three >8.00 m) in what towards the end of the nineteenth century was still 'Hudewald' (pasture woodland) with many large oaks; but Ivenacker was afforested in the twentieth century, leaving only some small sections as parkland and destroying many of the ancient oaks. During a visit in April 2015, I concluded that Ivenacker would not meet my criteria for a 'most important site' (see Chapter 9) had it been situated in England.

France

The Association A.R.B.R.E.S. in its *Guide des arbres remarquables de France* of 2009 has listed around 20 ancient oaks, seven of these with a girth of over 9.00 m although measurements are not given for all trees. From additional website sources, I have come to a total of nine such large ancient oaks in France. The inventory for France is probably incomplete, yet the numbers suggest that there are not many ancient sessile or pedunculate oaks in France. Most are in the northern and western parts of the country and none are recorded in southern France. The Chapel Oak in Alouville-Bellefosse near Rouen in Normandy was turned into a Catholic shrine in the seventeenth century. The ancient oak pollard near Bulat-Pestivien in Brittany has a girth of *c.* 12.60 m (it has split in two parts), the largest recorded oak in France.

Hungary

Only one oak tree has been measured as over 9.00 m, with a girth of 9.80 m at 1.30 m above ground, in 2004.

Another large oak in the same location in Zsennye was measured at 7.26 m. The location is a small wooded park around a country house surrounded by arable fields.

Ireland

The largest oak in Ireland is identified as a pedunculate oak with a girth of 9.36 m at 1.30 m above ground. It is doubtfully a pollard as its first major division is only about 2.00 m above the ground and it has very large and tall limbs (crown height *c.* 32 m) that seem to have been formed on the tree at a young age. If a pollard, the tree could have been cut only once as a young tree. It stands in a walled garden behind a house named Oak House, but is undoubtedly older than these buildings. There are over 200 oaks with girths of more than 6.00 m in Ireland, quite a large figure for a country that is historically poor in tree numbers.

Latvia

The ancient oak in Kaive was measured with a girth of 10.20 m at 1.30 m above the ground in 2010. Its present situation is in a meadow on the edge of open woodland, but there are no records of its historical landscape context. Two oaks are just over 9.00 m at 1.30 above ground, but could be smaller at 1.50 m. A fourth oak with 9.60 m girth seems to be a natural pollard through storm damage.

Lithuania

The ancient pollard oak in Stelmužė Manor Park stands in parkland near some buildings and a road. The tree is the only one in the country measured greater than 9.00 m in girth, and in 2005, it had a circumference of 9.58 m at 1.30 m above the ground.

Netherlands

In the Netherlands, just one oak has a reported circumference above 9.00 m; this tree stands along the Borculose Weg in the village of Ruurlo in Gelderland and had a girth of 10.21 m measured in 2009. This oak, locally known as the Kroezeboom, may in fact be either an ancient coppice tree or even the result of multiple plantings (J. Philippona, pers. comm., 2013). A maiden oak at Verwolde Castle in Laren, Gelderland has been measured with a girth of 7.70 m; all other recorded native oaks (*Q. robur*) are smaller. The oak at Ruurlo does not appear in the 'Lijst van dikste bomen in Nederland' because of doubts about its status as a single tree (www.bomeninfo.nl/lijst.pdf), but it can be found on the map of the website on monumental trees (www.monumentaltrees.com). It is excluded from the list of oaks in Europe >9.00 m in girth in this chapter.

Norway

Three ancient oaks in southern Norway are recorded with girths of around 10.00 m. The Mollestad Eik near Grimstad on the south coast and the Brureik near Ullensvang on the Hardanger Fjord are both ancient pollards; the Tordenskjold Eik in Horten is presumed to be a pollard, as inferred from its estimated height, but no image was available and I have not visited it.

Poland

There are 11 ancient oaks greater than 9.00 m in circumference recorded in Poland (www.monumentaltrees.com). None of these are in the forest of Białowieża, which is famous as the "last primeval forest of Europe", and has many large oak trees (three above 7.00 m in circumference). These are maiden trees growing in more or less closed forest and probably up to 500 years old. The 11 larger oaks are scattered across lowland Poland. The ancient oak at Bakowo along road 391 is recorded with a girth of 10.28 m at 1.30 m above ground, making it the thickest oak in Poland. Two other large oaks measuring between 7.00 m and 8.00 m also occur here. All oaks >9.00 m in circumference in Poland are maidens.

Romania

There are several locations with ancient pasture woodland dominated by oaks in Transylvania, these landscapes are thought to have originated with Saxon immigration around the thirteenth century. The largest oak in the village of Cajvana has been measured with a girth of 11.01 m around a broad-based bole and the oak at Mercheasa has been measured with a 9.24 m circumference; a third oak with a girth >9.00 m died recently. An oak in Botosana has been measured with a girth of 8.96 m, nearly in our class. Recording in another reserve, the Breite Ancient Oak Tree Reserve near Sighisoara, established the largest oaks at around 6.00 m in girth. Two other localities near this town, Langa Fiser and Ticusu Vechi, are of a similar nature, pasture with solitary trees and woodland, and have greater numbers of oaks. At Langa Fiser, the largest recorded oak was 7.91 m in girth. Most oaks at these sites are maidens and range from 4.00 to 6.00 m in girth. Recording in Romania has only begun recently, and it is likely that more oaks in the larger size classes will be found.

Spain

The list of remarkable trees in Spain, 'Árboles Singulares' (www.dgmontes.org/arboles.htm), gives only a few oaks with girths over 9.00 m, but additional large oaks are recorded on the website www.monumentaltrees.com. In all, six large ancient oaks are known, all growing in the north of Spain. From the (authentic) photographs that accompany these records, it seems all but one of these trees are pollards. The tree with the greatest girth, measured as 12.20 m at 1.30 m above the ground, is located on the Monte Rio los Vados in Ruente (Ucieda). The maiden oak at Pazo de Cartelos appears to stand with a few similar large oaks on a boundary bank/ditch.

Sweden

The largest oak in Sweden and, claimed by some, in Europe is the Kvilleken (Kvill Oak) (Figure 6-5) near Vimmerby in Kalmar County. It is an extreme natural pollard with a quickly tapering, very hollow and fragmented bolling. Its measurement of 14.75 m girth at 'breast height' (1.30 m) seems rather arbitrary as d.b.h. has little true meaning in a tree like this, given that this convention in forestry practice is meant to avoid root buttresses and other basal irregularities in moderately sized trees (see also Chapter 2). The tree stands in an area of pasture woodland with several mature oaks of small dimensions. On a visit to this great oak 21 August 2015, I found it to be in serious decline, with thin yellowish foliage and recent die-back in the upper branches. Another large but very different oak tree stands on the verge of a small road in Gällstaö (in the municipality of Ekerö); its circumference was measured as 10.25 m in 2007. A recent listing of ancient oaks in Sweden can be found at www.tradportalen.se/, where the Kvill Oak is mentioned. In total, I have found 14 recorded oaks with a girth >9.00 m in Sweden, all are in the southern part of the country; in August 2015, I visited several of these and all were found to be maidens. According to the same website source, there are 1,259 living ancient or veteran oaks measured between 6.00 m and 15.00 m in girth in Southern Sweden, indicating that this region may be second only to England in importance for the preservation of ancient oaks. Most of the big oaks stand alone in an agricultural landscape and not in parkland. Most sites with pasture oak woodland that I have seen, often still grazed by cattle, had few if any veteran oaks.

Figure 6-6. The ancient pollard oak Kaba Mese near Karacaozu, Anatolia, Turkey

Figure 6-7. The pollard oak Kaba Mese is, with a girth of 12.05 m, the largest oak in Turkey

Turkey

No oaks >9.00 m have been recorded in the European part of Turkey. There seems to be just one in western Anatolia, in the Black Sea Region, and I have included it here as it stands within the natural range of *Q. robur*. On a visit to this great pollard (12.05 m in girth) on 7 May 2015, I was informed by an ecologist of the Turkish Forestry Department that it is the sole survivor of oak pasture woodland that still existed here 150 years ago.

Ukraine

The only oak tree with a girth of more than 9.00 m in Ukraine is situated in woodland near the village of Stuzhytsia in the far west of the country. It has a girth of 9.60 m at 1.30 m above the ground. It seems unlikely to be a pollard, although it branches low above the ground and then divides into two stems. It was estimated to be 30 m tall, which also indicates a maiden tree, probably grown in more open terrain for much of its life.

United Kingdom outside England

The list of 10 ancient oaks with a girth >9.00 m presented here includes only those records that have been confirmed by the ATI verifiers; there are reports of some '10.00 m' oaks in Wales that are unconfirmed. In Scotland, the large coppice oaks recorded with a girth over 9.00 m have not been included as they are in a different category. In April 2013, the well-known Pontfadog Oak in Wales was felled and killed in a storm (Miles, 2013), so it is not included. This leaves only two eligible oaks in Scotland but eight in Wales, which is a relatively high number considering the size of Wales and its mountainous terrain, which limits where oaks could grow large. The two oaks in Scotland have been identified as pedunculate oaks and are maiden trees; all the listed oaks in Wales are pollards.

Summary of the distribution of oaks in Europe

The inventory given here has identified 96 ancient oaks with a circumference >9.00 m outside England; 85 of these occur in continental Europe and one in Turkey. A total of 115 oaks with similar dimensions have been found in England. England thus has more very large ancient native oaks (*Q. robur*, *Q. petraea* and hybrids) than all other

European countries combined, even though these species are widespread and abundant across most of Europe and grow as well there as in England. In relative terms, considering land area, the density of these largest ancient oaks in England is vastly greater than in continental Europe. Outside England, only southern Sweden has substantial numbers of oaks with girths ≥6.00 m; in an area similar to England, there are about one third as many ancient oaks as there are in England.

There seem to be no consistent environmental factors that could be responsible for these figures. England has areas with as little or less precipitation than much of lowland northern Europe, especially in parts of East Anglia and around London, where large ancient oaks are numerous. Windsor Forest and Great Park, partly on poor Bagshot Sands, receive an average of less than 650 mm precipitation annually, yet this area has one of the greatest concentrations of ancient oaks in Europe (Alexander & Green, 2013; see also Chapter 9). Staverton Park has many ancient and veteran oak pollards but is in the driest part of England, with only 500 mm precipitation, and on windblown sand. Oaks in England occur on as many soil types as in continental Europe and may grow at different rates accordingly. Winters are generally milder in England than in countries such as Germany, but the growing season in England is not significantly longer than in other parts of Western Europe. The widths of annual increments in ancient oaks found in Ivenack, NE Germany (Blank, 1996) are very similar to those found in Sherwood Forest, England (Watkins *et al.*, 2003). Scotland has as few large ancient oaks as most parts of Europe, yet apart from the highlands, its climate is very suitable. In Sweden, the oaks reach their northern climatic limit, yet after England, this is the most important country for ancient and veteran oaks. The causes of this difference in the abundance of ancient oaks have to do with the history of land ownership, land use, forestry, social stratification in society, economic developments, even revolutions and wars that set England apart from other nations. An attempt to unravel this historic 'miracle' of abundant ancient oaks in England is presented in Chapter 7.

Of the 96 largest ancient oaks in Europe (outside England) listed here, 51 are maiden oaks, 28 are pollards and 17 remain undetermined, largely because there are no photographs in the public domain that I could use in lieu of visiting all these oaks. It is notable that maidens prevail in Eastern Europe, whereas pollards are more common in the west. All oaks >9.00 m in Poland are maidens; all oaks >9.00 m in Wales are pollards; Germany has 13 maidens, four pollards and seven undetermined oaks >9.00 m. In England, the ratio of pollards to maidens is also higher than in continental Europe, with 66 pollards, 45 maidens and four >9.00 m ancient oaks undetermined. It is, of course, more difficult to obtain sufficient reliable data on pollarding for smaller oaks, especially where this has not been recorded and no photographs are available. In Chapter 3, a spatial analysis of pollarding for all ancient and veteran oaks ≥6.00 m in girth in England is presented and discussed.

7

Why England has most of the ancient oaks

In Chapters 4 and 5 we investigated the medieval landscape in which the oaks that are now ancient occurred. The landscape was almost exclusively lowland pasture woodland and its uses were as pasture commons, deer parks, chases and Royal Forests. An unknown number of ancient oaks were boundary oaks or meeting oaks, which in all likelihood stood solitarily in hedgerows or at crossroads or on a village green. The number of ancient oaks ascribed to deer parks outnumber those attributed all other historical land uses combined; 47.5% of the total of ≥6.00 m girth oaks (which includes 14.5% with no determined association with a historical land use) are associated with medieval or Tudor deer parks. Of the 23 most important sites for ancient oaks in England described in Chapter 9, 20 were deer parks, 16 of them dating back to the Middle Ages. Of 115 ancient oaks with a >9.00 m circumference, 60 stood in medieval deer parks. Without its medieval deer parks, England would now have far fewer ancient oaks. Four of the most important sites were part of a medieval Royal Forest or a chase, and 14 ancient oaks with a girth >9.00 m are associated with Royal Forests and chases. Just one of the 23 most important sites described in Chapter 9 was a common. Of the largest ancient oaks in England (those >9.00 m in girth), 32 are associated with other or unknown uses of the land during the Middle Ages. We therefore have to look at the history of deer parks and Royal Forests to find an answer to the question of why

Figure 7-1. A herd of fallow deer in Parham Park, West Sussex

there are so many ancient oaks in England and so few in other European countries, as documented in Chapter 6.

Deer parks

Deer parks have a long history that begins well before their introduction in northwest Europe (Fletcher, 2011), and in England they date at least from Anglo-Saxon times (Hooke, 1998; Liddiard, 2003). Their English origin is probably linked with 'hayes' or 'haga', which have been interpreted as more or less temporary enclosures ('deer folds'). These ranged from long nets to hedges into which wild deer were either driven or lured to be killed (Fletcher, 2011) and, on occasion, may have become permanent enclosures in which deer were kept alive for later killing. Domesday Book mentions only 31 parks (Shirley, 1867), several of which had Anglo-Saxon owners before they were taken over by the conquering Normans. It is unlikely that this list is complete (Liddiard, 2003), but it is difficult to know how many permanent deer parks there were in England by 1086; probably not many, for various reasons.

The Anglo-Saxon kings, like their counterparts throughout northwest Europe, considered themselves owners of the wild deer — first and foremost the large red deer and second the much smaller roe deer (*Capreolus capreolus*) — wherever they occurred. There were specially designated hunting forests in which the kings hunted deer and wild boar, and these were not enclosed, at least not during the early and high Middle Ages. The Franks were the first monarchs in Europe to develop the legal concept of a hunting forest or *forestis* (Buis, 1985) that restricted access to the deer and the trees. The 'beasts of the forest', rather than the timber, were the greatest concern of the Crown; the trees or 'vert' primarily served the deer. The existence of a deer park, if not owned by the king, under such forest laws would infringe on royal prerogatives concerning the hunt. It is therefore understandable that deer parks did not proliferate under such regimes. However, by excluding not only the peasants but also the landholding nobility from access to the deer, which were considered a noble quarry providing noble food (venison), the kings created a potential for friction if not downright rebellion.

After they had appropriated most of England's estates and manors that had been in the hands of Anglo-Saxon nobility, the Normans found a solution to this quandary. They had already conquered southernmost Italy, and from there, they introduced fallow deer to stock English parks. These animals have several advantages over the indigenous species as park deer. Although red deer can be and are kept in parks (and mix well with fallow deer), they are less easily kept in numbers in relatively small enclosures (say up to 80

ha) than fallow deer. Being half the size of red deer, fallow deer can be kept in twice the number on a given area of park. The greater gregariousness and tameness of fallow deer are other factors in their favour as park animals. At least in the Middle Ages, fallow deer never occurred in the wild in England; in fact, if escaped from a park they would try to return to their kind within the security of the park (Fletcher, 2011). Escaped red deer would more likely stay outside the park and join other red deer still roaming wild, at least until these became rare in England. Fashion and aesthetic appearance may also have played a role in the popularity of fallow deer for parks (Fletcher, 2011), which were often envisaged as enhancing the status of the manor house (Mileson, 2009). Roe deer are unsuitable for any but the very largest parks (in which they probably occurred unintended) because of their very different habits and need not be considered further here. It was therefore possible to grant the nobility a (small) park stocked with fallow deer, often as a present from the king, thereby avoiding having to share the wild red deer of the forests. As a consequence, the hunting of deer developed in two very different directions without and within the parks (Fletcher, 2011).

From the late twelfth century onwards, numerous parks were established in lowland England, where the relatively mild winter climate (we are still in the Medieval Warming Period at this time) suited the southern fallow deer best. An upsurge in imparking occurred between 1250 and 1350 (Figure 7-2), with a peak in the 1330s (Mileson, 2009: 128), after the Royal Forests had been in decline for about a century. Indeed, many parks were granted within Royal Forest lands or in chases to the most powerful barons. Parks had become a status symbol, which every nobleman aspired to own, and the wealthiest magnates and prelates owned 10–20 or more, while the king had "many scores of them at any one time" (Mileson, 2009). The exact number remains elusive, but an estimate of 3,000+ is commonly given for the period between 1100 and 1400 CE (e.g. Rackham, 1976); Cantor (1983) listed 1,900 for which he found historical records. Archaeology and searches of more local archives have discovered many more medieval deer parks since Cantor's gazetteer was published (Mileson, 2009). Depopulation and the associated lack of cheap labour after the Black Death (1348–50) caused a decline in the number of new parks (Marren, 1990: 67; see also Figure 7-2), but this does not mean that there was a marked decline in the number of existing parks (Mileson, 2009). Abandoned fields and depopulated villages also led to the enlargement of deer parks.

The early sixteenth century saw a brief revival in the creation of new deer parks (Fletcher, 2011), followed by the

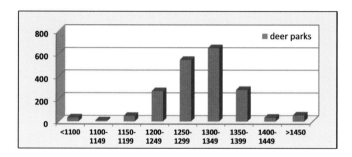

Figure 7-2. The number of medieval deer parks established within 50-year periods during medieval times. Data from Cantor (1983). All parks first mentioned in the historical records before 1100 and all those first mentioned after 1450 are also represented.

transformation of many of the parks into pleasure grounds around a mansion with less emphasis on deer hunting. The deer park kept its age-old allure among the nobility and gentry: "Whilst you have ... Dispark'd my parks and fell'd my forest woods, ... leaving me no sign, save men's opinions and my living blood to show the world I am a gentleman." laments Henry Bolingbroke in Shakespeare's *Richard II* Act Three Scene I. The number of deer parks in England had become so much greater than in any other European country that the English deer park remained a common feature of the countryside well into modern times.

Fallow deer from England were given to the Danish king in the twelfth century and became popular there too (Andrén, 1997). From Denmark, deer parks spread into Sweden and south into Schleswig-Holstein, but the first introduction of fallow deer into Germany occurred only in 1577, when the King of Denmark gave 30 animals to Landgraf Ludwig IV of Kurhessen (Ueckermann & Hansen, 1968). The first fallow deer park in Prussia was established in 1681 and the first in Brandenburg towards the end of the seventeenth century. Earlier, during feudal times (Feudalzeit), the high nobility of Germany kept red deer in large parks (Gehege, also known as Tiergärten). These were often very large, an example being the Saarkohlenwald in Saarland at 8,000 ha. They were reserved as hunting forests for the ruler (Landesherr) and therefore were more like Royal Forests in England except for the enclosing of the area. Apart from Artois and Normandy, which had deer parks in the Middle Ages, deer hunting in France remained mostly hunting 'par force', i.e. the chase on horseback and with hounds of red deer in the forests, based on Frankish concepts and traditions. The situation in the Low Countries was similar, and very few deer parks existed there until much later in the nineteenth century.

In Chapter 4, we analysed the losses in England and found that about 75% of medieval deer parks in 12 sample counties have completely disappeared. Even if losses of deer parks in other European countries had been at a similar or lower rate, this would still leave England with many more parks. We have also found that the presence of ancient and veteran oaks in the landscape is most strongly correlated with the (former) presence of (medieval) deer parks. The fact that England has a greater number of large ancient native oaks than the rest of Europe (Chapter 6) is therefore correlated with the abundance of deer parks in that country during a 350-year period in the Middle Ages. Ancient oaks that could now be 600–900 years old often stood in those parks as young or mature trees. But the preservation of parks alone is not a sufficient condition for the existence of ancient trees. We have also seen in Chapter 4 that in many locations where a medieval park is still a park, there are no ancient oaks. Medieval deer parks had trees of necessity to shelter the deer (Fletcher, 2011) and these were almost invariably, or at least predominantly, native oaks. Before 1550, these were the unplanted trees of pasture woodland with its glades and launds on the less fertile parts of the manor or in sections of the Royal Forests given by the king for the establishment of deer parks. Therefore, if a park was a medieval park but has no ancient oaks now, they have most likely gone. We have to investigate why so many more trees were preserved in England than elsewhere (if that is indeed the case).

In Chapter 4 we observed from the analysis of the fate of medieval deer parks that conversion to agriculture was the main cause for their disappearance. This is an economic cause: the land, even though usually less fertile, was put to more productive use when deer keeping went out of fashion, or when the deer herd was destroyed together with the park pale. When in the 18th century agricultural science discovered how to make the poorer sites productive there was a drive to abandon the low intensity use of pasture woodland and commons and sell the land off to entrepreneurial farmers to grow crops. This process continued well into the 20th century as my scrutiny of the two editions (19th century and 21st century) of OS maps for several hundred parks has shown. But a number of differences between the history of continental Europe and Britain as well as the conservative attitudes of the English landowning classes are probably relevant to the preservation of ancient oaks in (former) deer parks. The first factor is the comparative absence of violent conflict (wars, revolutions and revolts) in England after the end of the Wars of the Roses and the rise of the Tudor monarchy in 1485. The single most destructive episode after this time

was the English Civil War (1642–51), in which royalist and parliamentarian armies destroyed each other's castles and ransacked numerous country estates and parks. The sequestrated estates of royalists and Catholics were often plundered in the aftermath of military hostilities, with the deer killed or let free and the trees often cut. However, the Wars of Religion, the Thirty Years' War, the Revolutionary and Napoleonic Wars and the two World Wars, to name

just the most destructive of many conflicts on the European continent, were incomparably more ruinous, and the destruction happened time and again over centuries.

During the Restoration in England many parks were restored and restocked with deer, often coming as gifts from abroad (Fletcher, 2011). But changes were already apparent in the sixteenth century and these accelerated during the Commonwealth: many parks were transformed by agricultural uses, such as grazing cattle and sheep or even growing turnips on sandy soil for winter cattle feed, often at the expense of the park trees. Despite this, a conservative attitude prevailed among many park owners in England, no doubt strengthened by the knowledge that income could and did come from elsewhere: the park was an indulgence that a gentleman could afford (Mileson, 2009).

It is the combination of these two factors that allowed many English parks to be preserved until the realization dawned that the ancient oaks were actually 'beautiful' or 'picturesque', and still later, that they are living historical monuments supporting extraordinary biodiversity. In other European countries, these notions usually came much too late. In southern Sweden, millions of oak trees were felled in the eighteenth century and first half of the nineteenth century, many of these in park-like landscapes such as hay meadows with scattered oaks. At first, the felling was driven by the need for oak timber for warships, but it escalated into a conflict between the peasants and the Crown over ownership of the oak trees. The peasants' interests conflicted with the Crown's, not only over how to cut the branches but over their very existence. In Sweden's climate, haymaking for winter fodder was of the utmost importance and the shade of the oak trees limited the growth of grass. Most oaks not already felled for the Navy — those with partly rotten boles and/or crooked shapes, i.e. ancient trees — stood on farmlands owned by smallholders. When the ban on cutting these 'unfit' oaks on peasants' lands was finally lifted in the 1830s, many were quickly destroyed because they interfered with haymaking. More than 1 million oaks were cut and it is estimated that only about 2% of the 'oak landscape' existing until the nineteenth century has survived until the present (Eliasson & Nilsson, 2002). Despite this devastation, southern Sweden still comes next to England in numbers of ancient oaks with girths ≥6.00 m (see Chapter 6).

Forestry

There seems to have been a long-standing reluctance among English private landowners to undertake forestry planting. In the 'Notes' to the 1776 edition of John Evelyn's book *Sylva*, a Dr Hunter observed: "How many thousand acres of waste lands are there in this kingdom, that at this present time produce nothing, but may be profitably improved by planting! Did men of large possessions but rightly consider this, they would carefully look over their estates, search

out every useless bog, and plant it with poplars and other aquatics. They would examine all the waste grounds and set apart some for the cottagers, and apply the most barren and useless for plantations." Evelyn's book, first published in 1664, lamented the same attitudes as well as the destruction of the forests in past centuries up to and including the recent English Civil War (1642–51) and offered a remedy: planting forest trees. The Navy needed the (oak) timber and the resources within the country were becoming scarce. At least in Royal Forests, such as Dean and the New Forest, and on some large estates owned by magnates close to the Crown, the planting of oaks now increased.

According to Stebbing (1916), this action, instigated by Evelyn's book, "saved the country a hundred years or so later; since sufficient oak timber was forthcoming to build the 'wooden walls' which gave us the command of the sea." But tree planting soon lapsed again, and Stebbing's book, published half-way through the First World War, still complained about too little forest planting and the ongoing scarcity of timber, even though ships were by then built of iron. In 1946, W. L. Taylor could write: "the fact remains that at no time in our history has a concerted public effort to reafforest been sustained long enough to matter." (Taylor, 1946). (Note that afforestation now means planting with trees, not creating an area where the king could go hunting.) Britain was the largest importer of timber in the world, mainly from Scandinavia and Russia but also from Canada and the U.S.A. Especially "waste lands with useless scrub", of which Britain still had 1.5 million acres around 1916 (this figure also included lands clear-cut during the war), were to be afforested as a priority. One wonders what types of vegetation were actually included under this derogatory appellation; could it have included pasture woodlands and wooded commons?

It was not until the creation of the Forestry Commission as part of the Forestry Act in 1919, when the war had ended, that the planting of trees (now mostly conifers) on 'waste lands' commenced on a large scale and Stebbing got what he wanted, at least to a degree. As we have seen in Chapter 5, this has transformed what was left of the Royal Forests, destroying the ancient oaks that were still present in all but a few sites. Increasingly, Stebbing's recommendation that existing oak plantations, such as those in the Forest of Dean, should not be replaced was ignored. Conifers not only replaced many of these, but penetrated what was left of the oak pasture woodland here and elsewhere. Woods of less than 500 acres (c. 200 ha) were considered too small for commercial purposes, but many landowners began to plant conifers in woods much smaller than this as the value and use of native trees diminished and as government subsidies for planting with conifers were made available. Rotation time played a major role in this: it takes a century or more for oaks to become large while most conifers take only a few decades.

The presence of ancient trees was, for a long time, no barrier to conifer planting as many remnants of ancient pasture woodland still testify. The ecological and historical value of ancient trees was underestimated: when writing of the New Forest, the forester Stebbing (1928) could come up with no better descriptors than 'beauty spot' and 'playground', and was convinced that, if planned 'carefully', commercialisation of that forest with plantations would not be in conflict with these uses. Only when the utilitarian view of woodland began to recede from primary or even sole consideration in recent decades did conifer planting among ancient trees or in ancient woodland diminish, no doubt helped by changing economics. It is now cheaper to import timber than to grow it in the UK, unless on a scale proposed by Stebbing a century ago. Land is no longer as easily or cheaply available as it seemed to be in his time. Indeed, the bare deforested hills and mountains of Britain have been afforested to a much lesser extent than he envisaged, but where planting has occurred it has been with trees, mainly Sitka spruce (*Picea sitchensis*), that are an ecological mistake and an eyesore to all but the most mercantilist observers.

From Early Modern time, when deforestation and degradation of forested lands in Britain had reached maximum levels, re-afforestation through plantation has had a chequered history with ups and downs, never reaching the scale it attained in large parts of continental Europe (James, 1981). The State's efforts were largely confined to the remnants of medieval Royal Forests and a few other Crown lands, and only took off in earnest after the First World War around 1920. Private owners were mostly reluctant either to change traditional land use (amenity, sporting and pleasure grounds) or to make the investment necessary to commercialise the woods. Britain continued to import most of the timber it required, as it had done since at least the sixteenth century.

How does this brief history compare with that of mainland Europe?

Stebbing (1928) compared the Forêt de Tronçais in the Auvergne in France (10,530 ha) with the Forest of Dean in Gloucestershire, England (9,300 ha). He emphasized that plantations of predominantly sessile oak, with beech and hornbeam as subsidiary species, were undertaken in this French Royal Forest (forêt domaniale) under Louis XIV by his minister Colbert from 1660 onwards. By contrast,

the Forest of Dean was steadily depleted of its oak timber to supply the Navy until "the supplies gave out before the replacements of wood by iron in the construction of ships." What Stebbing overlooked was the fact that the Forêt de Tronçais was a newly planted forest on land that had been cultivated since Roman times, whereas the Forest of Dean was a remnant of 'wildwood' turned into a hunting forest in the Middle Ages. But the principal point that Stebbing made is valid and relevant to our question: forestry as a technique to grow trees was developed much earlier in France than in Britain, resulting in rational forest

management with the aim of producing large quantities of timber. Under such management, undertaken on a large scale in the many state forests of France for several centuries, there is little chance that the ancient oaks that must have been part of the French medieval hunting forests have survived. By the eighteenth-century, France had already begun to import naval oak timber, especially the curved 'compass timber' from abroad. France has 25 large state forests (forêts domaniales) among a total of 1,300 state forests with a total area of 1.8 million ha; 0.69 million ha originate from Royal Forests. During the French Revolution, 0.34 million ha of forest were confiscated from monasteries (forêts abbatiales). The examples of French forests below will illustrate the situation regarding ancient and veteran oaks.

The 34,700-ha Forêt d'Orléans is the largest state forest in France and is situated to the east of Orléans north of the Loire Valley. It is probably the core of a much larger forest dating back to the Roman period; according to Julius Caesar, the Druids had their annual gatherings here. In the ninth century, it became a Royal Forest under Hughes Capet, with purposes and uses similar to those of the later

Figure 7-4. Oak woodland in Savernake Forest, Wiltshire

Norman Royal Forests in England. The forest is on marls, clay and sand, on almost level terrain between 107 and 174 m above sea level. Before drainage with 5,000 km of ditches that began in the early nineteenth century, it was in many places boggy with numerous ponds. About one half of forest trees are pedunculate oak, a third Scots pine, and the remainder beech, hornbeam, birch, lime and hazel and Corsican pine. The entire forest consists of young to mature trees formed by plantations and managed for forestry, amenity and recreation. There are no ancient oaks.

The 14,360-ha Forêt de Compiègne in Picardy, about 60 km north of Paris, is the third largest state forest in France. It was already designated as a hunting forest under the Frankish kings in the sixth century. It remained a favourite hunting forest for royal parties up to the Second Empire (under Napoleon III) ending in 1870. It was opened up by roads and rides from the sixteenth century onwards. The dominant tree species are pedunculate oak, beech and hornbeam, mostly replanted since the nineteenth century after heavy exploitation threatened deforestation. The invasive American species black cherry (*Prunus serotina*) is widespread and an ecological sign of soil disturbance and enrichment; the original soils were poor in nutrients. This forest is managed for timber, amenity and recreation, with the latter function becoming increasingly important. There are no ancient oaks.

The Forêt de Notre-Dame is a forest of 2,200 ha just east of Paris. It was a common belonging to 10 different municipalities, not a Royal Forest. It consists mainly of oak-birch woodland (*Betulo-Quercetum roboris*) on poor sandy soil. Management is mainly aimed at amenity and recreation, with forestry secondary in importance. Roads radiating from a central point open up the forest to traffic and tourism. The forest is designated as the French equivalent of an SSSI. There is one large veteran oak in this forest, the Chêne Notre-Dame, which has a broad crown and a girth of around 6.00 m. It stands in semi-open terrain with rough grass and bracken, indicating a clearing in former pasture woodland. Presumably most, if not all, of this wooded common was over-exploited and turned into heath and scrub before trees could grow up again.

The Forêt de Fontainebleau in the Département Seine-et-Marne, *c.* 60 km southeast of Paris, is an ancient forest of 28,000 ha. It is situated on a sandstone plateau and there are many boulder-like outcrops of rock; the soil is sandy in most areas. Assembled from a variety of large and small landholdings, it was united as a hunting forest under the Capetian King Philippe I in 1067, but it was also exploited for timber. This resulted in a tree cover of only 20% in the seventeenth century, with 80% degraded to heath or even

sand dunes. Re-afforestation began in 1664 under Colbert. Plantings of oak in the eighteenth century often failed, but beech and Scots pine plantings (the first in 1786) were more successful. Pedunculate oak, mostly as more or less degraded oak-birch woodland, and Scots pine cover 44% and 40%, respectively; beech occupies 10% of the forest and the remainder is heath invaded by mainly birch. Some parts with mostly over-mature beeches and fewer oaks have been designated as strict reserves in which no management or logging takes place; other parts still bear the marks of pasture woodland in which occur several mature to veteran oaks such as Chêne Sully, but the greatest part of this forest has been and is managed as plantation forestry, with amenity and recreation important secondary functions. There are no ancient oaks; they have been outcompeted by beech and hornbeam following the banning of grazing about 300 years ago.

Since the 1660s, the State has taken the initiative to develop forestry in France on a rational and economic basis. It soon applied the knowledge obtained in the Royal Forests and other forêts domaniales on a large scale. Oak timber for the Navy was the main objective at first, but beech and Scots pine were planted from the eighteenth century onwards. With such an early start in most state forests, several rotations of planting or seeding and felling had occurred before nature conservation issues drew the attention of foresters in the late twentieth century. The ancient oaks had gone to make space for 'scientific forestry' aimed at timber production. France has one of the lowest numbers of surviving large ancient oaks in Europe (Chapter 6).

Of the 11.1 million ha of German forests, 4.8 million ha are in private ownership, 2.2 million ha are owned by corporations and 4.1 million ha are state owned; together, these forests cover 30% of the country (de. wikipedia.org/wiki/Forstwirtschaft). German law (§11 Bundeswaldgesetz) requires all forest owners to manage their forests on a multiple-purpose principle, providing not only timber but also non-commercial services such as eco-services, amenity and recreation. "Dazu erfordert die heutigen Forstwirtschaft ein ständiges Abwägen zwischen wirtschaftlichen und ökologischen Interessen, um die unterschiedlichen Ansprüche an den Wald berücksichtigen zu können." ["In order to satisfy the various demands upon the forest, today's forestry requires a continuous weighing up of economic and ecological interests."] As in other European countries, this is a relatively recent viewpoint, even though some of the non-economic functions of the forest were recognised much longer ago.

The Frankish kings appropriated large tracts of 'unclaimed' forest (possibly including tracts of 'wildwood')

Figure 7-5. An old oak among conifers in South Forest, Windsor Forest, Berkshire

distant from settlements and made them into hunting forests (*forestis*), which later in the Middle Ages became Royal Forests and then state forests (Reichswald). Princes and prince-bishops also acquired forests and many forest lands were given to monasteries. Royal Forests were often imparked as hunting reserves, perhaps to keep the deer and wild boars in as much as to keep the people out. More often than in England or France, forests were held in common, usually by village associations (Markgenossenschaften); these included woodland for timber and coppice, as well as pasture woodland and pasture (heath), that was managed by free agreement among the villagers, independent from overlordship. In other regions, villagers had user rights in manorial forests owned by the nobility, as was the situation in England. These forests were usually endowed by kings granting lands from the commons to their vassals, with hereditary ownership (Klose, 1985). When forest resources were gradually depleted through exploitation by an increasing population, the owners began to impose restrictions that led to various revolts and eventually to the Peasant's Wars (1524–25). Massive destruction ensued during the Thirty

Years War (1618–48), and by the middle of the seventeenth century, many forests in Germany were considered to be in a deplorable state. Timber was exported to the Netherlands and Britain in huge rafts floating down the Rhine and its tributaries, denuding the hills and the slopes of mountain valleys as private owners sold out their forests. As their power increased, the sovereign princes of the different Länder began to impose more restrictions and sanctions on forest use. Also during this time, mercantilist attitudes led to the beginnings of forestry economics.

Germany is considered to be the cradle of the science of sustainable forestry (Klose, 1985), which originated with Hans Carl von Carlowitz's book *Silvicultura oeconomika* (1713). Re-afforestation was undertaken on this basis and took off on a large scale after the destruction of the Napoleonic Wars (1808–15). The heaths on the North German Plain were planted with Scots pine, if not ploughed for agriculture. In more mountainous regions,

Norway spruce became the preferred timber tree. Oak and beech woods, if not replaced by conifers, were felled and replanted, and virtually all forest was managed with economic interests prevailing over common interests. Forestry became a business enterprise which saw restrictions by the State as a hindrance and advocated the principle 'laisser faire, laisser passer', as advocated by Adam Smith in his book *The Wealth of Nations* (1776). Perhaps even more than in France, two centuries of combining the science of forestry with economic principles ensured the replacement of "areas largely devastated by destruction of the forest, utilization of forest litter and grazing" with "highly productive forests" (Klose, 1985). As in France, the modern weighing up of economic and ecological interests came too late for many forest areas with ancient oak trees in pasture woodland. Prohibition of woodland grazing certainly has not helped ecological interests, but instead has contributed to the decline of ancient oaks, which are succumbing to beech and hornbeam suppression.

By the end of the eighteenth century, the situation in the Netherlands was even more serious than that in Germany, with almost no forest left other than a few isolated patches owned privately, such as the woodlands near Rheden in Gelderland and near Dieren in Noord-Brabant, and some common woods such as the Eder Bos and the Speulder-en Sprielder Bos in Gelderland on which timber trees could still be found (Buis, 1985: 521). From the seventeenth century onward, the Netherlands imported all its timber. On the sandy soils of Gelderland, Utrecht, Brabant, Overijssel and Drenthe, heath had taken the place of pasture woodland as it had in the north of Germany. A traveller by coach service between Amsterdam and Hamburg, a distance of 367 km by modern roads, would see open heath almost all the way. Re-afforestation was mostly on this heath and on the extensive inland sand dunes that had resulted from the abuse of even this sparse vegetation. Early attempts to plant oak produced low grade coppice woods at best; it was only in the nineteenth century that Scots pine was introduced successfully to tame the drifting sands, which in several places had swallowed villages. The commons were mostly privatised in order to plant the pines; the royal domains near Apeldoorn in the

Veluwe (Gelderland) became state forests. The Netherlands has almost no ancient and only a few veteran oaks.

The science of sustainable forestry that was developed in eighteenth-century Germany soon spread across central and north-western Europe but hardly reached Britain (Stebbing, 1928; Taylor, 1946). France had made its own start in the seventeenth century but adopted the same methods, using faster growing pines to replace beech and oak from the late eighteenth century onwards. Three centuries of modern economic forestry have destroyed the ancient trees

Figure 7-6. Oak woodland with cattle at Nora Kvill, Kalmar, Sweden

that developed in Germany and France, where the emphasis was on plantation. Southern Sweden, and especially Kalmar, Jönköping and Östergötland, were rich in pasture woodlands as well as in meadows, often along streams, where oaks grew in a parkland setting. In the Middle Ages, village lands were divided into 'innmarken', enclosed lands for arable crops and hay making near the village, and 'utmarken', the often wooded pasture lands beyond. Pasture woodlands, several times larger than the inner fields, were used collectively by the community. When agricultural land use was scattered and the associated population limited, there was not much conflict between trees and farming, but this intensified with a growing population. Oaks were cut for timber and grazing inhibited their regeneration, so that eventually, most of the remaining mature oaks grew in the 'innmarken'. Here, they interfered especially with meadows for hay making by casting shade and dropping slowly decomposing leaves.

In 1558, King Gustav Vasa declared all oaks growing on Crown and taxed land the property of the Crown. The nobility owned oaks on private estates and adopted the same attitude. The peasantry was prohibited from cutting these trees, which became increasingly important for the Swedish navy in a time of Swedish expansionism during the seventeenth and early eighteenth centuries. A conflict of interests between the nobility and the peasantry now became acute, with farmers at the very least lopping off lower branches (shredding, but not pollarding, which was prohibited) to reduce the shading of crops and grass for hay making, a crucially important resource for livestock in the long winters. Excessive branch cutting often slowly killed the tree. Selective cutting for the navy, foremost in Kalmar but later also elsewhere, had left mainly poorly

in both state and privately owned forests on the continent; in Britain, such forestry has prevailed for just one century, allowing the preservation of some ancient trees in England.

Sweden

The history of the oaks in Sweden differs from that in England as well as from that in Germany and France. The abundance of ancient and veteran oaks in Sweden today, second only to that in England (Chapter 6), can be attributed to 'forestry', but not the kind of forestry

shaped, old, hollow and senescent oaks standing, the timber still 'untouchable' to the peasants.

The protectionism enforced by the State caused much ancient 'oak habitat' to remain: 30–50% of 'oak habitat' present in the eighteenth–nineteenth century in Östergötland is still extant. However, a royal decree of 1789 gave freehold peasant farmers the same rights to their lands as the nobility, eventually reducing the efficacy of the ban on cutting oaks. In the nineteenth century, the ban was further lifted, beginning in 1805 when peasants were allowed to exploit, with legal permission, only 'old and bad' oaks; the good oaks fit for naval timber had been marked with two crowns stamped into the bark. The royal oaks were finally relinquished by lifting the ban completely in 1832. The State had surrendered its ownership and the oak became the peasant's hated tree. This led to a drastic reduction in the number of oaks in Sweden, not only mature and old oaks, but even saplings and young oaks. With this more permissive policy, however, attitudes began to change and the hatred that seems to have been felt towards nobility and oaks alike began to de-escalate.

Nevertheless, the ancient 'oak habitat', of which perhaps as much as 80% was situated in ancient meadows, had become fragmented and seriously reduced. Between 1790 and 1825, two surveys showed a total reduction of 82% in the number of oaks fit for naval use in southern Sweden. Much of this was cut for the navy, while the peasants took care of the older, 'rotten' oaks, destroying large numbers between 1810 and 1850. Perhaps over a million old oaks were felled in this effort to rid the meadows of their trees. Much of what remains now consists of younger oaks because full protection only happened in the twentieth century. Locally, the former hay fields, with glacial erratic boulders collected in heaps or dry stone walls and with a scattering of mature oak trees, are now grazed by cattle. The greatest threat to ancient oaks in the remaining fragments of historic 'oak habitat' is now the gradual infilling with a secondary growth of trees, caused by the cessation of hay making as well as grazing. The conservation issues affecting ancient and veteran oaks in Sweden are generally well understood, with emphasis on their value as habitat for biodiversity. [Information obtained from Naturvårdsverket, 2006 and from Eliasson & Nilson, 2002.]

Conclusions

The answer to the question implied in the title of this chapter is of a historical nature. When I pose the question to usually well-informed persons, such as my colleagues at the Royal Botanic Gardens, Kew, they tend to invoke

Figure 7-7. Veteran oaks at Burley, New Forest; drawing by J.G. Strutt in Sylva Britannica (1826)

climate or ecology, not history and people. Perhaps the climate and soils in Britain are more suitable to its native oaks than are those in other European countries. But oaks of the same species occur in similar abundance in most of lowland Europe north of the Alps.

The climate in Britain is relatively mild due to the surrounding seas, but both native species of oak are well adapted to greater climatic amplitude than that found in lowland England, where most of the ancient and veteran oaks occur. The most important region outside England for ancient oaks is southern Sweden, where pedunculate oak reaches its northern limit. Britain has diverse geology and many soil types, but all of these are present on the continent too. Soils that favour woodland types with pedunculate oak in particular cover much larger areas in France and Germany than in England.

Human interference favouring oaks in the process of woodland succession has happened for millennia in England and on the continent; grazing animals are likely to have had similar effects before the advent of husbandry (Vera, 2000). We owe it to English cultural, social and political history that this country has more ancient oaks than the rest of Europe. It is not that destruction did not happen in England. In chapter 4, we found that 75% of medieval deer parks have been lost without leaving a single ancient oak. Even more acreage of Royal Forests disappeared into oblivion or was so altered that no ancient oaks survived. Yet the enduring legacy of the Royal Forests

Figure 7-8. Veteran oaks in Bear's Rails, Windsor Great Park

is that they slowed down the rate of woodland clearance at a time of strong population growth and increasing land hunger in Western Europe (Marren, 1990).

If we can believe the extraordinary numbers of ancient oaks recorded in the late sixteenth to early seventeenth century in Duffield Frith Forest (Chapter 5) and Sheriff Hutton Park (Chapter 4), almost all of which are gone, we might conclude that not many oaks are left in England either. But what these figures really demonstrate, even if exaggerated, is how many ancient oaks England must have had before *c.* 1650, when the destruction began. There were too many for all of them to have been destroyed without a concerted effort by all landowners to do so. Fortunately, the landowners made no such effort; instead, they preferred to enjoy pheasant shooting, driving around in carriages and similar pastimes on their estates.

Sixteenth and seventeenth-century surveys of forests and parks often concluded that almost all oaks were 'decayed' and unfit for navy timber, indicating that they were ancient oaks. Many were maidens, especially in deer parks (Chapter 8), so previous owners had not been interested in taking the timber when the trees were still young and sound. The growing wealth of the British Empire meant that most

owners obtained other sources of income. English land ownership, despite attempts to demonstrate the contrary in the *Return of Owners of Land, 1873* (also known as the 'Modern Domesday'), remained concentrated in the hands of the traditionally privileged few. The principle of primogeniture has done much to keep estates within the same families. In 1879, according to a compilation by John Bateman (1883) based on the *Return of Owners of Land, 1873*, there were 3,672 'peers', 'great landowners' and 'squires' in England. The first two categories, the true great landowners, amounted to 1,521 individuals. The three categories, or the 'landed gentry', together held 16,260,452 acres (6,504,180 ha) or about half of England (land area 12,778,700 ha). Many families owned multiple estates. The big sell-off occurred only in the twentieth century, but many large estates remained, albeit often reduced in acreage. This private ownership was an important factor in the preservation of the ancient rural landscape, parks and country houses, as it still is today. In this, as often stated by the Historic Houses Association in its policy documents, the United Kingdom (and especially England) is uniquely rich in Europe. This has undoubtedly contributed to the preservation of the ancient oaks of England.

8

Ancient oaks in a pasture woodland context

The surveys of ancient and veteran oaks in England presented and analysed in Chapters 4 and 5 have demonstrated a strong association between the present distribution of these trees and the (past) occurrence of medieval deer parks and to a lesser extent Tudor deer parks, Royal Forests, chases, and ancient wooded commons. In all of these categories of ancient land tenure and land use, wood pasture as defined by Rackham (1986, 2003b) was the prevailing management method[1]. This method of land management attempted to combine the growing of trees with the concurrent grazing of animals.

The animals could be deer, wild boar (until these became extinct in England in the thirteenth century) or domestic animals, mostly cattle. All of these animals have feeding habits that are, in various degrees, detrimental to tree regeneration, but the relationships between tree rejuvenation from seedlings or suckers (e.g. elm) and the browsing effects of mammalian herbivores are complex and differ from species to species, both of tree and animal (Vera, 2000). Density of animals per unit area influences the gradient from successful regeneration to no regeneration at all, while site

Figure 8-1. Pollard oaks in Epping Forest, Greater London and Essex

---

1 Following the definitions of Harding & Rose (1986) 'wood pasture' is the management regime resulting in 'pasture woodland', the physical presence. Pasture woodland is also the correct translation of the Latin *silva pastilis* used in the Domesday Book survey. It is the term used in this book.

characteristics such as soil type, drainage, precipitation, composition of vegetation and fluctuations in seed production ('mast years' versus low-production years) can all influence the outcome in terms of regeneration success for trees.

The management of trees in pasture woodland was usually to create pollards, i.e. by cutting the tree above the reach of browsing animals on a periodic basis. Such management seems to indicate that the natural regeneration of trees was seriously hampered in pasture woodland. Trees needed protection from the incessant browsing by the animals if they were to be cut, either by enclosures or by cutting them above the reach of browsing animals; but how common were pollarded oaks in pasture woodland? Further on in this chapter, I shall demonstrate that pollards are not everywhere the most abundant tree form of oaks in ancient pasture woodland. Indeed, other tree species, such as beech or hornbeam, may often have been more abundant among the pollards than oak. Owing to its value for heavy timber, there was more incentive to leave the oaks uncut and only pollard the other trees. The cutting of oak, beech and hornbeam pollards stopped in conjunction with abandonment of the woodland pastures, mostly at the end of the eighteenth and the beginning of the nineteenth century, but perhaps continuing in some parts of the country such as Essex into the early twentieth century (Owen Johnson, by email 20th February 2015). But how and when did wood pasture begin? Were there perhaps natural precursors to this type of woodland use? And how did the oak become the predominant pasture woodland tree in most parts of the country? To answer these questions, we need to go back to prehistoric times and we have to investigate the ecology of the oak in a landscape context. We need to investigate whether medieval oak pasture woodland constituted a significant change in the natural vegetation, or whether it perhaps was merely a continuation of an ecosystem that already existed before human influence became substantial.

Prehistoric woodland with oaks

The native oaks of England first appear in abundance in the pollen record of the Boreal phase (pollen zones V and VI) of the Flandrian or post-glacial Holocene, beginning about 8,700 years BP. The two species cannot be distinguished by pollen, so we do not know from this evidence if they appeared simultaneously or one after the other. Low frequencies of oak pollen in much older samples may have resulted from long-distance dispersal (Godwin & Deacon in Morris & Perring, eds., 1974), but a foothold may have been gained in Cornwall by about 9,500 years BP, followed by entry into East Anglia across the then dry southern North

Sea a few centuries later (Bennett, 1988; Ferris et al., 1995; Brewer et al., 2002; Harris et al., 2003). Hazel followed by elm and oak first replaced Scots pine and to some extent birch in the south and southeast of England, which would be severed from the mainland by marine transgression across swampy lowlands that now form the English Channel only about a millennium later. Oaks remained an important component of the developing and eventually prevalent deciduous angiosperm woodland[2]; later, in pollen zone VI around 7,500 years BP, they were joined in co-dominance by lime, which perhaps became the dominant tree in some parts of lowland England during the Atlantic (Pollen zone VIIa) from c. 6,500–5,000 years BP. A predominance of oak occurred mainly in the uplands and in the northern and western parts of England (Rackham, 1976).

Ash first appeared around 8,500 years BP but remained more local for a long time. The climate was then warmer (c. 2° C warmer in summer and 1° C warmer in winter) and wetter than at present (Climatic Optimum). Oaks spread far north to Scotland, reaching the central highlands around 6,000 years BP (Harris et al., 2003) and onto the uplands at higher altitudes than at present, although the current distribution of trees in these areas is artificially reduced. Around 5,600 years BP there was a marked decline in elm, perhaps caused by an early outbreak of elm disease; this decline was soon followed by a more gradual decline of lime and ash, the latter especially in the south. The Sub-boreal phase (pollen zone VIIb) from c. 5,000–2,500 years BP remained relatively warm but became drier, this period may have seen the spread of pedunculate oak across most of England at the expense of sessile oak, which largely retreated to the wetter west and to the uplands. From around 2,500 years BP, the climate cooled during the Sub-Atlantic (pollen zone VIII) and it was only then that beech and hornbeam, which had first appeared more than three

---

2 It will be useful to define the terms 'forest', 'wood' and 'woodland' as used here to clarify what follows. These terms are often used for different structural types of woody vegetation or interchangeably by different authors. 'Forest' here will mean tree-dominated vegetation, where tree canopies will be in contact in an optimal undisturbed phase. Elsewhere in this book, the term may also be used in the medieval sense of a royal hunting area. 'Wood' will be tree-managed forest, either as coppice woods with or without standard trees or as plantations. 'Woodland' will be a broader term, which on a landscape scale may contain in a mosaic pattern a gradient from closed-canopy forest to park-like pasture woodland or even grassland with shrubs and single trees. In all three of these landscape types, trees are present and usually the dominant growth form. (See, for example, Ellenberg (1988) for a broad concept of 'woodland'.)

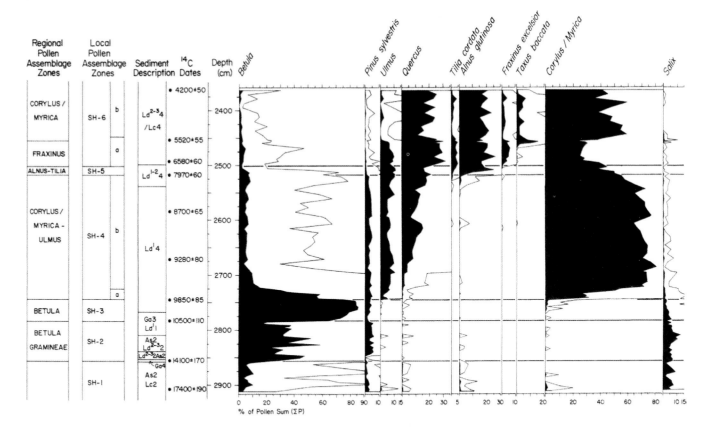

Figure 8-2. Pollen percentage diagram from Saham Mere, Norfolk, showing the arrival and increase of oak (*Quercus*) from around 9,400 years BP to 4,200 years BP. Amended from Bennett (1988) Figure 3 with non-arboreal taxa omitted.

millennia earlier (Rackham, 2003b), spread across the south and east and became abundant. Hornbeam is still common only in the southeast; beech has a wider distribution but has also been much planted.

The Roman Warm Period (*c*. 2,050–1,480 years BP) and the Medieval Warm Period (*c*. 1,120–710 years BP) were followed by the 'Little Ice Age' (*c*. 710–160 years BP) and then by the most recent warming phase; but human impact on the landscape, as well as the relatively short durations of some of these climate fluctuations, make it more difficult to separate climatic factors from other influences on tree distributions. In the context of the development and spread of pasture woodland in England, in which oaks played a key role, climatic factors are also less relevant than other influences. We can be certain that wherever wood pasture was practiced from its earliest beginnings, perhaps by Neolithic farmers and pastoralists during the Atlantic, oaks were usually present.

What did the lowland deciduous woodland look like during the Climate Optimum of the Atlantic, *c*. 6,500

years BP? Of what trees was it composed before the first farmers made their inroads? From pollen distribution maps for this period compiled by Birks *et al.* (1975) and some other sources, Rackham (1976) presented a map of Great Britain and Ireland showing 'wildwood provinces', each of which is characterised by one or two dominant (most abundant in % of total pollen) trees. Alder (*Alnus glutinosa*) was ignored because its abundance in the pollen records is caused by the circumstance that alders grow in fens and on the margins of shallow lakes, from which the peat deposits with preserved pollen are also taken. Most pollen grains do not travel far, even from trees such as alder that spread their pollen by wind, and the samples are therefore biased towards alder. We would only be interested in the composition of the woodland further from the peat source.

Tree species vary greatly in the amounts of pollen that they produce and disperse in spring. Wind-pollinated trees, such as birch and oak, produce much greater volumes than insect-pollinated trees, such as lime. The pollen diagrams that show the pollen counts as a percentage of the totals at depositional intervals in the peat column must somehow be corrected for these differences in pollen production if they are to be used as proxies for the relative abundance of tree species in the landscape. We can obtain data for existing mixed woodlands by catching the pollen rain and comparing the counts with the true abundance of the source trees. When adjustments are made (for lime with a correction factor of 8 for instance), it turns out that most of the lowland woodlands in England during the Atlantic were diverse. Much of lowland England had abundant lime, for the most part presumably small-leaved lime, but this is not separable in the pollen record from large-leaved lime which is the other, but less common, indigenous species. This type of tree vegetation is mapped as the 'lime province' by Rackham (1976). The other province, comprising northern England from Yorkshire into Scotland and western England into Wales and Devon and Cornwall, is the oak-hazel province. Hazel has become an underwood shrub or tree as the result of many centuries of coppicing.

In this situation, under a canopy of oak, hazel rarely flowers, but in the 'wildwood' 7,000 years ago, it was often a medium-sized canopy tree.

The actual composition of the woodlands would have varied regionally and also locally with soil type, moisture availability and aspect on slopes. Oak would have been more abundant or dominant on poor sandy or loamy soils, whereas lime, elm and hazel would have dominated on richer soils or on soils that would retain precipitation longer. Ash usually occurs on moister sites too; it is also a pioneer, as can be seen on the South Downs where beech was levelled in the hurricane of October 1987 and where the National Trust decided to let succession run its course on parts of the Slindon Estate in West Sussex. Beech and hornbeam had not yet arrived in the Atlantic or were very rare in the southeast. Birch, another pioneer, would be combined with oak on poor sandy soils, regenerating abundantly after disturbance as we can observe today.

Figure 8-3. Oak tree broken by storm in High Park, Blenheim Park, Oxfordshire

This disturbance would rarely have been caused by fire; as Rackham pointedly remarked: indigenous deciduous woodland burns like wet asbestos. The preoccupation of the Forestry Commission with the danger of forest fires stems from their widespread conifer plantations, especially those planted with the resinous Scots pine, which is indeed highly flammable during a dry midsummer spell. An increase of bracken has made un-grazed pasture woodland more flammable in dry conditions in spring, when the fronds are brown and dead. Wind, and possibly also death caused by pathogens, could create the openings into which the pioneer species would invade. The natural woodland in lowland England would thus have become a patchwork of a few tree species, with one or the other dominant in different places, and with a mosaic of succession stages from pioneer growth of seedlings and saplings to old and senescent trees of the longest-lived species. Glades may have formed in places where trees had died from some cause and where browsing and grazing animals prevented regeneration; often these are now associated with a rising water table or flooding. There would have been much dead wood in the landscape, both from fallen and still-standing dead trees.

Apart from the evergreen yew, which would have been locally abundant on chalk slopes, oaks are the longest living indigenous trees. Succession involves competition, and one of the driving factors in this is shade tolerance. More light-demanding trees tend to be succeeded by more shade-tolerant trees until the next disturbance re-sets the clock. Pioneers such as birch are the most light-demanding species, but oaks also need much light to succeed from seedlings. Lime, elm, ash and also the later arrivals, beech and hornbeam, are all more shade-tolerant than oak when young (Ellenberg, 1988). Oak does not regenerate well under a canopy, even under its own canopy, but its longevity would guarantee that when opportunities arose from the natural removal of shade-tolerant competitors, its acorns were still available for regeneration. Oaks would also have been favoured when wood pasture activities by the earliest farmers began to create more openings, resulting in a more parkland-like woodland around settlements. Or was this parkland perhaps already present before agriculture arrived?

Researchers who have an interest in and knowledge of bryophytes, lichens, non-lichenised fungi and invertebrate fauna have observed that the various types of grazed and formerly grazed woodland today are the strongholds in terms of species diversity for these tree-associated organisms. Many species are virtually restricted to such habitats, with lichens becoming richer also where the air is unpolluted, and with invertebrate fauna and fungi becoming richer with

an increase of large dead and decaying trunks and limbs (see Chapter 10). Factors associated with this biodiversity are a) a wide range of age classes of trees, b) a mosaic landscape of open glades, open and dense woodland, and c) tree boles that are well lit, not shaded as in un-grazed woods. Requirements for (half) light, microclimate and age of substrate are crucial for epiphytes, which often tolerate only a narrow band of variation in these factors; these requirements are all more likely to be met in more or less open woodland or parkland with ancient trees. It is therefore argued, for example by Harding & Rose (1986) and by Rose (1993), that as these tree-associated organisms, many of which are slow dispersers, must have existed in Britain well before pasture woodland and parks were created by agricultural people from the Neolithic onward, such park-like landscapes must have been already widespread during the Boreal and Atlantic periods. Ancient English woods without ancient trees are, in most cases, former coppice woods. Although coppicing was discontinued a century or longer ago, the trees are still all young[3] (relative to potential age) except for birch, wild cherry (*Prunus avium*), alder and willow. Other woods, even if containing only native broadleaved trees, have usually been managed as high forest for timber; these have no ancient trees either. Comparisons with these, as made by Rose (1993) and by Chatters & Sanderson (1994), show the paucity or absence of the tree-associated organisms mentioned above in high forest, demonstrating that much of the 'wildwood' was not tall closed-canopy forest.

There are very few examples of woodlands (or forests) that have not been grazed by livestock for more than a century and that have been protected as non-intervention reserves for about the same length of time, and none of these occur in Britain. Sections of Fontainebleau in France, Hasbruch and Neuenburg in Germany, and Białowieża in Poland and Belarus are examples in the northern lowland plain of the continent. The former three were heavily grazed from medieval times until the nineteenth century and there are still pollards at these locations, especially in Hasbruch and Neuenburg, but a non-intervention policy has been applied for such a long time that natural processes rather than a wood pasture history have begun to determine the forest structure (Koop, 1981). All three of the factors associated with this biodiversity listed above are still present, but these woodlands are becoming forests, and glades, if present, are small. Gaps are created by fallen trees but are

---

3 With 'young' I refer to the stems, not necessarily to the coppice stools, which can be ancient.

not perpetuated by grazing animals: the wild deer that are present do not usually keep them open for very long.

When compared with the other examples of non-intervention forests, Białowieźa has been a non-intervention reserve for longer and is by far the largest forest, but its climate is more continental; it has not only broadleaves but also abundant Norway spruce and locally Scots pine. There was grazing of cattle, sheep and pigs in what is now the national park on both sides of the international border, but this has been abandoned (and forbidden) in the Polish section since 1923 (Vera, 2000). The strict forest reserve in Poland of 4,700 ha (in 1996 expanded to 10,500 ha) is surrounded by a wide buffer zone of managed forest. There is much standing and fallen dead wood and there are 21 living large, partly hollow oaks with girths of between 6.00 m and 7.50 m, all of which are tall maidens, in the Polish National Park area of this World Heritage Site. These oaks, as tall canopy trees with relatively small crown/bole ratios, could be up to 500 years old. Mature oaks (mostly pedunculate oak) have a significant presence in the lime-hornbeam (*Tilio-Carpinetum*) forest, which covers about 45% of the area. Vera (2000) wrote that there are only sporadic oak seedlings and a few young trees in this forest, based on observations made in the 1970s. More recently, however, it was found that regeneration of oaks does occur in this forest type (Jeroen Philippona, by email 7th March 2016).

Oaks could be declining in mixed spruce-broadleaved forest types within the Białowieźa forest, with only old trees present, while spruce, lime and hornbeam are increasing. Competition from the latter two species does inhibit oak regeneration. In the absence of beech, however, the situation with direct competition between mature trees in Białowieźa is very different from that in the old pasture woodlands in Germany (Hudewälder) and Fontainebleau in France. Areas in the National Park where oaks are (co-)dominant (thermophile oak forest) appear to be associated with pasture woodland of the past (Falinski, 1986); however, the largest-girth oaks are not of an open grown architecture like those in pasture woodland but indicate a forest environment that is very different from parkland. In several English high-forest woods now managed as nature reserves with little or no interference, such as Savernake Forest in Wiltshire and The Mens in Sussex, natural processes are beginning to alter the structure of the forest, with increasing amounts of dead wood and gaps being created and filled again. However, even in the forest of Białowieźa, these processes have not yet been given the centuries needed to develop into a resemblance of the hypothetical prehistoric 'wildwood'. It is possible that the result will never be the same as it was six or seven thousand years ago.

The Dutch ecologist Francis (Frans) Vera has become widely known as the principal proponent of a hypothesis that envisions a pre-Neolithic central and west European lowland landscape that had more similarity with savannah than with closed forest (Vera, 2000). Savannah *sensu stricto* is grassland with scattered trees, as in East Africa or Australia, but what Vera has in mind would have been more like half-open parkland with expanses of grass and/or heather and scattered bits of forest, surrounded by margins ('mantles') of thorny or prickly shrubs such as blackthorn (*Prunus spinosa*), hawthorn, holly, brambles (*Rubus* spp.) and locally gorse and common juniper (*Juniperus communis*). There may have been solitary (free-standing) trees, but these would also be surrounded by shrubs. Wet places would also have been more open, but with rushes, reeds and willow thickets and with alder or black poplar (*Populus nigra*) as the dominant trees. Perhaps the closest to this 'savannah' landscape in England at present would be the New Forest, which has heath, grassland, marsh and open woodland where grazing is allowed (closed forest only occurs within enclosures).

The drivers of this dynamic landscape mosaic would have been large grazing and browsing animals such as deer, elk (*Alces alces*), aurochs (*Bos taurus*), horse (*Equus caballus*) and wisent (*Bison bonasus*), as well as beavers (*Castor fiber*) along streams. Each different animal has

different feeding habits and they live in groups or herds or as solitary individuals, thus diversifying their impact on the vegetation. Tree regeneration would mostly have occurred under the protection of thorns, therefore in the margins around the forest patches which then tended to expand. Towards the centre of the forest, the thorns would become out-shaded and the succession from light-demanding tree species such as oak to shade-tolerant trees such as lime took place. Eventually these trees would die and fall, creating openings in which the animals could browse once more (in a dense forest they would find little to eat) and so expand the openings to pasture. Here, thorns could have started the cycle again, first giving opportunity for light-demanding trees such as hazel and oak to regenerate.

There are places in northwest Europe where this model seems to operate, but now with domestic animals; these include Hatfield Forest in Essex (Rackham, 2003b) and the small nature reserve Borkener Paradies in Lower Saxony (Vera, 2000). The New Forest in Hampshire is often mentioned as another example, but here the composition and density of the animal population are also quite artificial and high, with a predominance of horses (Tubbs, 1986). Most of these traditional pasture woodlands are or have been overgrazed and may not be a true model for the past. Often, nature conservation management using grazing in woodland settings has as yet not lasted long enough to observe full cycles from grassland to decaying woodland, especially not in relation to oaks, but some reserves have been established in pasture woodland where ancient and veteran oaks are already present, such as Cranbourne Park in Windsor Great Park, Grimsthorpe Park in Lincolnshire

and Birklands in Sherwood Forest. The current campaign for 'rewilding' could create other examples where these processes can be observed in due course.

Much has been written about this already and this is not the right place to review all the evidence or the arguments, I shall only touch on a few of the more pertinent points. Hodder *et al.* (2005) and Rackham (2003b, 2006) have discussed evidence in support as well as against Vera's model (the alternative is sometimes known as the Tansley model), particularly for Britain and Ireland. The entomological evidence allows us to infer the past presence of large ancient oaks similar to those present today, which are associated with pasture woodland, not with high forest with a closed canopy. Some of the evidence from pollen samples, such as the presence of light-demanding herbs, indicates that there were open areas in the forest and this supports the evidence from beetles for open areas or parkland. Other flowering herbs occur most abundantly or in some cases exclusively in closed canopy forest, and these plants are still indicators of ancient woodland, some of which may have been wooded since pre-Neolithic time (Rackham, 1986, 2003b, 2006). The pollen of such species,

Figure 8-5. Ponies grazing in Bolderwood, New Forest, Hampshire

usually present in very low numbers, has often been omitted from diagrams on which conclusions about the vegetation were based, or was lumped as 'grasses' and 'herbs' because the researchers were interested in the tree composition of the ancient woodland.

The most serious objection to Vera's model for vegetative succession in the pre-Neolithic landscape, according to Rackham (2006), is its general application across much of lowland Europe. Vera's model predicts that oak and hazel did not regenerate without grazing animals' effect on the vegetation, which turned the landscape into parkland. During the Atlantic, only aurochs, red and roe deer, elk (which became extinct towards the end of the period), beaver and wild boar were left in Great Britain. The (sub) fossil record is always incomplete, but provides the only evidence from which we can infer the prehistoric fauna. Ireland had only wild boar in abundance and probably only a few red deer, yet in Ireland, oak and hazel are as abundant in the pollen record as they are in Great Britain. According to Vera's model, with the exception of aurochs, which like cattle were predominantly grass eaters, the large herbivores that were present during the pre-Neolithic period were unspecialized grazers or feeders and did not create a park landscape with grassy glades and fields.

Vera juxtaposed two extremes, closed canopy high forest (the 'null hypothesis') and more or less open parkland (the 'alternative hypothesis') as the two possible models for the pre-Neolithic lowland landscape in central and western Europe. Rackham (2003b, 2006) suggested that Vera's model may have applied in a gradient from east to west across cool temperate lowland Europe, with half-open pasture woodland prevailing in the east, where most of the large grazing animals occurred in the wild for the longest, but less widespread in the west. Locally, climate and soil, with help of light animal browsing and grazing, may have favoured parkland landscapes rather than closed forest. Rackham (2003b) explains the common death of 'infilled' oaks and other trees among the much older oak pollards in Staverton Park in this way. No large herbivores other than deer are present, nor are they needed to maintain the park woodland of Staverton, which grows in pure sand and receives only 550–600 mm of rain per year. But, as elsewhere in pasture woodland that has long been neglected, bracken has spread and inhibits tree seedlings.

The correlation between oak regeneration and grazing can be observed in pasture woodlands, as Vera (2000) emphasised, but often after grazing by domestic animals has ceased and the only browsers and grazers are some deer and rabbits. His assertion that oaks fail to regenerate in pasture woodlands in England after grazing stops (Vera,

2000, p. 274) is untrue as a general statement. Examples I can give and have seen are at Ashtead Common in Surrey, Aspal Close in Suffolk, Enfield Chase in Greater London and Kenley Common in Greater London. The first two locations have many veteran oak pollards and a new generation of self-seeded oaks filling in gaps between these. In some localities, birch is the pioneer, as in several larger areas with few oak pollards in Ashtead Common, whereas elsewhere, oak is the first to colonize the grassy glades, as in Aspal Close, Enfield Chase and Kenley, where the oak trees are accompanied by a few thorns. Bramble thickets can also enable oak seedlings to progress to saplings and young trees in the presence of deer. No evidence of common death of these new oaks was observed in these localities in 2013–16. On Ashtead Common and Aspal Close beech, lime and hornbeam are absent or rare, which probably makes all the difference. On Kenley Common, hornbeam is present, but not in the half-open grassland areas where oak is regenerating. Hornbeam spreads much slower than oak, but it may follow it. Whether these competing species would arrive eventually is now a moot point in locations where they are absent from the woodlands, given that these locations are now often surrounded by treeless farmlands or urbanisation.

On poorer acid soils especially, oaks can regenerate in managed oak woods, albeit sporadically and probably in bursts correlated to mast years. Despite this, spontaneous regeneration in oak woods has been poor or absent for more than a century (Rackham, 2003b), perhaps because of the introduction of the oak mildew fungus early in the twentieth century. A long-living tree like oak does not need many surviving offspring to maintain the species in a particular locality. Nevertheless, competition from beech is observed in many oak woods on slightly heavier soils, and if undisturbed for a long time, oaks in such locations will die, often even killed by overtopping beeches, and fail to regenerate. In the non-intervention forest reserves of Fontainebleau, Hasbruch and Neuenburg, the former oak-dominated pasture woodland is slowly but inexorably being replaced by a beech-dominated forest with virtual exclusion of oak (Koop, 1981; Vera, 2000). However, beech, although established in a few localities in the south by 4,300 years BP (Oxford, 1975 cited in Harding & Rose, 1986) only spread in Great Britain during the sub-Atlantic from c. 2,500 years BP and remained confined to the south of the island for a considerable time. In the current oceanic climate of much of Britain, beech is favoured in the succession, but if drier and warmer summers are indeed the future, this may change.

Lime, mostly small-leaved lime, as the other serious competitor of oak was sometimes dominant but usually

Figure 8-6. Pollard oak in pasture woodland, Burnham Beeches, Buckinghamshire

co-dominant with oak in the south and east of England, but it was by no means ubiquitous and sometimes only a minor component of the Atlantic 'wildwood' in the east of England (figure 8.3 in Rackham, 2003b), with oak always present. Elm would have been another competitor, but it apparently suffers periodic setbacks from elm disease (Rackham, 2006, p. 107) and was not common everywhere. Hazel, since prehistoric times a coppice tree, is also present in most Atlantic period pollen diagrams; this species can sometimes become a medium-sized canopy tree if left undisturbed. The pollen diagrams show the 'wildwood' to have been of variable composition even within a restricted area such as East Anglia.

Under Vera's model, it is difficult to envisage how an oak in a closed-canopy forest could ever become a large tree, because when under a canopy, more shade-tolerant trees would soon out-compete it. Yet large 'bog oaks' (Reddington, 1996; Pilcher, 1998), as well as remains of forest submerged by the rising ocean that include large stumps of oaks, are evidence that oaks grew large in Wales and East Anglia between 8,000 and 5,000 years BP. A 13 m long trunk of an oak found in the Norfolk Fens (Miles, 2013) is of a tall maiden without low branches, that is, a tree that grew in high forest 5,000 years BP. At Ramsay in Cambridgeshire, a Bronze Age fen-oak was found in 1976 and measured to be 20 m long below the first branch and nearly a metre thick 6 m above the ground (Marren, 1990). This oak trunk had numerous large tunnels made by the great capricorn beetle (*Cerambyx cerdo*), which only lives in very large oaks; it is extinct in Britain. Even longer oaks, up to 27 m below the first branch, have been found in the older clay deposit under the peat (Rackham, 2003b); oaks of such size are rarely found in managed oak forests today. Surely such oaks grew in high forests with a closed canopy that were drowned by a rapidly rising sea level, not in open woodland or parkland. While it is true that open-grown oaks eventually have wider trunks than oaks in closed-canopy forest (Green, 2010b), oaks could have been much larger in prehistoric forests with natural canopy gaps and forest margins than in the present situation. Primeval virgin forest no longer exists in Europe (Peterken, 1996), but where it remains in temperate regions of the world, an abundance of very large dead wood is present.

There is, at present, no conclusive evidence that the presence of oak and hazel in the Atlantic forests in Britain and Ireland was dependent on the activities of large grazing animals. The ecological evidence, especially that from invertebrates associated with ancient oaks, but also that from lichens and epiphytes (Chapter 10), points at the existence of large, open-grown trees that are very different from the tall bog oaks. Whatever the precise causes, and contrary to the conclusions of Hodder *et al.* (2005) and Mitchell (2005), the 'wildwood' in England must have been made up of a mosaic of closed canopy forest patches and more or less open parkland, closely connected to varying and probably shifting local conditions. It could be informative for this discussion to consider another dynamic which occasionally creates large gaps in forests in this part of Europe: severe hurricanes such as the one that hit the south of England in October 1987.

Historic pasture woodland with oaks

The pasturing of domestic animals in the natural forests of Europe goes back at least to Antiquity (Ellenberg, 1988; Vera, 2000) and was probably common in prehistoric times before that. The Romans distinguished two types of pasture woodland. *Silvae glandiferae* were those woodlands in which oaks dominated, providing animals (mostly pigs) with their nutritious acorns in season and in good mast years. These woodlands were the more valuable and their use was usually strictly regulated as the crop was variable and often limited. *Silvae vulgaris pascuale* were the pasture woodlands without such fruits, in which animals would often be grazed without any regulation for at least as long as woodland still covered large parts of the land surrounding settlements. During this period, there was probably not much need to pollard trees to protect the new growth from the animals. Explanations for the Elm Decline that invoke pollarding the trees for animal fodder not only fall short of the facts (Rackham, 1986, 1996) but are unconvincing in themselves: if pollarding caused the sudden decline of elm in the pollen record, then coppicing would have had the same effect and the pollen record could not tell us which method was used. The woodland resource

must have seemed inexhaustible in much of Europe, although the Romans established a dense network of villas in several regions where this would no longer have been the case. Agricultural produce was exported from England to other parts of the Roman Empire.

During the Middle Ages the woodland resource became scarcer, depending on region and related to population density. Rackham (1986, 1996) and Williamson (2013) have described for many areas of lowland England that around 1,000 years ago the rural landscape was not essentially different from the present. Forests and woodlands were mostly turned to heath and scrubs with some wooded areas; most woods were, as now, isolated patches of trees, with abrupt transitions to the surrounding farmland. The distinction between 'ancient countryside' and 'planned countryside' was already evident. After the Statute of Merton (1235) most medieval woods were intensively managed as coppice, often with young oak standards as the only (planned) trees of any stature. Domesday Book (in 1086, see Hinde, ed. 1995) records many woods as *silvae pastilis* or pasture woodlands, and their value is usually expressed as the number of pigs that these woodlands could sustain or that were brought in as a form of rent; this use was more valuable to the manor than the timber they could yield. The use of woodland for pannage or the feeding of pigs by turning them out into oak- or beech-dominated lands (usually Royal Forests) was strictly regulated in the Middle Ages. It was seasonal, coinciding with the crop of acorns and beech nuts, and fell roughly between the end of September and 20th November; the numbers of pigs allowed depended on the richness of the crop. In England, mast years with abundant acorns occur irregularly, averaging only one in three years. This traditional use came to an end in England during the later Middle Ages (Rackham, 1986) and in other parts of Europe around 1500 or later (Buis, 1985), but it still continues in the New Forest (Tubbs, 1986).

When did the practice of pollarding trees in pasture woodland begin? Archaeological evidence of woodland management during the Neolithic is preserved in the Somerset Levels, where the remains of extensive trackways are laid across the peat so as to connect islands (Rackham, 2003b). One of these, the Sweet Track, dated *c.* 5,920 BP (Coles & Orme, 1976) is made of various species of tree, with definite evidence of woodmanship indicating the practice of coppicing of ash, hazel and oak. This shows us that coppice woods already existed in England 6,000 years ago. The quantities of small and similarly sized poles seen in these tracks cannot be obtained easily from 'wildwood', except in extraordinary circumstances, for example, subsequent to massive regeneration of birch or ash after a

Figure 8-7. Ancient oaks and Longhorn cattle in Cranbourne Park, Windsor Great Park, Berkshire

storm disaster. The high diversity of tree species used in the construction of the Sweet Track rules this out. But could some of the poles not have come from pollards?

Obtaining wood from pollards requires climbing a ladder or the tree itself, a much more laborious effort than coppicing. It is unlikely that this was done unless necessary, and it became necessary only if the grazing pressure of animals was too high for coppicing *and* if it was uneconomical or otherwise undesirable to exclude animals by hedging or fencing around the coppice wood. This only became the case when woodland had become a scarce commodity. Woods did not become scarce across all of England at the same time; the intensive Roman farming of the 'planned countryside', which stretched roughly across central England, would have already caused woodland to be scarce in these areas while it was still abundant elsewhere. Pasture woodland in Roman Britain, especially if of the more valuable *silva glandifera* kind and if situated in the agricultural belt, is likely to have been managed as oak pollard or beech pollard woodland. Certainly in later Anglo-Saxon time, when the population again increased and agriculture expanded, the necessity of pollarding trees would have become more prevalent; the 'wildwood' had virtually disappeared in lowland England by the seventh or eighth century (Rackham, 1986, 2003b). Nevertheless, extensive areas of woodland still existed, such as the Forest of Dean, Wychwood in Oxfordshire, the Weald in the southeast and large parts of Essex and Middlesex. These were essentially commons, which meant

that they were in use by the community, including pasture. Kings and privileged nobility would have hunted there, imposing restrictions on common use mainly after the start of the second millennium CE. The pollarding of trees was introduced in these larger wooded areas too, perhaps as a consequence of the prohibition of cutting the trees which now became incorporated into Forest Law. But as we shall see below, the practice was far from universal.

Vera (2000) has suggested that the pollarding of oaks may also have been carried out to increase the crop of acorns produced by each tree, especially when the oaks were still young trees. By the time of Domesday Book (1086), the area of England covered in woodlands of any kind (pasture or coppice mainly) was down to between 10% and 15% (Rackham, 1994). With the establishment of Royal Forests, chases and deer parks, many wooded areas dating from Anglo-Saxon time or earlier became regulated by law or, in the case of deer parks, enclosed with a park pale to keep the deer inside the area. Despite these restrictions, many of the commoners rights remained and were exercised, often under licences or against payment in kind or otherwise. Uses of the woodland for pannage or the taking of firewood could not be denied because they were subsistence for the population, they could only be regulated. While the king or his officers regulated these uses (or tried to) in Royal Forests and royal parks, the landed nobility and also the Church and monasteries owned most of the chases and deer parks, which fell under the respective manors belonging to them. In England,

Categories	Maiden	Pollard	Other
CHA+RF	175 (43.7%)	179 (44.8%)	46 (11.5%)
COM	79 (36.4%)	114 (52.5%)	24 (11%)
MDP	716 (63%)	365 (32.1%)	55 (4.8%)
TDP	324 (74%)	89 (20.3%)	25 (5.7%)
Manor	349 (53.9%)	225 (34.8%)	73 (11.3%)

Table 8-1. Numbers and percentages of maiden oaks, pollard oaks and other tree forms with a circumference ≥6.00 m in chases and Royal Forests (CHA+RF), commons (COM), medieval deer parks (MDP), Tudor deer parks (TDP) and manors. 'Pollard form (natural)' in the database is here interpreted as a maiden tree 'pollarded' by natural causes and therefore combined with maiden oak numbers. 'Other' includes 'multi-stem' and 'unknown' trees plus four 'stumps'.

few pasture woodlands remained under jurisdiction of the commoners, who then attempted to regulate uses by common assent. With so little of the land being woodland, trees were a valuable asset that were used in some way almost everywhere, at least in the lowlands. Oaks were more valuable than most trees and were often not left to grow into an ancient stage; fallen dead wood was often removed for firewood (Kirby et al., 1995). Damage by gales resulting in wind-blown trees led to the regulated removal of wood and timber; the regulations stipulated who could have the spoils and nothing of use was left. Under these circumstances, the most likely way for an oak to become old was for it to be cut as a pollard. Its trunk (bolling) would become hollow and worthless, but new branches kept coming after each cut as long as the tree remained alive and standing.

Nevertheless, five of the most important sites for ancient and veteran oaks in England (Chapter 9) have few pollards: Cranbourne Park in Windsor Great Park (formerly Cranbourne Chase), Birklands in Sherwood Forest, Brampton Bryan Park in Herefordshire and High Park (formerly Woodstock Park) in Blenheim, both medieval deer parks, and Grimsthorpe Park in Lincolnshire, a Tudor deer park. Other sites where the ancient oaks are mostly maidens are Bear's Rails in Windsor Great Park and Cornbury Park in Oxfordshire. In Savernake Forest, the largest (and oldest?) oaks are pollards but many of the smaller ones <7.00 m in girth are maidens. Several of these locations were forests or parks that had been owned by the Crown since medieval times and must have been excluded from pollarding at some point in time, whereas grazing may have been limited to the king's deer. Some of the maiden oaks on these sites exceed 9.00 m in circumference, several

more are 8.00–9.00 m, and these trees almost certainly date back to medieval times.

In Table 8-1 maiden and pollard oaks in England with a circumference ≥6.00 m are divided between five categories of former land use (excluding undetermined records). We can see a different pattern here: commons, chases and Royal Forests have a higher proportion of pollards than medieval and Tudor deer parks. It therefore appears that the above-mentioned sites that were once a chase or Royal Forest and have mostly maidens are the exception. The use of the oaks in chases and Royal Forests was often similar to that on commons and pollarding was common, although by no means universal. In deer parks, which were enclosed and therefore more private, pollarding of oaks was less prevalent. This is especially true in the later Tudor parks, where 74% of oaks with ≥6.00 m girth are maidens. Perhaps we may conclude from this that the private owners during Tudor times became even less permissive of the exploitation of the trees by the peasants and tenants of their estates than their medieval ancestors had been. Perhaps they preferred to keep the oaks uncut to provide the timber needed for work on the demesne, or even for sale, giving the peasantry access to trees outside the park (Muir, 2005). But then it seems that, in the end, many trees were not cut after all.

Contrary to what is sometimes stated (e.g. Rackham, 1989), maiden oaks can provide the special habitats associated with pollard oaks; at this size and age, they are hardly different from ancient pollards. In fact, some maidens may have lost much or all of their upper trunk and branches and become a 'natural' pollard; it can be difficult to determine the former state of very large and old oaks of this habit.

In earlier times, pasture woodland may often have been composed of a variety of tree species, all naturally generated. Pollarding could have been practiced mostly on beech, hornbeam and even holly and hawthorn, and less on oak. But because natural tree regeneration was hampered, the long-lived oaks were promoted and may have eventually become dominant. Unless the use of woodland as pasture has continued until modern times, beech has tended to regain dominance where present — as we can see in many abandoned pasture woodlands such as Epping Forest in

(as wood pasture) reached in the mid-nineteenth century and lasting well into the second half of the twentieth century. The Enclosure Acts between 1750 and 1860 removed much of what had remained up to that time from common use, leading to transformations ranging from agricultural fields to plantation forestry (Pryor, 2011). While individual ancient oaks were often venerated and depicted in art, poetry and folklore (Rooke, 1790; Strutt, 1826; for other examples see Pakenham, 2003; Hight, 2011; Miles, 2013), their ecological context — ancient pasture woodland — was often seen as both backward and useless and systematically destroyed (Rackham, 2003b, 2006). A revival is now taking place within several former pasture woodlands that have become nature reserves.

In only a few instances, the New Forest in Hampshire being the largest and best-known example, have common grazing rights been exercised without interruption since the Middle Ages (Tubbs, 1986). The same is probably true for a few medieval deer parks in which deer remain the principal, but not always the only large herbivores, several of which are mentioned and/or described in Chapters 4 and 9. Where grazing was (temporarily) discontinued, the former pasture woodland may have remained recognisably similar to the presumed situation in the Middle Ages, except for the filling in of parts of the open parkland with spontaneous growth of younger trees (and sometimes with exotic invasive trees like rhododendrons in Eridge Old Park); examples of such pasture woodlands include Ashtead Common in Surrey, Birklands in Sherwood Forest, Cranbourne Park in Berkshire, Eridge Old Park in Sussex, High Park in Blenheim (Oxfordshire), Moccas Park in Herefordshire and Aspal Close in Suffolk. In such places, the (pollard) oaks or

Essex/Greater London and The Mens in Sussex — and few if any of the earlier pollards would have survived. In some areas, we find beech or hornbeam among the most common ancient and veteran pollards. Examples can be found in Epping Forest and at Burnham Beeches, both near London, where wood was harvested from pollarded trees to be burnt in the bread ovens of the city, and in the New Forest, where oaks were cut but beeches were left because they were less useful for ship building.

Since the medieval period, pasture woodland has steadily declined, with a low point in terms of active use

Figure 8-9. Natural pollard oak in Sherwood
Forest, Nottinghamshire

other trees remain present, as do some
of the open glades, although these
are mostly maintained by mowing
the invading bracken. In four of these
examples, grazing with cattle or sheep
has been (re)introduced, but whether
this in itself will lead to a restoration
of the medieval situation is very
uncertain. Another likely difference
with the situation in the Middle
Ages is the far greater presence of
fallen dead wood in many of these
sites, especially since they have been
managed as nature reserves in which
dead wood has been recognised as
important for biodiversity (Harding
& Rose, 1986; Green, 2001, 2010a;
Alexander & Green, 2013).

In other ancient pasture
woodlands, there are few or no
remaining ancient pollards or
maidens and a high forest of younger
trees dominates. This situation is
probably much more common than
are pasture woodland with ancient
oaks and/or beech and hornbeam
pollards. Examples include The Mens
(a common) and nearby Ebernoe
Common (which is partially grazed
again) in Sussex, as well as large
sections of former Royal Forests
such as Epping Forest (Essex/Greater
London), the Forest of Dean and
Savernake Forest, where indigenous
trees still dominate but are nearly all
tall maidens or lapsed pollards. Especially in the southeast
of England, beech has often become the dominant tree,
outcompeting oak. It is therefore imperative for the long-
term preservation of ancient oaks in England that pasture
woodland is maintained or restored in those locations
where these trees still occur in numbers or, where they have
mostly gone, that future veteran oaks are nurtured from
spontaneous or planted young native oaks. The grazing of
woodlands now has many advocates; resuming the cutting
of pollards seems to be subject to more hesitation. This is
in no small part due to the perceived risks of drastic cutting
of ancient pollards which have suffered no such treatment
for 200 years (see contributions in Read, ed. 1996).

Experimental pollarding of oaks, especially re-pollarding
of old trees, is still rare. More experience has been gained
with beech and especially hornbeam, for example in
Burnham Beeches and Epping Forest, both managed by
the City of London Corporation (see e.g. Read, ed. 1996),
and in Bockhanger Wood, an SSSI in Kent that has ancient
hornbeam pollards. The work on oak pollards in Burnham
Beeches suggests that a cautious approach, taking off a few
boughs at a time over several years, is the safest strategy.

However, the restorative power of some of the oak pollards and maiden oaks after the 1987 and 1990 storms should be taken as encouragement to be not too shy (Green in Read, ed. 1996). I have seen many 'natural pollard oaks' vigorously growing new crowns from a bolling created by storm breakage. Perhaps we should imitate this rather than applying a clean chainsaw cut. Success may also be dependent on the site, especially its soils and climate, with ancient oaks on poor sandy soil in areas of low precipitation being much more vulnerable than those on richer bedrock sites in the rainier west. An attempt at re-pollarding of oaks in Epping Forest in the 1980s turned out to be largely unsuccessful (Dagley & Burman in Read, ed. 1996), but leaving some larger limbs intact appeared to have a positive effect on tree survival. One alternative to re-pollarding seems to be to prop up the too heavy limbs that are supported by a hollow bolling, but this strategy does not succeed for very long and merely postpones the demise seen in numerous historical examples. These pollarding strategies leave aesthetic considerations aside, although they are not unimportant for iconic trees such as the Major Oak in Sherwood Forest that serve as flagships for the conservation of ancient trees.

Finally, with a long view into the future, new pollards should be made much more often (Atkinson in Read, ed. 1996; Rackham, 2003b). In Richmond Park (Surrey), which has many ancient and veteran oak pollards but continues to lose them slowly but steadily, new oaks that are increasingly

Figure 8-10. Natural pollard oak at Manor Farm, Middle Chinnock, Somerset

Figure 8-11. Natural pollard oak in Ickworth Park, Suffolk

raised from English acorns to preserve native gene pools are planted under a continuing restorative programme. A programme of pollarding some of these young oaks has recently been cautiously started (Adam Curtis, pers. comm., February 2014). New oak pollards have also been created in Moccas Park, Herefordshire at a rate of about 5–10 per year since 1990 (Wall in Read, ed., 1996). The risks are likely to be less than with re-pollarding or even crown-reducing and there is a greater stock of available young oak trees in many parks, allowing contingencies to be spread over many different sites. It is encourageing to notice the extensive planting of native oaks from own-park-tree acorns, such as that in Brampton Bryan Park, Herefordshire, and even additional protection of spontaneous oak seedlings by the use of herbivore-excluding guards (for example at Grimsthorpe Park, Lincolnshire), which is perhaps closest to the medieval situation.

9

The most important oak sites

In this chapter, I list and discuss those sites in England which have substantial numbers of ancient and veteran native oaks, and in which the landscape has retained elements of the (late) medieval pasture woodland in which these trees were a prominent feature (Figure 9-1). I consider them to be the most important sites in England for ancient oaks. These sites have already been briefly mentioned (among others) in Chapters 4 and 5, but here they are described in more detail. As far as is possible, I give a summary of the history of these sites up to the present, in particular in relation to the presence of the trees, which includes the current status of the land and its management. The oaks are described in terms of their estimated numbers, sizes (girth) and possible ages, past management and present conservation, and in terms of the landscape and vegetation in which they have occurred in the past and occur at present. Recommendations for the conservation of the oaks and the ancient landscape are given where appropriate.

Which sites to include and exclude cannot be determined easily because the criteria used, including the numbers of oaks that can be considered ancient, are difficult to quantify. My criteria for defining ancient oaks here differ somewhat from those used by the Woodland Trust in its Ancient Tree Inventory (ATI) and are more restrictive. The aim here is to list sites with oaks that may go back to the Middle Ages, or at least to the Tudor period, and were mature living trees around 1600 CE or earlier and that have at least 10 such trees still alive. Natural England (formerly English Nature) (Castle & Mileto, 2005) have suggested that the presence of >15 ancient trees in an area

with trees of special interest (TSI) should qualify the area as 'high value', so setting a minimum of 10 ancient oaks may approach a similar threshold if we assume that other ancient trees are also present on most such sites. Oaks with a girth ≥6.00 m are here assumed to be older than 400 years, but depending on site and other factors, such as pollarding history, some oaks of between 4.00 m and 6.00 m girth may also be that old. The number of oaks on a site that fit my concept of ancient may therefore be a subset of these two size classes; oaks with girths ≥6.00 m recorded in the ATI as 'veteran' are also included. In addition, Natural England's criteria for 'high value' TSI sites include the presence of 100 or more veteran trees, as well as 15 or more trees with a stem diameter (d.b.h.) ≥1.5 m. Many sites have more veteran oaks with girths <6.00 m than oaks with

Figure 9-1. Map of England showing the 23 most important sites for ancient oaks described in Chapter 9. 1 – Chatsworth Old Park, Derbyshire; 2 – Birklands (Sherwood Forest), Nottinghamshire; 3 – Calke Park, Derbyshire; 4 – Grimsthorpe Park, Lincolnshire; 5 – Bradgate Park, Leicestershire; 6 – Kimberley Park, Norfolk; 7 – Aspal Close, Suffolk; 8 – Brampton Bryan Park, Herefordshire; 9 – Ickworth Park, Suffolk; 10 – Staverton Park, Suffolk; 11 – Moccas Park, Herefordshire; 12 – High Park (Blenheim Park), Oxfordshire; 13 – Hatfield Park, Hertfordshire; 14 – Ashton Court Park (near Bristol); 15 – Windsor Great Park, Berkshire/Surrey; 16 – Richmond Park, Surrey; 17 – Cranbourne Chase or Park, Berkshire; 18 – Hamstead Marshall Park, Berkshire; 19 – Savernake Forest, Wiltshire; 20 – Ashtead Common, Surrey; 21 – Parham Park, West Sussex; 22 – New Forest, Hampshire; 23 – Woodend (Shute) Park, Devon

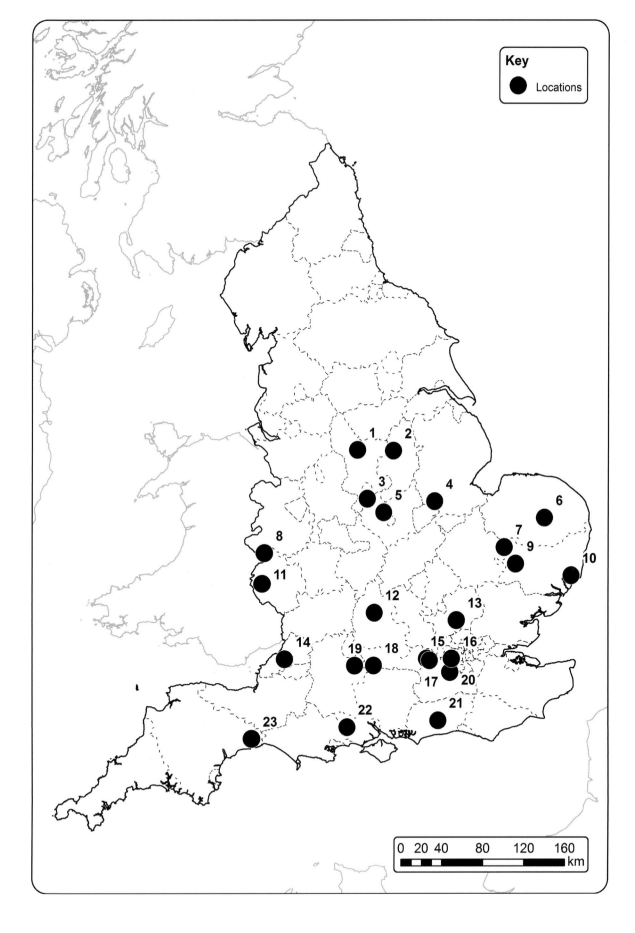

girths ≥6.00 m, but to qualify as a 'most important site', I have given extra weight to oaks with girths ≥6.00 m, or on some poor-soil sites ≥4.00 m or ≥4.50 m. If sites have many recorded veteran oaks but few or none exceed 5.99 m in girth, these sites usually do not meet the size (age) criterion.

The landscape, although having undergone changes over 500 years or more, should not have been deliberately altered so as to obliterate the pasture woodland with its unimproved grassland glades and plains, possibly alongside heath, bracken, shrubberies especially of hawthorn, or sometimes enclosed coppices. Archaeological evidence should also have survived together with the ancient trees themselves, which are the native, unplanted trees of England. For a site to be eligible, any alterations that have taken place should only cover a relatively small part of the site or be of a 'light touch'. There should be documentary or other evidence of historical habitat continuity since the Middle Ages. Standing and fallen dead oak wood should be present in some abundance, although this is more a modern ecological requirement than a reflection of the situation in the past. Landscaping should have been carried out in such a way that the earlier characteristics of the medieval pasture woodland are largely retained or at least recognisable, and can include ponds or small lakes and avenues planted with native trees (beech or lime) or sweet chestnut, the latter already present in England before the Middle Ages but often planted after.

I have not taken a site's size as a criterion for inclusion, but there is evidently a relationship between the number of hectares and the number of trees present. Sites differ much in size and in some cases, such as the New Forest, Sherwood Forest, Blenheim Park, Grimsthorpe Park, Richmond Park, Savernake Forest and Windsor Great Park/ Windsor Forest, ancient oaks are abundantly present in only parts of these domains. For some of these sites, I have limited the description to the parts where many ancient and veteran oaks occur and have omitted, apart from a contextual mention, the remainder. The New Forest, where the ancient and veteran oaks have a scattered distribution, is described in its entirety, but Table 9-2 includes only the open pasture woodland where these trees occur.

Aspal Close in Suffolk

Aspal Close at Beck Row near Mildenhall is the remnant of a small and probably medieval park; it has the remains of a moat on its eastern edge which once surrounded Aspallgate Manor, built in the fourteenth century. It became common pasture woodland later but was enclosed again by 1836. This site is situated on nutrient-poor wind-blown Pleistocene sand at 5–10 m altitude. Now completely

surrounded by urban development, in the nineteenth century there were only farm fields, the village of Beck Row, the hamlets of Holmsey Green and Cake Street Green, and a house known as Aspall Hall near the moat, now replaced by smaller houses. At around 12 ha, the current nature reserve is much smaller than the area of pasture woodland present in the nineteenth century and has even been reduced since 1940 for further housing development. A village playing field, Beck Row Recreation Ground, is within the

Figure 9-2. Pollard oaks in Aspal Close, Suffolk

park, but otherwise much of it remains in a medieval state with open acid grassland, partly overgrown with young oaks and thorn scrub, as well as more wooded parkland that contains around 160 old pollard oaks (Figure 9-2). None of the pollard oaks are large and only 27 are 4.00–5.50 m in circumference, but they are growing on poor sand in a low precipitation region so some could still be late medieval. A larger oak pollard of 7.00 m girth stands amongst houses in the southeast corner on ground that

was undoubtedly part of the park. They are perhaps up to 400–500 years old, but certainly not 800 as claimed on the Forest Heath District Council website for the reserve (for the probable ages of ancient oaks see Chapter 2).

Aerial photography proves that in 1940 the site was essentially open unimproved grassland with scattered oak pollards; since then, it has mostly become woodland with much infill from secondary tree growth, most of which is hawthorn and oak. Grazing with livestock, lastly as part of a local farm, ceased completely in 1960, leaving rabbits as the only grazing animals in the park. In the southern part, there are still some open areas with Breckland Grassland, a varied vegetation type specific to the Brecklands in East Anglia. The wooded two-thirds of the site can be characterised as an impoverished form of oak-birch woodland with few birch trees and without species such as bilberry (*Vaccinium myrtillus*) and broom fork-moss (*Dicranum scoparium*). Instead, there are patches of gorse and, in the northern 'panhandle' part, thickets of blackthorn; both could be remnants of open pasture with scrub islands. The ancient or veteran oaks are an important habitat for fungi and invertebrates; at least six species of bat have also been recorded here. Due to the surrounding housing estates, visitor pressure on the park is heavy. It is a local nature reserve managed by the Forest Heath District Council, which is currently embarking on a 50-year restorative programme for the oak pollards and the ancient pasture woodland landscape in general.

Ashtead Common in Surrey

Ashtead Common in Ashtead is an ancient pasture woodland common of *c.* 200 ha. The site was occupied in the Roman period and archaeological excavations have revealed the remains of a Roman villa and tile works, with the course of a Roman road to it. A triangular earthwork of unknown origin, but possibly Iron Age, also lies in the middle of the woodland common. These remains suggest that this location was occupied at least until the Romans left and that the ancient woodland was managed and utilized. The presence of the tile works implies that wood burning was taking place, so it is likely that some woodland was coppiced while other parts were grazed (*silva pastilis*). There is no evidence of farming on the hill, but fields may have been present in lower areas, as they are now. The medieval history of Ashtead Common is unrecorded, but the land became part of the manor of Ashtead and was probably 'waste' used to graze livestock (Forbes & Warnock in Read, ed. 1996 and references therein). This common is situated between Epsom Common to the east and Leatherhead Common (now mostly a golf course and an upmarket housing estate) to the west.

Ashtead Common has many veteran oak pollards, but most of these trees are not very large in terms of trunk circumference and the largest are perhaps only *c.* 400 years old. The veteran oaks were mapped and tagged in 1993 by the City of London Corporation and again in 2007–8 by volunteers (Ashtead Common National Nature Reserve (NNR) local plan 2011–2020, City of London Open Spaces Department). A survey carried out on behalf of the City of

Figure 9-3. Pollard oak in Ashtead Common, Surrey

London Corporation by Treework Environmental Practice between January and March 2009 found 1,195 living veteran oaks (Bengtsson & Fay, 2009). Concentrations of these oaks are mainly located in the northern part known as The Forest, the south-western part, and along the northwest margin, where the wooded common borders on arable fields, totalling *c.* 130 ha. The tree with the largest circumference recorded in the ATI was again measured by me on 18th October 2014 with a 7.10 m girth but most are much smaller, with girths of between 4.00 m and 6.30 m for the larger oaks. Numerous fires which occurred in the latter part of the twentieth century killed or seriously damaged up to 40% of the oak pollards in affected areas, causing birches to invade the burnt areas; 425 still-living oaks were found to have some fire damage in the survey of 2009.

Many pollards have large crowns with uncut massive limbs, indicating that the last time these oaks were pollarded may have been around 150 years ago. Re-pollarding of some of the oaks is desirable as they are relatively young, still strong and relatively sound, but work has mainly been limited to the 'crown reduction' of trees that presented a perceived health and safety risk (Forbes & Warnock in Read, ed., 1996). For the past few years, an active work programme of restoration pruning has been established at Ashtead, with 70–100 trees per year being cut. Many are host to various fungi (beefsteak fungus is particularly common) and important for invertebrate fauna associated with dead and decaying oak wood, with more than 1,000 species of beetle (Coleoptera) identified. In several areas there is coppiced hazel as underwood among the oak pollards, so perhaps most of these are the last crop of oak standards to be pollarded when the common became intensively grazed with cattle in the seventeenth century. There is no clear evidence of compartmentisation with wood banks in the areas with hazel coppice.

Newton Wood, in the northeast of the common, is a large enclosure bordering on Epsom Common in which there are a few oak pollards and much secondary woodland, so it is likely that most of the old oaks were felled here. The current boundary of the wood is the same as that marked on the nineteenth-century OS map; it has a vague bank and ditch separating it from the present commons. What were farm fields bordering the commons to the south in the nineteenth century is now almost all built up, but no further encroachment onto the commons has occurred since then, at least in part thanks to the intervention of the City of London Corporation. On the common there are four permanent ponds dug into the Eocene London Clay which underlies most of the site, two

of these are possibly medieval and must have served the grazing livestock on this east-to-west hill ridge site (max. 88 m a.s.l.). Other small ponds are now silted up. Ancient clay pits also attest to intensive use in the past. There are some areas without pollard oaks which would have been more open grassland, or perhaps on sandy soil some heath, with clumps of gorse and thorn, but as with all neglected commons, there is much young tree growth (here mostly birch) that is slowly turning the common into a forest. On the nineteenth-century OS map, these areas, mainly in the central part of the common, are indicated as much more open pasture. Grazing with Sussex cattle has been reintroduced in a small area in the south of the common, where there are few pollard oaks, and has now been extended northwards to include an area where there are more pollards. Some hazel coppicing has also been resumed and, since 1992, bracken has been mowed or rolled where possible to create and maintain open glades and rides and to reduce the fire hazard. Grazing should ideally be extended to the entire common and at least to the pasture woodland areas with oak pollards.

Owing to the clay soil, the woodland vegetation is on the richer side of oak-birch woodland with, besides pedunculate oak, birch, rowan and hazel (common), some holly, much bracken and brambles, and here and there hawthorn, grey willow or sallow (*Salix cinerea*) and even gorse, indicating the more open past. In moist hollows tufted hair grass (*Deschampsia caespitosa*) dominates; elsewhere the glades and rides have various rough grasses but no heath. Indicators of ancient woodland, such as wood anemone (*Anemone nemorosa*) and wood sorrel (*Oxalis acetosella*) are present to a limited extent in the areas with hazel coppice. Most of the common is a National Nature Reserve (NNR) and all of it is within a Site of Special Scientific Interest (SSSI).

Ashton Court Park in Somerset

Ashton Court Park near Bristol is a landscaped park of 340 ha owned and managed by the City of Bristol. There has been a manor house here since the eleventh century; in Domesday Book (1086) it is recorded as a substantial estate owned by Geoffrey de Montbray, Bishop of Coutances. A park for deer was created by Thomas de Lyons in 1392, this became known as Ashton Lyon Park (Bond, 1998). In some areas of the present park, evidence of medieval ridge and furrow and strip fields indicates that some of the imparked land was not ancient pasture woodland but arable fields. The estate was bought by a Bristol merchant in 1495 and Ashton Court, a Tudor mansion, was built near the centre of the park in the late fifteenth century and subsequently enlarged. In the sixteenth and seventeenth centuries, the park was extended, presumably mostly to the southwest where no ancient or veteran oaks have been recorded.

The underlying geology in the park is quite complex, consisting mainly of Carboniferous sandstone or gritstone on the plateau, Early Carboniferous (Tournaisian) 'bands' of limestone on the slopes in the north and west, and dolomitic conglomerate or Mercia mudstones of Late Triassic age overlaid with reddish brown sandstone and stony loam in the lower parts. The topography is also varied, with a plateau at 130–135 m a.s.l. in the west and escarpments falling off into coombes and sloping down to the entrance of the Avon Gorge in the east at around 10 m a.s.l. Extensive landscaping was planned by Humphry Repton in the late eighteenth century but not all of this was executed, the main features from this period that are still present are plantation enclosures and remnants of avenues planted with non-native trees, as well as groups of parkland trees, including sweet chestnuts, to the northwest of the mansion. Cedars (*Cedrus*) and giant sequoias (*Sequoiadendron giganteum*) were planted nearer the

Figure 9-4. The Domesday Oak at Ashton Court, Somerset

Figure 9-5 (left). An ancient maiden oak in Sherwood Forest, Nottinghamshire

courses are now laid out, but since 1970, there have also been two enclosed deer parks on the estate. There are also tarmac roads, access for cars and car parks, a miniature railway and some other attractions to please the 1.6 million people that visit per year.

Ashton Court Park has 42 ancient and veteran oaks with a girth ≥5.00 m, among which is the Domesday Oak (9.19 m girth) (Figure 9-4), which is supported by props to prevent its heavy branches from further splitting the trunk. Six ancient oaks (recorded as 'veteran') exceed 7.00 m circumference and most of these are pollards. Besides oaks, in some areas ash, beech, wych elm (*Ulmus glabra*) and field maple occur in the parkland or woods and hawthorns are common on some of the more open slopes. The grassland is unimproved in most areas, except on the golf courses. An area of 210 ha in the park has been notified as an SSSI primarily on the grounds of high invertebrate diversity and records of rare beetles (Coleoptera) associated with the ancient trees. A management plan has been in place since 2009 which seeks to protect ecological assets but also has to allow and manage high visitor densities linked to the proximity of Bristol. Enlargement of the enclosed deer parks and removal of planted conifers in the parkland are among the actions proposed.

Birklands (Sherwood Forest) in Nottinghamshire

Birklands (Birkland Hay in medieval time) near Edwinstowe is the last surviving major fragment of Sherwood Forest which retains its medieval character of pasture woodland, albeit with the usual infill of secondary woodland, here dominated by birch and oak, since grazing came to an end. As was most of this large Royal Forest, Birklands is situated on dry acid sands of the Triassic Sherwood Sandstone Plateau. By the eleventh century, it was already surrounded by extensive lowland heath and acid grassland, in which there were few similar patches of oak-birch woodland. One of these may have been a nearby area known as Bilhaugh (Bilhaugh Hay), which has many veteran oaks in a section called Buck Gates (now part of the Thoresby Estate) but elsewhere the woodland has been mostly coniferised and much altered for a military training camp. Birklands, as circumscribed here, contains the area of Sherwood Forest Country Park and of Birklands East managed by Forest Enterprise, an executive agency of the Forestry Commission. These areas, totalling 228 ha, are areas A and B in Clifton (2000).

Conifer plantations surround Birklands on the north and west sides, with arable fields, the village of Edwinstowe and a coal mining site to the south and east. The plantations on the north side of Birklands, known as Assarts Wood and

mansion in the 1880s. Other woods, such as Church Wood (now Pill Grove), are presumably older. Adjacent to this wood is Clarken Coombe, a wood in which there are some veteran oaks, as well as a small-leaved lime with 7.34 m girth. There were extensive areas of undeveloped parkland and woods well into the nineteenth century. Two golf

Figure 9-6 (right). An oak tree split by lightning in Sherwood Forest, Nottinghamshire

Plantation. The ancient wooded area of Birklands and its surrounds is recorded on maps of Sherwood Forest going back to the fourteenth century; detailed maps have existed since 1735 (Clifton, 2000). On the nineteenth-century OS map an open corridor connects the heath in the northeast section with another, triangular section that ends at the northern end of Church Street in Edwinstowe. Older images of the Major Oak show it to stand in that corridor amid open heath or grass with a few scattered small trees. These open areas have now filled in with secondary oak-birch woodland and only a few small glades with heath and coarse grass remain.

By 1608, the royal commissioners had already reported that 54% of oaks in the wooded northern part of Sherwood Forest were unsuitable as timber for the English Navy, in other words they were ancient or veteran oaks. By 1680, that proportion had risen to 96% but we do not know whether this was the result of progression into the ancient stage or of the cutting of most of the remaining 'sound' oaks. I suspect it was the latter, because oaks do not normally decline at such a rate (see Chapters 1 and 2). A progressive and steep decline in the numbers of oaks reported to exist between 1608 and 1788 reflects deforestation through 'mismanagement' (Clifton, 2000). A survey carried out during the period 1996–99 identified 954 standing 'veteran' oaks in Birklands, of which only 537 were still alive (Clifton, 2000); significant numbers of veteran oaks were also found in the plantations surrounding Birklands. Within Birklands, 58% of standing veterans (alive or dead) were found on 28.6% of the surveyed area. This indicates that although Birklands is the most important area for ancient or veteran oaks at present, a much larger area has been pasture woodland in the not so distant past. This recent loss is mostly due to forestry plantations and, to a lesser extent, to military use of the land. The famous Major Oak (10.66 m in girth) (Figures 2-12 and 2-13) is much visited by tourists (Palmer, 2003); there is a Robin-Hood-oriented visitor attraction area nearby.

In the Ancient Tree Inventory (ATI), several hundred oaks at Birklands are recorded as veteran and 24 oaks are recorded as ancient, having girths ≥6.00 m and possibly dating back to the Middle Ages. A dendrochronology study (Watkins *et al.*, 2003) in nearby Bilhaugh (Buck Gates section) found the earliest datable tree ring in an existing oak to be from 1415. Two oaks were 8.00 m and three were 7.00–7.30 m in girth. The Major Oak is thus by far the biggest oak now alive at this location; an ancient oak of similar girth blew down in 1961. It is an obvious ancient pollard, while most of the ancient and veteran oaks in Birklands are not (Clifton, 2000). They are often short and

Seymour Grove, separate it from the heath area known as Budby South Forest on the OS map. In the nineteenth century Seymour Grove was an arable field. A small area in the northeast of Birklands is marked as heath on the nineteenth-century OS map and was planted with native oak later to form Queen Oak Plantation and Robert's squat, and evidently grew in more or less open parkland, but they are not pollards that have a clear distinction between the bolling and branches that grow from about the same height on the tree. This situation occurs in a few other remnants of Royal Forests and in several ancient parks; it could have significance for how these trees were utilized in the past. Pollards undoubtedly existed, as a 1680 report by royal commissioners stated that 8,060 trees (24%) in Sherwood Forest were subject to "oft lopping", a term Rackham (2003b) identified with pollarding.

The proportion of standing dead ancient or veteran oaks in Birklands (43.7%) is much higher than that in any of the other 22 sites described in this chapter, and far exceeds the national average of 5.5% (Chapter 11). In the conifer plantations, this demise of veteran oaks will be mostly due to shading and possibly to root damage that occurs during ground preparation for planting or even in deliberate attempts to kill the oaks. These causes can be excluded in most of Birklands, where the oak plantations were established on heath where only a few trees remained and there are no conifer plantations. A lowering of the groundwater table is a possible cause, perhaps related to large pits created by the Thoresby Colliery immediately to the east. That mining site was still a part of Bilhaugh and was under (planted?) woodland in the nineteenth century. Birklands' relief, though gentle, slopes downwards towards the north and east in the direction of these pits, from 90 m a.s.l. to an unknown level below 65 m.

Birklands is an SSSI and, for the most part, also an NNR. Grazing with cattle has been resumed (it may have been sheep in the past on these poor sandy soils) in a limited area of Birklands. The vegetation is predominantly that of oak-birch woodland, with patches that had been degraded to grassy heath and scrub now again succeeding to this woodland. Apart from oak and birch, few other trees are present, mostly only rowan in the understorey. Bracken is ubiquitous, perhaps a result of abandoned grazing as cattle can suppress it but deer will not touch it. Bilberry and, in shady locations, broad buckler fern (*Dryopteris dilatata*) occur in patches. More open areas are dominated by tufted hair grass and heather; in moister areas, purple moor-grass (*Molinia caerulea*) can dominate. Mosses, such as broom fork-moss and hair moss (*Polytrichum* spp.) as well as lichens are very common on oak trunks and dead wood, but are still less well developed and diverse than elsewhere in this type of woodland because of air pollution in the past (see Chapter 10). Birklands is a nationally important site for fungi, both mycorrhizal and saproxylic, that are associated with birch and oak and for invertebrates that are associated with dead oak wood.

Bradgate Park in Leicestershire

Bradgate Park was first mentioned as a deer park in 1241, when it belonged to the manor of Groby which was owned by the Earl of Winchester. It was imparked some time before that date from Charnwood Forest and enclosed by a ditch and wooden park pale. The medieval park was initially much smaller than the present one; the first substantial enlargement occurred in the late fifteenth century and included farmland of the lost village of Bradgate. The current dry-stone wall, sections of which date from the seventeenth and eighteenth centuries, encloses *c.* 332 ha of acid grassland much encroached upon by bracken, a few patches of ancient pasture woodland and eight wooded enclosures, called either plantations or 'spinneys'. A spinney used to be a wooded area dominated by thorns, but the present woods are either a mixture of broadleaved trees (oaks and sweet chestnut) and pines or, in the case of the Coppice Plantation, nearly all Scots pines. These woods were enclosed by dry-stone walls in the early nineteenth century and planted as coverts for pheasant shooting. The ruins of the Tudor mansion, Bradgate House, built by the Grey family who had owned the park since 1445, stand in the southern part of the park. In addition to the Old John Tower, an eighteenth-century folly on the highest point (220 m), there is now a modern visitor centre and cafe in the park. The landscape is mostly open and undulating. In the southwest corner is the incised valley of the River Lin, a small stream that has exposed much of the late pre-Cambrian intrusive diorite rock formations in the park; other and slightly older rock outcrops, mostly Charnian pyroclasts, are on the hill around Old John Tower. In the Lin valley is a string of dammed ponds and there are a few mature exotic conifers, but no other landscaping; native oak woodland dominates. Upstream are some marshy fields with rushes. The soils are sandy and acid, mostly derived from Palaeozoic slates and sandstones. A hard surface footpath runs through the southern end of the park, otherwise there is a network of sand paths, many unplanned.

Ancient and veteran oaks occur in two main patches of pasture woodland. One is situated in the southwest corner of the park above the incised valley of the Lin and on its south-facing slope; the other is located between Coppice Plantation and Dale Spinney. A few other ancient oaks

Figure 9-7. Veteran oaks and drystone wall in Bradgate Park, Leicestershire

occur scattered between these locations, with a fairly large pollard (not yet measured) on the south side of Bowling Green Spinney (inside its stone wall). The largest oaks (7.00–8.00 m in girth) are few; my estimate from a visit in October 2011 was that there are some 80 ancient or veteran oaks, mostly pollards and all in the southern half of the park. On 23 May 2016, I recorded 25 ancient or veteran oaks with girths from 4.40 m to 6.88 m in the area between Coppice Plantation and Dale Spinney. Several ancient oaks in this area are remarkable for their 'repair' capacity, evident in the strong lateral growth of callus wood around huge gaps in the trunks, which often supports large new crowns of foliage. Native oaks have naturally regenerated in these areas in the past, but in the last 30 years, they have

mostly been planted. The southern section is the oldest part of Bradgate Park; later additions have mostly been treeless moorland, turned to acid grassland and bracken by the grazing of the (now *c.* 370) red and fallow deer; there are only a few patches of marshy heath with cross-leaved heath (*Erica tetralix*), purple moor-grass and *Sphagnum* pools. Bradgate Country Park is now administered by the Bradgate Park and Swithland Wood Charitable Trust and, except for the wood enclosures, is open to the public year-round. It is notified as (part of) an SSSI on account of its ancient oaks and acid grassland.

Brampton Bryan Park in Herefordshire

Brampton Bryan Park is situated in the extreme northwest corner of Herefordshire on the Welsh border and is a private park of 191 ha. The castle of 'Brantune' held by Ralph de Mortimer is mentioned in Domesday Book (1086) but the earliest reference to a stone 'tower with curtilage' dates from 1295; the castle came into the ownership of Robert Harley in 1309 through marriage and is still owned by a descendant. The Norman castle probably stood within the present park but was later abandoned and rebuilt closer to the River Teme at the site of the present village (Whitehead, 2001); the re-built medieval castle was ruined during the Civil War. Evidence of a moated house and arable fields in the form of ridge and furrow indicate that the eastern part of the present park was established on agricultural land that belonged to the Norman manor as its demesne land. The park, which is not mentioned in Cantor (1983) but which was probably created in the fifteenth century before the start of the Tudor reign (Whitehead, 1996), is situated between the village at 135 m and the highest point of the hills to the west and southwest at 305 m a.s.l.

The hills form part of the Pedwardine Hills escarpment; the middle part of the park lies on Upper Silurian Wenlock shales while the lowest part of the park is on much younger sediments overlying Precambrian rocks. The Silurian formations are overlain by glacial drift material forming acidic soils. The boundaries of the park on the hills are well marked on both nineteenth-century and recent OS maps; they are visible from the air in part as hedgerows and as the demarcation between parkland and (conifer) plantations on the hills. In the valley, agricultural fields have separated the park from the road (A4113) and the village since at least the nineteenth century and probably earlier, as indicated

Figure 9-8. Ancient oaks in Brampton Bryan Park, Herefordshire

by a few hedgerow oaks >5.00 m in girth, but the roundish outline of the park may indicate that these fields were once part of it. The park is shown to have a round shape on Christopher Saxton's map of Herefordshire and Shropshire 1577, the earliest unambiguous reference to a park here.

The uphill western parts were probably a wooded common through most of the Middle Ages, but there is some archaeological evidence of an early enclosure which could have been a small deer park (David Whitehead in Harding & Wall, eds., 2000). Remains of a nineteenth-century iron fence mostly follow the present boundary of the park; deer were kept in the park until the 1920s. The remains of planted rows of ancient sweet chestnuts, dating from the seventeenth century, could indicate earlier internal divisions or mark major rides. The nineteenth-century OS map shows sections of tree rows, presumably formed of sweet chestnuts, that are now partly gone, so it appears that there may have been a division between the eastern and the western halves of the park. These rows are not, as in parks such as Croft Castle (Herefordshire), the remains of avenues and do not follow straight lines. A few trees in them are oaks and limes. The nineteenth-century OS map shows several enclosed woods, and most of these still exist while some have been expanded and others added. Some were already planted with conifers in the nineteenth century, but the later woods or plantations are conifer plantations only. In addition, in the lower part of the park are several mostly solitary planted non-native park trees dating from the nineteenth century or later; they include various oaks, maples, cedars and giant sequoias. These trees are signs of 'park improvement' on a modest scale that did not have as much impact as the conifer forestry plantations. The present owners have planted many new parkland oaks from acorns gathered in the park, a positive move towards restoration of the ancient park landscape.

Apart from 'Park Cottage' (a substantial mid-nineteenth-century lodge) and its garden, there is no other habitation in the park at present. The park grassland is mostly unimproved, especially on the hills, with some areas dominated by bracken. Other areas show traces of past cultivation, especially in the 'launds' or meadows near the park keeper's house, which is in the vicinity of the Norman manor site. A small section on the highest hill in the south is heath dominated by heather and billberry, presumably the upland western part of the park was more widely covered in this ground vegetation before the modern increase of sheep grazing, as suggested by the scattered presence of some indicator species. Nearby is an oak coppice wood just outside the park in Pedwardine Wood. Several small ponds are present in the lower eastern part of the park and some of

these are marked as fish ponds on nineteenth-century maps; all date from the nineteenth century (Whitehead, 1996). Aside from the obviously planted rows of sweet chestnuts, there are some large small-leaved limes and some old ash and alder trees, as well as an old row of planted beeches on the top of a hill (which is marked on the nineteenth-century OS map). On the summits and steeper hill slopes that are not afforested, there are scattered old shrubs of hawthorn and elder (*Sambucus nigra*), patches of secondary woodland with birch and rowan, as well as some ancient and veteran and a larger number of mature native oaks.

The ancient and veteran oaks are scattered throughout most of the park; there seems to be about an equal division between sessile oak and pedunculate oak. One of the best-preserved areas of medieval parkland is on a gentle westward slope west of Broomy Hill Plantation and north of the presumed park pale remains. This area of parkland is now being expanded by the removal of planted Scots pine and the planting of native oaks. A total of 53 oaks with girths ≥5.00 m had been recorded by 2014; six of these exceed 7.00 m and one has a circumference of 9.20 m. Nearly all of these oaks are maidens and typical parkland trees; one 8.65 m oak that may have been a pollard fell over recently and is now dead. Recording was incomplete as observed on a visit on 3–4 September 2014, with at least 4–5 more oaks >5.00 m in girth present near the ancient remains of the park pale. In April 2016, an additional eight living oaks with >6.00 m girth were found, including one possibly >9.00 m in girth plus a large oak coppice stool. Brampton Bryan Park has been notified as an SSSI mainly because of its ancient and veteran trees and their associated lichens and varied entomofauna, mostly beetles associated with dead wood.

Calke Park in Derbyshire

Calke Abbey was not a monastery but an Augustinian priory, founded between 1115 and 1120 by the Second Earl of Chester. The canons moved to Repton Priory in 1172 and made Calke a subsidiary cell. It is likely that Calke was managed as an agricultural estate rather than as a religious house during the fourteenth and fifteenth centuries, but documentary evidence is lacking. It is possible that a deer park was established in this period, but none is listed in Cantor (1983) for this location. A sketched reconstruction of the surroundings of Calke House in around 1590 (National Trust Archives) shows much of the present park as ploughland, but it also shows the rounded contours of a park including 'Roche Wood', 'Bowley Wood' and 'Castle Close', in which nearly all of the largest ancient oaks now present occurred. It seems that an earlier park had existed but was disparked and given over to Calke Village. A wooded area just south of Ticknall shown on Speed's map of Derbyshire (1610) could indicate just that; this map would not mark coppice woods and the oaks are evidence that it wasn't coppice. The outline of the old park is still clearly visible, marked by hedges and lanes, but no park pale remains are indicated on the current OS map. The outline of the park on the nineteenth-century OS map is very similar to that of the present park of 243 ha — more or less a heart shape with the point in the south and Calke Abbey in its centre. In the nineteenth century, the medieval pasture woodland landscape was still much in evidence throughout the park, but some of it must have been 'reparked' from the arable fields and thus is secondary. The ancient and veteran oaks are smaller there.

The underlying geology of Calke Park is Triassic red sandstones, overlain by disconform Pleistocene till including sands and gravel, resulting in acidic soils. Apart from a valley incision, the terrain undulates gently to a height above sea level of 116 m in two localities. Apart from the creation of a string of ponds in the west-east oriented valley by damming a brook, we see fields dotted with occasional trees alternating randomly with more densely wooded parkland. The more cultivated parts of the park are concentrated around Calke Abbey, an eighteenth-century mansion that replaced an earlier house. Mainly owing to the owners' lack of financial means, no substantial changes have been made since, and the house and park are now in the care of the National Trust. Planted avenues of beech, oak and sweet chestnut have mostly disintegrated. The more open parts of the park are grazed by cattle and sheep and the deer are enclosed in a smaller part of the park; the present deer park hosts around 200 fallow deer and some red deer. Most of the ancient pasture woodland that remains lies north of the valley with ponds and another section is situated along the western side of Calke Park. Here, too, some improved grassland occurs, but for the most part, this grassland is in its semi-natural state: acid grassland with much bracken and with scattered trees and shrubs.

The park has many ancient and veteran oaks, among which is the 'Old Man of Calke' with a circumference of 10.01 m, which is a short, hollow and senescent pollard oak. Four oaks are recorded as ancient with a girth >7.00 m, 12 oaks with a girth >6.00 m and c. 70 veteran oaks with girths between 5.00 m and 6.00 m in the Ancient Tree Inventory (ATI). Most of the veteran trees are native oaks, but some beeches, sweet chestnut, small-leaved and large-leaved lime, ash, hornbeam and hawthorn are also recorded. Birches, hawthorn, locally hazel and some oaks form secondary woodland infill and alder dominates the

Figure 9-9. The Old Man of Calke, the oldest oak in Calke Park, Derbyshire

shores of the ponds. Non-native trees (introduced after 1500 or neophytes, here mostly horse chestnut) are uncommon, which is another sign of a lack of 'park improvement' following eighteenth and nineteenth-century fashions. This sets Calke Park apart, despite its obscure medieval history which makes it difficult to ascertain that this is a medieval deer park. Part of Calke Park is an SSSI and its pasture woodland, about one third of the park, is an NNR, mainly on the grounds of rare invertebrates associated with dead wood, for which it is one of the most important sites in Derbyshire. Calke Park is heavily visited because of its proximity to major urban centres, which limits the options for its management as a nature reserve.

Chatsworth (Old) Park in Derbyshire
In Domesday Book (1086) the manor of 'Chetesuorde' is listed as the property of the Crown in the custody of William de Peverel. It had been part of a much larger estate in late Anglo-Saxon time but this was divided by the invading Normans. The manor of Chatsworth was acquired by the Leche family in the fifteenth century; they already owned property nearby. They enclosed the first park during their tenure, perhaps in the late fifteenth century as it is not listed as a medieval park by Cantor (1983). Chatsworth Park was originally situated on the east bank of the River Derwent, between the river and the escarpment known as Dobb Edge leading up onto East Moor. Although we do not know its exact extent, it must have included the two areas with ancient oaks north and south of Chatsworth House, which lie on gentle slopes above the river floodplain and below the steeper escarpment. In the early seventeenth century, the park extended onto the moor, but William Senior's survey map of 1617 shows an internal division separating the wooded park from the moorland park. A

hunting tower was erected on a high point overlooking both areas.

The earliest house stood on high ground in what is now known as Old Park, formerly in the southeast corner of the late medieval park. In 1553, Bess of Hardwick built a new house on the site of the present house, more towards the middle of the park. Formal gardens were laid out in its vicinity and fish ponds and lakes were created on the slope behind the house. By the end of the seventeenth century, extensive formal gardens had separated and reduced the two areas with ancient oak parkland. The landscape park was much enlarged in the eighteenth and nineteenth centuries; the village of Edensor was moved so that the park could extend to the west bank of the Derwent. It now encompasses 400 ha within a 5,000 ha estate. To the east, above the park, are the moors of the Derbyshire Dales rising to 323 m, to the west, the hills are lower (to 200 m) and farmed. The underlying geology is Carboniferous sandstone, mudstone and siltstone of the Millstone Grit Group on the moors and similar sedimentary rocks of the Bowland High Group to the west. The soils in the park are therefore acidic except in young fluvial deposits in the floodplain of the Derwent; they are nutrient-poor, especially on the higher slopes.

The escarpment is densely wooded. In the nineteenth century, this was all still deciduous woods, but in the twentieth-century, conifer plantations replaced this woodland. The only parts of the escarpment that are not now planted with conifers are an area at the northern end (within the park) and another section behind the gardens, where exotic trees are mixed with natives and almost all of the trees will have been planted. Plantations with conifers also predominate in Stand Wood on top of the escarpment. Much tree planting, both of solitary parkland trees and in enclosed woods and along roads, also took place from the mid-eighteenth century onwards on lower parkland on either side of the river. This has left two parkland areas north and south of Chatsworth House undeveloped or only lightly altered with a few plantings. In the southern area, Old Park, there are two plantations; in the northern section, trees have been planted in the late twentieth century on both sides of a road in avenue style and also in a circle. These landscape plantings did not exist in the nineteenth century and some are indeed recent alterations; they are now seen as inappropriate and are being removed (Parkland Management Plan, Chatsworth Estate, February 2013). The northern parkland section with ancient oaks has public access; the southern section (Old Park) is private and managed as a deer park with a small herd of fallow and some red deer.

In the parkland north of the house there is an area on the slope with unimproved but coarse grassland, which is grazed by sheep and lacks bracken. Here, 41 ancient and veteran oaks with >5.00 m girth have been recorded; three of these exceed 8.00 m girth. Younger and smaller oaks are also present and, in the northern part of this section, there is an oak wood with a few sweet chestnuts and elms surrounded by acid grassland and/or bracken heath. Old Park has many more ancient and veteran oaks; this is now effectively a separate area and is included as such in Table 9-2. The best location for old oak trees is in the area south and east of Old Park Plantation, there are also a good number in the northeast corner between plantations. This is the least altered section of medieval parkland in Chatsworth Park. Here, acid grassland is much invaded by bracken apart from a system of rides which have been mown and are kept open by deer grazing (Steve Porter, pers. comm. September 2014). A few indicator species, such as mat-grass (*Nardus stricta*) and heath bedstaw (*Galium saxatile*), may be remnants of heather vegetation. Other areas of unimproved parkland in Old Park have mostly mature oaks (mostly planted) and veteran oaks and less bracken, while lawns with improved grassland are situated between these areas and the east bank of the river. Apart

from a small-leaved lime, some birches and a few alders in seepage locations, there are no other trees among the oaks. The terrain on the slopes is strewn with large sandstone boulders and many ancient oaks have grown upon and over these. The ancient and veteran trees of Chatsworth were comprehensively surveyed in 2012 (Parkland Management Plan, Chatsworth Estate, February 2013). The largest oaks here are between 7.00 and 9.00 m in girth[1] and most are maidens; some of the pollard forms are probably natural (Figure 9-10). The number of ancient and veteran oaks

---

1 The veteran tree survey 2012 by the Chatsworth Estate records two oaks with circumferences >10.00 m in the Old Park. Tree No. 390 is actually two trees fused by extensive root bases over rocks, and I measured tree No. 306 on 16 June 2015 at 8.10 m circumference (trees on steep slopes can be difficult to measure).

Figure 9-10. A natural pollard oak with a girth of 8.10 m in Old Park, Chatsworth, Derbyshire

≥5.00 m girth in Chatsworth is 180 (139 in Old Park), of which 74 (63) are ≥6.00 m circumference and eight (five) are ≥8.00 m circumference (measurement was generally at 1.30 m). There is an active programme of planting new oaks in Old Park, but several were seen to be affected by oak mildew (September 2014). The site of Old Park (deer park), with an area of 72.5 ha, is an SSSI because of the presence of many ancient and veteran oaks and areas of unimproved acid grassland, and the concomitant biodiversity.

Cranbourne Chase or Park in Berkshire

The designation 'Cranbourne Park' appears on the nineteenth-century OS map for pasture woodland and fields north and east of Cranbourne Tower. John Norden's map of Windsor Great Park from 1607 calls this area 'Crambourne Woode'. On the map by Menzies Sr. (1864), this area was a part of Cranbourne Park, a park for deer that was separate from Windsor Great Park. Menzies Jr. (1904) described the area northeast of Forest Gate, including Cranbourne Tower and the woodland to the east, as lying within Windsor Great Park, and this is the present administrative situation. I here use Cranbourne Chase primarily for the enclosed area of pasture woodland that is situated between Cranbourne Tower and the A332, even though the current OS map places Cranbourne Chase in the plantation forest further west. The chase was probably more extensive in the past, as a map of 1752 in the Royal Collection still indicates.

To the west of Cranbourne Chase or Park is the current Cranbourne Wood, part of the plantations of Windsor Forest. Between the plantations and Cranbourne Chase or Park remain four notable ancient oaks, two of these along Forest Road and a third the dead stump of the 'Conqueror's Oak' on the defunct boundary of the 'White Deer Enclosure'; these are on the edge of the forestry plantations. The fourth large ancient oak stands in a meadow south of Cranbourne Tower. These four oaks are likely to be the last remaining medieval oaks of a greater number in this area, and one is indeed a mere rotten stump. On the 1752 map, the area now mostly grassland with a scattering of (ancient) oaks was still covered in trees. Remnants of prominent avenues radiating from Cranbourne Tower and shown on that map are also still present. The pasture woodland here described extends from the road to Cranbourne Tower off the A332 up to the ancient park pale line and the fields of Flemish Farm. To the east of the A332, a strip of land from the Cranbourne Gate to Ranger's Lodge historically also belongs here, as shown on the map of 1752. The ancient oaks on this strip are here also considered to belong to the Cranbourne Chase pasture woodland, and so the A332 runs through this pasture woodland; indeed, Sheet Street Road (now the busy A332) was only a sand path through this area until well into the nineteenth century. I also include a section of woodland behind an old fence immediately north of the reservoir at Cranbourne Tower; the entire area is about 40 ha. Historical evidence indicates that it was part of the Royal Forest of Windsor since the eleventh century, and that there may have been an Anglo-Saxon hunting forest there since *c.* 700 CE; the site then became part of Cranbourne Chase when the Forest was divided in the thirteenth century, and a park in the seventeenth century. The Keeper of the Chase lived in Cranbourne Lodge, on

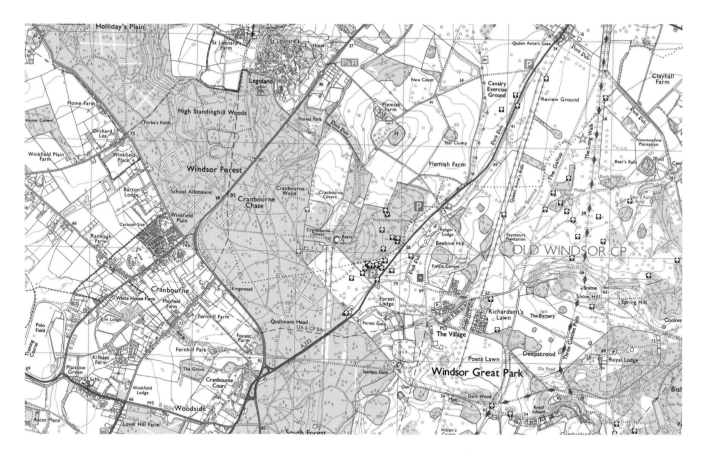

Figure 9-11. Surroundings of Cranbourne Tower on the Ordnance Survey map with the ancient oaks recorded under the Ancient Tree Inventory (ATI) project of the Woodland Trust. © Crown copyright and database rights 2011 Ordnance Survey 100021607

the site of the later tower. Much of this chase, as well as the remaining Royal Forest beyond, must have been heath as it is on poor Bagshot Sand, but the part with the later park roughly east of the line of the Forest Road–White Deer Enclosure boundary has a loam or clay soil resulting from exposure by erosion of the underlying London Clay Beds. This area retained most of the trees, hence perhaps the use of the name 'Crambourne Woode' for it in 1607.

The terrain slopes from 90 m a.s.l. near Cranbourne Tower to 50 m at the Flemish Farm boundary, with some small erosion valleys or gullies draining towards Pickleherring Pond. This ancient pasture woodland, most of it now again grazed with longhorn cattle, is the richest site for ancient and veteran oaks in the Windsor area. There are some 250–300 ancient and veteran oaks on this site; 90 of these exceed 5.00 m in circumference. One oak has a 9.20 m girth, seven are of 8.00–9.00 m in girth and another has been measured at 7.96 m. Remarkably, most

of the ancient or veteran oaks here are maidens, some appear to be 'natural' pollards from storm damage, and only a few seem to be genuine pollards. Some of the latter stand on the edge of or just outside Cranbourne Chase as here defined; the two great oaks along Forest Road (one known locally as King Offa's Oak with an estimated girth of 11.18 m) are also pollards. This unusual 'management' has been observed in other Royal Forest areas, including Birklands (see above), and could mean that commoners did not have access to the living trees in certain parts.

The pasture woodland of Cranbourne Chase has some open grassland, many grassy glades and also areas of closed tree canopy, this mainly due to secondary tree growth from a period when grazing was discontinued. Among the trees, oaks are dominant, but in the northernmost part, beech is co-dominant in places. Both ancient and veteran and younger oaks occur; there is no evidence of substantial planting; for example, there are no even-aged groups of trees. Nevertheless, it seems to have been around here that Lord Burghley's 1580 order to seed acorns was carried out in order to replenish the stock of oaks that had been greatly reduced to supply oak timber for Elizabeth I's navy (Menzies, 1864 cited in Smith, 2004). Hawthorn is common in the more open areas, and domestic apple (*Malus*

Figure 9-12. Ancient maiden oak of 8.20 m girth in summer, Cranbourne Park, Windsor Great Park, Berkshire

Figure 9-13. Ancient maiden oak of 8.20 m girth in winter, Cranbourne Park, Windsor Great Park, Berkshire

pumila), some sycamore and a few sweet chestnut trees also occur. Willows are common in the stream gullies and around Pickleherring Pond. The ground flora is poor, mainly grassland species, bracken and brambles, with sensitive herbaceous indicators of ancient woodland (e.g. bluebells) absent or sparse due to grazing, which may have been the main land use for more than 1,000 years. There is no evidence of coppicing. Fungi are abundant and diverse, with a hotspot for the rare species oak polypore (*Buglossoporus* (*Piptoporus*) *quercinus*). There is a very rich invertebrate fauna, including for example the click beetles *Ampedus cardinalis* and *Lacon querceus*, due to the presence of ancient trees and abundant dead wood (Alexander & Green, 2013; Chapter 10, Figures 10-8 and 10-10). The site forms part of Unit 4 of the Windsor Forest and Great Park SSSI and belongs to an SAC under the EU Habitats Directive. This area is developing, under careful management, back to a situation which approaches medieval pasture woodland. As few 'working' examples exist, it is also an experiment and should be carefully monitored.

Grimsthorpe Park in Lincolnshire

Grimsthorpe Park is a large landscaped park between Little Bytham and Grimsthorpe. Vaudey Abbey ('Vallis Dei') was founded here in 1147 as a gift to the Cistercians of Fountains Abbey. It acquired some more land in due course but remained a small religious estate. Some earthworks and 'stew ponds' mark the location of this religious house. Grimsthorpe Castle was given to Ranulf de Blondeville, Earl of Chester and First Earl of Lincolnshire, in 1217. At the southeast corner of the present castle, which is a Vanbrugh creation in part, is King John's Tower, a possible remainder of the medieval castle. This property was granted in the general area of the 'Forest of Kesteven', which apparently ceased to exist well before Kesteven Forest (north of Bourne) was created by Henry I and then disafforested early in the thirteenth century. There is no mention of a deer park here before 1485. Grimsthorpe Park was granted by Henry VIII to the Tenth Baron Willoughby de Eresby in 1516. Two deer parks were established in 1536, one for red deer and one for fallow deer (www.lincolnshiredeergroup.co.uk);

they appear on a map of 1576 ('Saxton Survey' in Shirley, 1867). By the early eighteenth century, a very large deer park almost equal to present-day Grimsthorpe Park was surrounded by a wooden park pale but without a bank and ditch. Deer were kept in a deer park until 1949.

The park is situated partly on glacial till with pockets of sand and gravel, but the underlying disconform Middle Jurassic Lincoln Limestone deposits ('Cornbrash') are exposed in several areas, including some disused quarries on which species-rich chalk grassland has developed. The terrain is mostly level in the southern part of the park, which reaches a maximum of 70 m a.s.l., with a more undulating landscape elsewhere. Some larger woods are present in the park and the oldest of these follow the contour of a ridge and slope. Two long rides through the parkland, Creeton Riding and Dark Riding, which are orientated towards the castle (vistas), and Four Mile Riding may date from the seventeenth century. There are shorter straight rides in different directions elsewhere through the southern half of Grimsthorpe Park. In the eighteenth century, the park was altered by Lancelot ('Capability') Brown. The most prominent features that remain from this period are further avenues laid out in a radial pattern from a high point on the southern plateau, with the long Carriage Road or Chestnut Avenue (with horse chestnut trees) directed towards the castle (vista), an artificial lake, and several small woods surrounded by open spaces under grass or arable crops. There was already much farmland in the park in the eighteenth century (the large Foal Field south of the abbey site was a medieval arable field) and arable fields have expanded since. Recently, management has reversed the trend towards arable cropping with some fields being turned back to grassland in which new trees are being planted. A section of the home park is still parkland but it has no ancient or veteran trees.

On the site called Hospital Field, there are a few ancient and veteran oaks which could date back to Tudor times after the monastery was demolished, a few may be even earlier hedgerow oaks. There is a 'Tudor oak park', which is the section of parkland with many oaks known as High Wood crossed by Steel's Riding. The oaks now growing there, with one or two exceptions, are too young to date from the Tudor period; Rackham (2009) estimated their maximum date of origin as the second quarter of the seventeenth century. Many oaks were felled during the Civil War, a troubled time for the estate. An ancient wood bank is present between the main section of High Wood and a southern spur, which is probably part of the boundary of a medieval wood (Rackham, 2009). If this is correct, then High Wood was perhaps a coppice wood in medieval time (most likely exploited by the

monks), which would explain the younger age of the oaks now growing here. East of the main avenue through the park, the grassland of this section is unimproved and partly dominated by bracken, especially north of Steel's Riding. A smaller parkland area west of the main avenue has more productive grassland on black sandy soil.

The largest oak in the park (of 9.32 m girth) is located near Steel's Riding, where it crosses open fields. This oak stands in an old and deep drainage ditch, as do several smaller oaks; they do not seem to be park trees. Most of the ancient and veteran oaks with a girth ≥5.00 m in Grimsthorpe Park are concentrated in a preserved section of unimproved ancient parkland on either side of Creeton Riding, known as Bracken Beds. Unlike most of High Wood, the parkland of Bracken Beds has much bracken as its name suggests and tufted hair grass is the dominant grass; the oaks that are present there are, on average, larger than those in the remainder of the park. Ancient field maples and hawthorns are also present and regenerating at Bracken Beds. The parklands of High Woods and Bracken Beds are separated by fields, some arable, between the radial avenues. Old maps show many trees in these fields and the two pasture woodlands appear to have once been continuous (Rackham, 2009); only a few oaks remain in the fields and some are dying or dead.

This large park has seen substantial alterations, but many oaks remain in its two main sections of pasture woodland, 90 of which are ≥5.00 m in circumference, with two oaks between 7.00 m and 8.00 m in circumference. Nearly all of the oaks in Grimsthorpe Park are maidens, in Bracken Beds a few are pollards, some not man-made but caused by storm damage. The oaks in Bracken Beds are nearly all wide-crowned parkland trees; the parkland landscape therefore dates back at least to the early growth of its oldest oaks, which I estimate to be 400–500 years. Rackham also concluded that Grimsthorpe was probably a Tudor deer park. Bracken Beds was not a coppice wood in the Middle Ages and may have been a wooded common. In the two ancient parkland sections, oaks predominate, but ash, field maple, wild service tree and hawthorn are also present. The pasture woodland areas, totalling 110 ha, are notified as an SSSI. The present management of these areas aims to maintain this landscape and its biodiversity, by opening up the bracken to create launds and connecting glades, grazing with longhorn cattle, protecting natural seedlings of oak and other native trees from browsing by cattle and wild deer, leaving all dead wood *in situ* and removing invading non-native tree saplings. New oaks are also planted, presumably to replace dead mature oaks, which are quite common in High Wood but less so in

Bracken Beds. Whilst post-medieval, these two sections of Grimsthorpe Park are among the best preserved oak pasture woodlands in England.

Hamstead Marshall Park in Berkshire

Hamstead Marshall Park (or Hampstead Park) was imparked in 1229 by William Marshal, Second Earl of Pembroke, son of William Marshal I, the famous (or notorious, depending on the chronicler) champion of tournaments and Marshal of England. In the fourteenth century, the park reverted to the Crown until it was sold in 1613. The medieval park is still mostly extant, with remnants of the old park pale visible in many places. Remains of three castle mottes exist in the northern perimeter and near St. Mary's Church. One of these, built on a natural hill, has a dry moat and was situated in a corner of the medieval park; it may have defended a park keeper's lodge of which nothing remains. Another motte is the possible site of Newbury Castle, first held by the Marshals. The River Kennet (and now also a canal) borders the park on the north side. The medieval park was enlarged in the sixteenth century by Sir Thomas Parry, who built an Elizabethan mansion near the church; a second and larger version of this house was destroyed during the Civil War. The church remained, but the village was moved to its present site, presumably so that the park could be enlarged. Later in the seventeenth century, a large palatial mansion was built in the northwest corner of the extended park. This house burnt down in 1718, but the stone pillars of four gates in a wall which surrounded a formal garden and forecourt are still standing. The present house inside the park dates from the eighteenth century and was remodelled in Regency style and enlarged from an earlier hunting lodge; it is situated more centrally in the medieval part of the park.

The park was first landscaped in the seventeenth or early eighteenth century with avenues of horse chestnuts and sycamores, some of these connecting the seventeenth-

Figure 9-14. Oak parkland in former deer park at Grimsthorpe Park, Lincolnshire

Figure 9-15. Oak with a huge burr in Hamstead Marshall Park

century house and eighteenth-century lodge. Most of these avenues have subsequently disappeared but some fragments remain. Later, in the nineteenth and early twentieth centuries, many trees were planted in the lawns and meadows; these trees include several Turkey oaks (*Quercus cerris*). The majority of the park trees are native, mainly oaks, but there is a scattering of exotic conifers as ornamental trees in the park outside of the woodland garden and parterres laid out around the present house. Three or four plantations of a mixture of conifers and broadleaves were created during the twentieth century in the northwest part of the park, mostly within the bounds of the medieval park pale. They serve as coverts for pheasant shooting. Enborne Copse extends from the park perimeter eastwards and is ancient woodland (and

an SSSI). The present park is 220 ha; the medieval park may have been about half that size and was enlarged on its two longest sides. It is situated mostly on London Clays with small exposures of Bagshot Sands on high ground in the southeast. A stream in a valley runs from southwest to northeast through the centre of the medieval park. At the lower end of the stream are three fish ponds dating from the Middle Ages; an ancient dam with a small bridge crosses the stream below the third pond before it cascades down into the river. Plans for a lake envisaged by Lancelot ('Capability') Brown, which could have drowned the lower part of the valley, were fortunately never realized.

There are 21 ancient and veteran oaks with girths of 6.01–8.35 m, but four of these stand outside the medieval park pale boundary; these oaks could date from the time of enlargement of the park in the Tudor period. Half of the largest eight oaks (>7.00 m in girth) are pollards; seven of these are within the medieval park and could date back to the Middle Ages. The oak recorded as having a girth of 8.35 m (Figure 9-15) has an extraordinary large burr around its base and its girth includes it. A large number of smaller oaks have been recorded throughout the park, many as 'ancient', but most of these are likely to date from the seventeenth century or later. Much of the park grassland is unimproved, with fields of rushes in wetter places and acid grassland on higher and drier sites. Bracken is confined to a few small localities. Ancient ash trees, several old field maples and copses of hawthorn are also present. Much of the park is rough pasture; a few sheep were seen on a visit in March 2016. Standing dead trees are retained if not a safety problem, but fallen dead wood is cleared from the park. There is an active programme of planting oaks and some ash to preserve the ancient parkland for the future. Natural England conducted a survey of the mature, veteran and ancient native trees in this park in early 2015.

Hatfield Park in Hertfordshire

Hatfield was given, with 5,000 acres of land, by the Saxon King Edgar to the monastery of Ely in 970. In Domesday Book (1086), it is recorded as being in the possession of the Abbey of Ely. In 1277, Hatfield had two parks, 'Great Park' (also known as Innyings Park) and 'Middle Park', owned by the Bishop of Ely. The palace of the bishops, built around 1485, was confiscated by Henry VIII and turned into a royal palace, of which only parts remain. A large Jacobean house was erected nearby in 1611 by Robert Cecil, First Earl of Salisbury, and large formal gardens were created around it. By the late seventeenth century, a long double avenue centred on the new house ran north-south through the two parks, now united. In the eighteenth century, further landscaping work created The Broadwater in the northern part of the park, a smaller lake east of the house, further avenues crossing the north-south one, and various woods and wide lawns. A map of 1786 shows these avenues but also a much more ancient line of trees, probably marking a boundary of one of the medieval deer parks. These trees were oaks of which some survive; one is possibly a multi-stem tree locally known as the Coppice Tree (Tag No. 0899), which may have been a coppice stool although now appearing to be three separate trees. On the nineteenth-century OS map there are 'Home Park' and 'Hatfield Park' which are adjacent; the former was then already mostly wooded. In the twentieth century, further plantations with conifers filled in more parkland enclaves, while other parts of the park were converted to farm fields.

Plantations with various conifers continue to be managed for timber inside the park, especially in the Home Park section. Hatfield Park is now a landscaped park with high

woods, conifer plantations (also ash near the lake), farm fields and some remaining parkland, a large part of which is open managed grassland mainly used for events. Nearly all of this grassland is improved and regularly mown. A small section of the park nearer Hatfield House has been managed as a deer park since around 1995 and has a herd of fallow deer, but this part has no ancient oaks. The soils of Hatfield are derived from proto-Thames gravelly sands and till deposited by Pleistocene (Anglian) glaciers over Palaeocene Reading Beds or Eocene London Clay. The terrain slopes gently from south to north and west, from *c.* 110 m a.s.l. down to 60 m around The Broadwater.

Remnants of the medieval deer parks remain in a few locations; the most notable of these, known as the Elephant Dell, is to the northwest of Hatfield House. This remnant includes a few areas of unimproved grassland, several veteran hawthorns and the largest concentration of ancient oaks. On a third visit to this location on 9 April 2014, I counted 22 living ancient oaks, 8–10 dead oaks that had been ancient or veteran when alive, and 10–11 large mature or veteran oaks (see also 'Hatfield's Grand Old Masters', a trail guide to the oaks published by Hatfield House in 2012). One of these ancient oaks, 'Old Stumpy' (Figure 9-16), is a pollard of 10.50 m circumference but with a large gap and some burrs; the strangely shaped 'Elephant Oak' is another large pollard. Another great oak at Hatfield, locally known as the 'Lion Oak', is near the stable yard at the boundary of the park, adjacent to the former yard of the White Lion Inn. This pollard oak was filled with concrete and bricked up in 1870 in an attempt to save it from collapse; it also has props. It was probably about 10.00 m in girth but is now impossible to measure due to the surrounding brickwork. Some ancient oaks are on the fields of the show grounds, standing in an irregular line; a few others stand elsewhere in these grounds but most oaks indicated there on the nineteenth-century OS map have gone. Scattered in the Home Park section, now mostly converted to plantation forestry, I found 36 living ancient oaks on my 9 April 2014 visit, confirming earlier sightings and adding a few not then seen. Several of these have suffered from crowding and shading, but work is now being done to 'halo' them. Yet there are also a number of dead stumps for which this rescue operation comes too late. With a total of *c.* 60 living ancient oaks, Hatfield Park is an important site, but its medieval pasture woodland and park landscape have mostly been modernised save for a small section near the house. This section, Elephant Dell, is now treasured and new oaks from ancient provenances at other important sites, such as Birklands and Windsor, are being planted to replace the dead trees. Measures to protect the grass and soil surrounding the ancient oaks from trampling have also been taken recently; the location is near the house and popular with visitors.

High Park (part of Blenheim Park) in Oxfordshire

Woodstock Park was a royal deer park imparked by a seven mile (11.2 km) long dry stone wall, which was begun around 1110 on the order of Henry I and completed a few years later; it was the first of its kind on such a scale. It was situated on the eastern margin of Wychwood Forest, a Royal Forest dating back to late Anglo-Saxon times (Schumer, 1999). Wodestok is Old English for 'stockaded clearance in the forest' and refers to a park with a lodge within Wychwood Forest (Banbury et al., eds., 2010); this was mentioned in Domesday Book (1086). The park of Henry I was therefore an enlargement of an existing park; earlier it may have been an Anglo-Saxon chase. A Roman road (Akeman Street) and the remains of Roman villas and other habitation in the northern part of the present park (known as Blenheim Park) are evidence of agricultural activity of considerable extent in this area centuries before the medieval park. 'Grim's Ditch', a possibly Iron Age earthwork defending a boundary and crossed by the Roman road, may be evidence for earlier habitation of farmers in the area. It is therefore likely that this area no longer had much woodland in Roman times. It is indeed striking that the northern half of Blenheim Park has no ancient oaks whatsoever; there could be a historical connection.

According to Schumer (1999), the medieval Wychwood Forest boundary ran diagonally through the middle of the present Blenheim Park, roughly from Old Woodstock to Park Farm. All extant ancient and veteran oaks occur south of this line, congruent with the Forest. The park's geology and geomorphology are mostly oolithic limestone and mudstone deposits dating from the Middle Jurassic, with Late Jurassic Oxford Clay capped by gravelly sand on the upper part of High Park. The landscape is gently undulating with the valleys of two converging small streams, mainly the Glyme, which runs through the park and discharges into the River Evenlode at its southern perimeter. Several disused quarries for limestone are situated in High Park on the slope towards the lake. The highest point in the park at 120 m a.s.l. is in High Park; the ground slopes down to c. 70 m at the confluence of the rivers. In the Middle Ages, the park was only slightly smaller than the present Blenheim Park; its extensions are mainly in the southeast towards Bladon and include The Lince, near the confluence of the Glyme and the Evenlode, and an area called New Park in the west. In the Lower Park, which was added in the late twelfth century, there are a number of ancient and veteran

oaks, and c. 50 of these exceed 5.50 m in circumference, with 15 ≥7.00 m in circumference and the largest measured at 8.92 m, a nearly 9.00 m girth oak. The largest of these are pollard oaks. In 1705, when landscaping features were first planned, the Lower Park was still "heavily wooded" (Baggs et al., 1990).

At first there was a hunting lodge in the park, most likely a wooden building with a thatched roof, but this gradually expanded to become Woodstock Palace. To allow access to this building, there was a causeway from the Oxford Road, located between Old and New Woodstock, that crossed the water meadows of the Glyme; this has now been drowned by the greatly enlarged Queen Pool. This palace was heavily damaged in the Civil War in the mid-seventeenth century, after which royal hunts no longer took place at Woodstock. The park was given to John Churchill, First Duke of Marlborough, who between 1705 and 1722 completed a vast English Baroque country house that was designed by John Vanbrugh and is known as Blenheim Palace. Formal gardens were laid out around it, as well as a vista in a straight line (as an avenue originally planted with elms) from the main entrance and courtyard across a large bridge spanning the valley of the Glyme to the Column of Victory (commemorating Churchill's victory at the Battle of Blenheim (1704)) and beyond to the Ditchley Gate at the north end of the park. The ruins of the old palace were demolished when the new park was developed, and from 1764, Lancelot ('Capability') Brown made great alterations to create one of England's largest and most famous 'landscape parks' by damming the Glyme and half drowning Vanbrugh's bridge in a lake. New trees, among which were many beeches but also conifers such as Lebanon cedars (Cedrus libani), were planted, either as solitary trees or in 'clumps', i.e. within enclosures as small woods. Earlier avenues were broken up or left to disintegrate. In the Lower Park, a number of ancient and veteran oaks were spared (integrated) but they came to stand in a wide-open landscape with improved grassland from which other ancient relicts were carefully erased.

The only area that escaped these alterations is High Park, in the seventeenth century known as 'The Straights'; this is situated on the high ground between the lake and the western boundary of the park near East End. The area of High Park began to be separated from the rest of Woodstock Park in the sixteenth century. The southernmost section of High Park, known in the sixteenth century as Bladon Wood, became part of the park some decades after 1576. High Park is a mostly untouched (in terms of landscaping and forestry) section of ancient pasture oak woodland. The present High Lodge stands near its

Figure 9-16. Ancient oak 'Old Stumpy' with a girth of 10.50 m at Hatfield House, Hertfordshire

highest point, surrounded by several spring-fed ponds, and the terrain slopes down towards the lake in the east and towards the River Evenlode in the south. High Park has secondary broadleaf woodland planted in peripheral areas ('boundary belts') and young to mature oaks filling in spaces between numerous ancient and veteran oaks. There are also grassy or bracken-dominated glades, including one forming a broad ride and a vista from the palace to the highest point near High Lodge. Most of the grassland is unimproved and with limited to moderate species diversity

(Thames Valley Environmental Records Centre, grassland survey 2015 High Park, Blenheim). There are a few plantations of conifers dating from the 1950s; the largest of these (mixed with oak and beech), is on a strip of land that is shown as open space to the south of the road from Springlock Gate to High Lodge on the nineteenth-century OS map. Sycamore was planted and is spreading in a few localities, but is generally not abundant. Ash and birch are the most common woodland trees after oak; willows also occur frequently, mostly in seepage areas indicated by tufted hair grass. Wild service (*Sorbus torminalis*), field maple (one a pollard of 3.29 m girth) and hawthorn occur occasionally or locally, blackthorn is rare and I have seen only one large broken beech and a few beech saplings and

Figure 9-17. Ancient oaks (of 7.60 m, 7.17 m and 6.72 m in girth) in High Park, Blenheim Park, Oxfordshire

young beeches in the core area of High Park. There are remains of a few large but felled beeches elsewhere in High Park. There is a giant natural coppice stool of hazel in an area with veteran oaks and glades, but I have seen no evidence of former coppice woods. Common indicators of ancient woodland are dog's mercury (*Mercurialis perennis*), false brome (*Brachypodium sylvaticum*), wood anemone and cowslip (*Primula veris*) as well as bluebells which are common in some parts.

There has as yet been no complete inventory by the ATI of the ancient and veteran oaks in High Park; up to the 21th April 2016, I have recorded a total of 97 out of an estimated >100 with girths of ≥6.00 m; however, most of the large oaks have metal number tags and were recorded in 2002 for the Blenheim Estate by English Nature (Natural England). The largest of these is a (possible) pollard oak that I measured on 5th March 2014 as having a girth of 10.37 m. The largest living maiden oak, which I measured on 28th April 2015, has a girth of 10.34 m. There is also a dead oak with 9.62 m girth and two living oaks with 9.47 m and 9.40 m girths. No other site in England has more than three living oaks >9.00 m in girth. Most of the ancient oaks here seem to be maidens, but several have low starting branches, indicating that they have grown in a parkland setting. Some are perhaps pollards or natural pollards through storm breakage.

High Park is an SSSI of 129 ha, notified in 1956 on account of its ancient oaks and associated biodiversity. It is managed by the Blenheim Estate to preserve the ancient woodland with minimal interference. Dead oak wood is left where it fell and there is much of it in all phases of decay. There are also large standing dead oaks. In a Management Plan for High Park March 2014 (www.wychwoodproject. org), long-term replacement of conifer and non-native broadleaf tree plantations, haloing of veteran and ancient oaks, and eventual grazing in some form are envisaged and the first two measures are gradually progressing. Work

is also underway to thin the stands of ash and sycamore where these interfere with ancient oaks or a parkland landscape. Glades and rides are mowed to suppress bracken and natural regeneration of oaks is stimulated but considered difficult because of deer browsing (but see under Grimsthorpe Park above); alternatively, oaks should be planted from acorns gathered from the ancient oaks present. Planting of new native oaks has been undertaken at least since the Ninth Duke of Marlborough started it on a substantial scale in the 1890s. His belief that many old oaks were dying and had to be replaced presumably came from a general misunderstanding of ancient oaks with their retrenching crowns, dead branches and hollow trunks.

Figure 9-18. Broken pollard oak in Ickworth Park, Suffolk

Ickworth Park in Suffolk

Ickworth ('Ixworth' in Cantor, 1983) was first mentioned as a deer park in 1313. It is a large park on the nineteenth-century OS map, containing extensive parkland and several woods; it was much enlarged in the eighteenth century by imparking farmland, as is still visible from remnants of hedgerows. The present park is *c.* 370 ha and has a 'herd' of fallow deer. From 1437, the estate was owned by the Hervey family, later Marquesses of Bristol, until the house and park were passed on to the National Trust in 1956. At its centre now stands a large Italianate-Georgian mansion with an impressive rotunda surrounded by formal gardens. Marriage of the Fourth Earl of Bristol to Elizabeth Davers in 1752 combined the Ickworth, Horringer and Rushbrooke estates, allowing expansion of the park. The park's extent in the nineteenth century was greater than at present, as parts of the parkland were later converted back to arable fields.

It seems possible to reconstruct the outline of the pre-eighteenth-century Ickworth Park from Google Earth imagery, and it is notable that nearly all the oaks recorded as ancient by the ATI are situated within that smaller area. In the west, this outline follows the River Linnet, a small stream originating in Lady Hervey's Wood at the southern end of the old park. In the north, it borders on Little Horringer Hall, where there are now mostly arable fields and woods, and in the east, the village of Horringer and a strip of woods form a more unclear boundary. On the nineteenth-century OS map, five tracks radiate out across the park from the lodge near the village of Horringer. Apart from enclosures near Ickworth House and west of Ickworth Lodge which still exist, the individual trees shown on the nineteenth-century OS map, which tends to be very accurate in these details, have a typically random parkland distribution and there are no trees apparent in lines of presumed derelict hedgerows. The Elizabeth Grove and other small copses are already present on the nineteenth-century OS map; there is a straight N-S running ditch (still present) skirting Elizabeth Grove, but otherwise few straight lines other than tracks in the 'old park' area are visible on this map in the 'old park' area. The few that exist now are (remains of) modern fences. No evidence of early agriculture is visible from the air in the 'old park' and I think it is reasonable to assume that this part of present-day Ickworth Park is more or less congruent with the medieval deer park. There are scattered small ponds which may also indicate an ancient park.

As in most parts of Suffolk, the soil in the park is formed from till deposited during the Anglian Glaciation, overlying disconform beds of Late Cretaceous chalk. The

Native oaks that have been planted often show past damage done by grey squirrels; this invasive animal is now rarely seen in High Park. The site also serves as a pheasant shoot, leading to seasonally high densities of these raised birds. Reducing the pheasant population outside the shooting season is recommended. Deer were kept in Blenheim Park until the First World War, but there are wild populations of fallow, roe and muntjac deer and a few badger setts. Several very rare invertebrates and fungi that are associated with ancient oaks have been recorded in recent years. In 2012–14, several mature oaks with symptoms of Acute Oak Decline (AOD) have been found in High Park (Blenheim High Park Management Plan, 2014).

park's terrain is mostly level, with the highest point at 83 m a.s.l. and gently sloping down to 60 m in the valley of the Linnet. Fallow and other deer are often sighted in the park but there is no confined or maintained deer herd; instead the park is mainly grazed by sheep. There are many ancient and veteran oaks in the parkland, nearly all in the 'old park'. The largest ancient oak is known as the Tea Party Oak, now a living wreck of a pollard with a circumference (estimated!) of 9.79 m. Another ancient oak has a girth of 8.29 m and there are eight oaks with girths of 7.00–8.00 m; in total, I have counted 67 ancient and veteran oaks in the ATI database ≥5.00 m circumference. There are also many ancient field maples, which seem to be a characteristic of some East Anglian parks. Only a few exotics, such as horse chestnut, sweet chestnut and sycamore, have been planted in the 'old park'. Although improvement of the grassland has meant that the landscape is somewhat sterile, with (mostly) oak trees on a green sward, most internal fences have been removed and a return to acid grassland with an extensive grazing regime seems possible. The park is managed by the National Trust and designated as a Suffolk County Wildlife Site.

Kimberley Park in Norfolk

Kimberley Park has been mentioned as a deer park since *c.* 1401 (Cantor, 1983). Around that time, John Wodehouse built Wodehouse Tower, a fortified manor house, "at the site of present-day Kimberley Hall", which was built in 1712. In 1480 the park was owned by Sir John Wodehouse, a descendant. However, there are the remains of a moat in Gelham's Wood within the northern perimeter of the park; these remains are marked as Gelham's Hall on the nineteenth-century OS map. At the western end of the park was another moated house, known as Kimberley Tower or Kimberley Hall, this is now surrounded by farmland. A third moat ruin is situated at the southern end of Falstoff's Wood and forms the archaeological remains of Falstoff Hall. Any of these may have had a deer park that is not mentioned in the surviving records. The house that preceded Kimberley House ('Hall') and stood on its hill was Downham Hall.

In the nineteenth century, this was still a large landscaped park, laid out in 1762–70 by Lancelot ('Capability') Brown, with parkland, a wood (then called Great Carr but now Falstoff's Wood), a lake and an avenue. The house stands on a slight elevation at 40 m a.s.l. and most of the park is on nearly level terrain at *c.* 25–40 m composed of disconform Pleistocene glacial till overlying Cretaceous limestone beds. The present park has been reduced in size, especially in the western part, by arable fields near Park Farm. Beyond the formal and walled gardens near the house is the 'deer park', to the west separated by a haha from the gardens. In this small park section between the house and the lake, three ancient or veteran oaks ≥5.00 m in girth are recorded, the largest of which was measured at 10.70 m circumference in 2010 by the Norfolk Heritage Tree Project. This measurement is incorrect, as on a visit accompanied by Mr. Robert Buxton on 25th April 2014, we found no oaks larger than 7.50 m here. There were more parkland trees in this part of the park in the nineteenth century than there are now, but from the OS map, it is impossible to say how many of these were oaks. Two of the recorded veteran trees here are limes (*Tilia* sp.) with ≥7.00 m circumference. The part of the park between the River Tiffey, a small lowland stream coming from the artificial but natural-looking lake, the Barnham Broom Road and the house was a deer park until sometime in the nineteenth century. According to the nineteenth-century OS map, this was unimproved grassland with many ancient and veteran oaks (and perhaps some other tree species). Much of this has been turned into farm fields in which most of the trees have been destroyed. These fields are no longer farmed but are now turned back to grassland. Several trees have survived in the north corner, none of which are ancient.

A few scattered ancient or veteran oaks remain elsewhere within the park; two of these stand on the edge of two ancient ponds which are interesting remains, possibly of the medieval deer park. Most of the existing ancient and veteran oaks are in a remnant of ancient parkland to the south of the driveway to Kimberley House. A total of 31 ancient and veteran oaks with girths ≥5.00 m have been recorded here, ten with girths from 7.00–8.50 m which are likely to be ancient. The inventory is incomplete; five ancient and veteran oaks were added on my visit on 25th April 2014. Plantations now surround this remnant of unimproved parkland. Kimberley Park is privately owned; the house is only open to booked weddings and corporate functions. No park management plan is available and no part of the park has protected status as an SSSI or NNR. Some parts, including the park fragments with ancient oaks, are designated County Wildlife Sites (CWS). Kimberley Park still has a number of ancient and veteran oaks in sections of parkland that retain their ancient character, but there have also been substantial losses of this parkland to farming since the nineteenth century. The present owner has a keen interest in the preservation of the ancient oaks and the remains of the medieval park.

Moccas Park in Herefordshire

Moccas Park is an ancient deer park, but there is no

historical record of it prior to the seventeenth century (David Whitehead in Harding & Wall, eds., 2000); it was first mentioned as 'Moccas Deer Park' in 1617. Remains of a Norman motte and bailey castle are situated at the eastern boundary of the park, as indicated on the nineteenth-century OS map, but this may never have had a stone building. Instead of belonging to this castle, Moccas Park may have been the park belonging to Geoffrey de Bella Fago in 1316 and listed by Cantor (1983) as part of the Parish of Dorstone. The castle there was about 2.5 km from the present park, but the wooded Dorstone Ridge, on which a large part of the present Moccas Park is situated, may have been the only land in the parish suitable for a medieval deer park. In the sixteenth century, it may have

been joined with the Moccas estate, then a part of the lordship of Bredwardine.

Moccas Park is situated on a northnortheast-facing slope from around the 280 m contour down to *c.* 80 m at Lawn Pool. Dorstone Hill forms the northeastern limit of the Black Mountains range and consists mainly of latest Silurian (Pridolian) Old Red Sandstone and associated mudstones. Pleistocene till and fluvio-glacial deposits mingle with floodplain sediments in the lowest parts of the deer park. The present deer park is 136.7 ha (inside the NNR/SSSI boundary), but a larger park extended to Moccas Court on the River Wye in the seventeenth century. Before 1772, the park included land to the south, around present Cross Lodge and Windy Ridge, that is now mostly farmland. The park was partly landscaped in the eighteenth century, when the present house Moccas Court was built. Some of the large sweet chestnuts in the park may have

Figure 9-19. Ancient pollard oak in Kimberley Park, Norfolk

Figure 9-20. A pollard oak in Moccas Park, Herefordshire

been planted around 1660. The park is commonly divided into an upper and lower park, referring to the more wooded part on the slope and the more open parkland down below. Parts of the lower park were established on medieval arable land, as shown by ridge and furrow in the vicinity of Lawn Pool. Most of the ancient and veteran oaks are in the lower park. Before 1950, what is now Moccas Hill Wood and mostly planted with larches belonged to the park, but this hilltop section was mostly moorland rather than pasture woodland according to a 1946 RAF aerial photograph. It is now being clear-cut in stages and grazed to restore it to coarse pasture.

Moccas Park is still a deer park with a herd of *c.* 300 fallow deer; it is surrounded by a dry-stone wall and along the B4352 road by an oak wood pale fence in traditional fashion. The smaller present park, now an SSSI and an NNR and privately owned, is one of the most important sites in England for ancient and veteran oaks. The Moccas Oak is an ancient pollard that has been recorded as having a 9.60 m circumference, but recently its bolling has broken and I re-measured it as 7.92 m in October 2015. It stands on the north bank of the Wye in a meadow. It is unlikely that this tree is related to the park, but its medieval context is not clear, perhaps it was an assembly tree. The name seems to have been transferred to this oak from an even larger oak pollard that stood in the deer park and was drawn or painted by various artists, among whom was J. G. Strutt in his *Sylva Britannica* (1822); these illustrations are reproduced in Harding & Wall, eds. (2000). In 1870, this older tree was measured as having a girth of 11.00 m (36 ft) at 1.50 m from the ground. It was again measured by members of the Woolhope Club in 1891; shortly thereafter it died and by 1930 was not heard of anymore. Another large oak pollard stands near Lower Moccas Farm, and several oaks that have girths >6.00 m were recorded near Moccas Court in the spring of 2016. Today, the stoutest oak in the deer park is known by entomologists as the Hypebaeus Oak (or Tree) as it was in this tree that the beetle *Hypebaeus flavipes* was first found. Moccas Park is to date the only known British location of this rare insect (see Chapter 10). The Hypebaeus Oak is a pollard that has a squat bolling with a girth of 9.29 m.

A total of seven Moccas Park pollard oaks have been recorded by the ATI as having girths of between 8.00 m and 8.60 m (one is just outside the park). In all, some 90 ancient oaks have been recorded in Moccas Park. An inventory with measurements in 1870 reported 68 ancient or veteran oaks with girths of between *c.* 3.50 m and 8.50 m, with one oak at 11.00 m, but this was not a complete count of the ancient oaks then present. Despite this, it is likely that the number of large ancient oaks in the park has increased since 1870, despite losses such as that of the largest oak known to have been there (Harding & Wall, eds., 2000). As elsewhere, trees older than 500 years are very unlikely to have been planted and are remnants of the medieval pasture woodland in which the park was established; pollard oaks in the above 7.00 m girth class are almost certainly medieval in origin. Many large oaks have spreading root bases partly above ground. There are also many large mature and veteran maiden oaks. Oaks and other trees have been planted since the seventeenth century and new oaks are still being planted. Active tree planting has added more than 1,000 trees, mostly oaks, to the park between 1985 and 2010. The upper park on the slope has become closed-canopy high forest except for one open area in the western part; on the nineteenth-century OS map, there is still more parkland with free-standing individual trees. Tree planting is largely responsible for this, as was already commented upon in 1870 (Harding & Wall, eds., 2000). It may also be the cause of the presence of fewer ancient oaks on the slopes than in the lower park, as oaks in the upper parkland may have been shaded out by beeches and sweet chestnuts, as well as by horse chestnuts planted since the 1780s. Moccas Park also has small numbers of old field maples and hawthorns, indicators of more open parkland, and very few lime and ash trees; a few exotic conifers are reminders of nineteenth-century park landscape fashions. Moccas Park is one of the most intensively inventoried sites of ancient trees in the country, and as a result, much knowledge is available about its fungi, lichens, flora, and invertebrate and vertebrate fauna. This information is summarised in Harding & Wall, eds. (2000). Management, under the guidance of Natural England, is increasingly focused on the conservation of this biodiversity, but other objectives issuing from private ownership also have to be met. Additional summer grazing is carried out with cattle and sheep on the slightly improved grassland in the lower park. Bracken is controlled by rolling in late spring or early summer and the removal of fallen timber is limited to branches of less than 15 cm diameter by an agreement with Natural England.

The New Forest in Hampshire and Wiltshire
The New Forest is the largest of the medieval Royal Forests that still exist, located west of Southampton between the Solent and Salisbury, with an area of 37,100 ha. It consists mostly of unenclosed pasture woodland and heath, grassland and valley bogs. The New Forest is drained to the south by three small rivers: from west to east, the Lymington River, the Beaulieu River and Avon Water. The highest point is Pipers Wait at 129 m a.s.l. The New Forest lies in the Hampshire Basin surrounded by low chalk downs on three sides. This basin is filled with flint gravels, sand and silt, which were eroded from the hills and deposited mainly as glacial outwash (gravels) on top of earlier sand deposits. Stream erosion of the gravels capping the higher parts has exposed underlying sands and clay. The gravelly soils now support only heath and acid grassland, but they may have been covered with oak-birch woodland before deforestation. The area was densely populated in the Bronze and Iron Ages, as attested by the presence of *c.* 250 round barrows. Much of the forest, especially areas on the gravel deposits, was already cleared and turned to heath in prehistoric times. In about 1079, William I ('The Conqueror') afforested the area; it was called New Forest (*Nova Foresta* in Domesday Book (1086)) because no Royal Forest rights had existed here in the Anglo-Saxon period. Within the New Forest are several archaeological remains of medieval forest lodges and park pales or other enclosing earthworks. Brockhurst Park was mentioned as a park belonging to the crown ('New Park') in 1484 (Cantor, 1983) and still exists within the New Forest. There are seven ancient or veteran oaks with >5.00 m girth in this park, including a multi-stemmed tree which is probably grown from a coppice stool and was recently measured as 9.10 m in circumference at 40 cm above the ground. The poor soil of the area makes it unlikely that farming was of much significance, and animal grazing and the exercise of other rights of common must have been the main use of the land by the inhabitants throughout the Middle Ages. Common rights are still exercised, especially grazing and pannage, and are regulated by the Verderers of the Forest. The rights of common are attached to property (houses) in the villages and hamlets in the area, thus limiting the number of commoners. There are several villages and hamlets with enclosed fields around the houses, forming cultivated enclaves. The commoners' animals (ponies, cattle and pigs) are now subsidised under a Department for Environment, Food and Rural Affairs (DEFRA)/Rural Payments Agency (RPA) arrangement and consequently grazing pressure is high.

Deer are common throughout the New Forest, although they mostly avoid the open heath during daylight hours. The impact of deer and domestic animals on the vegetation is profound; regeneration of trees is severely limited and generally occurs only within dense stands of holly or gorse or sometimes within the crown of a fallen tree, if light can reach the ground vegetation. Despite this, regeneration has taken place over time in about three-quarters of the 'open' woodlands and the total acreage of broadleaved forest and

Figure 9-21. A pollard oak near Rowbarrow in the New Forest, Hampshire

woodland has increased since 1867 (Small, ed., 1975). On the other hand, Tubbs (1986) concluded that "tree and shrub regeneration effectively ended by about 1970" because of the intensification of grazing and browsing. During the late Middle Ages and in early modern times, extensive areas were encoppiced, especially after an act allowing enclosures was passed in 1482 and followed by another act in 1698 allowing the enclosure of 6,000 acres (2,428 ha). The Deer Removal Act of 1841 allowed the Crown to enclose a further 10,000 acres (4,046 ha) for forestry plantations (almost needless to say, the deer could not be removed). Open pasture woodland now covers 14,600 ha, of which only 3,800 ha is classified as 'ancient pasture woodland' (Hayward in Read, ed., 1996). Heath and grassland cover 11,800 ha, 3,300 ha are wet heath and bogs, and 8,400 ha are enclosures with tree plantations, of which 8,000 ha of Crown land have been planted with conifers since 1920 by the Forestry Commission. The

earliest enclosures incorporated the existing wild-growing trees and created coppice with standard oaks; plantations in enclosures began with the planting of oaks in the eighteenth century. Some of the oldest enclosures are now open to grazing again, and few effectively prohibit deer from entering.

The wooded areas are now mostly dominated by beech because oak was felled extensively over several centuries, mainly to supply the Navy (Tubbs, 1986); beech is also more shade tolerant and outcompetes oak if left uncut. As a result, there are relatively few large pollard oaks, most around 7.00 m in circumference; the largest measured oak is tree No. 88534 near Woodgreen, with a girth of 8.78 m. The largest maiden oak is known as the Knightwood Oak and has a girth of 7.38 m. Eight oaks have girths of 7.00–8.00 m (up from four known by Tubbs, 1986) and 22 measure 6.00–7.00 m in circumference. Most numerous are the ancient and veteran oaks in the class 5.00–6.00 m. The total number of ancient and veteran oaks with ≥5.00 m circumference is 161 as assessed in November 2015 using ATI data shown on an Interactive Map (www.ancient-tree-hunt.org.uk/). This total omits dead trees or stumps and trees that looked to me like mature oaks in fields or hedgerows (almost all recorded trees have photographs attached). Maiden oaks are common in the veteran category. Given the much greater area of ancient pasture woodland in the New Forest, these numbers are significantly lower than those for many other important sites here described.

There are many old beech pollards in the New Forest which have not been cut for 150–200 years (Hayward in Read, ed., 1996). No trees are removed from the unenclosed ancient woodland and the old beeches are now often senescent or dead. Most of the ancient woodland (outside of plantations) is oak-beech, beech or oak-birch woodland. Small-leaved lime was present in the past but is now virtually gone (Tubbs, 1986) and hornbeam is rare. Birch is common, as is rowan, and there is a scattering of wild service. Holly is common and outside the woodland forms distinctive thickets ('holly holms') on the heath. Alder carr occurs alongside major streams, accompanied by willows or ash where flooding is limited. The New Forest, mainly because of its size, is the most important site for fungi and invertebrates associated with dead and decaying wood in England (see Chapter 10). The New Forest is an NNR, a Special Area of Conservation (SAC) and contains several SSSIs. It forms the greater part of the New Forest National Park and is administered by the New Forest National Park Authority, which has the difficult task of balancing nature conservation, rights of common and tourism in the densely populated south of England. Minimising the impact of tourism, mainly by regulating where people can park their cars, has been successful. Forestry managers are acknowledging ecological values besides timber, and their management is changing accordingly. It is now time to reduce the impact of grazing on the ancient woodland, which could be achieved by altering the subsidies for domestic animals and by a regular deer cull.

Parham Park in Sussex

Parham Park in West Sussex has an Elizabethan house built in 1577, after the manor of Parham was confiscated from the Abbey of Westminster by Henry VIII in 1540. It was granted to a London mercer called Robert Palmer, who is likely to have imparked part of the estate. The late medieval church of St. Peter near the house may have belonged to the small village which was moved to create the park. The deer park surrounding the house (presently 120 ha) may have been imparked in the sixteenth century, but the earliest mention of a herd of fallow deer is from 1628. It

is not mentioned in the gazetteer of Cantor (1983) and is, according to the evidence given here, a Tudor deer park. A stone wall remains in parts, but most of the park is now surrounded by a modern deer fence.

The larger park of Parham, at present comprising 354 ha, is situated at the foot of the South Downs. It is on cross-bedded sands of the Folkestone Formation of Aptian-Albian (Cretaceous) age, in many places overlying chalk with flints. Its highest elevation is Windmill Hill at 55 m a.s.l. The northern half of this park is mostly oak woods and conifer plantations (Northpark Wood), but there are some arable and pasture fields near Parham Farm. There are some small hazel coppice woods in this northern half, as well as high beech forest in the south and west. This part of the park has no recorded ancient or veteran oaks. At the northern-most end is a small raised bog and heath, partly planted with Scots pine and invaded by purple moor-grass; there is also some secondary woodland with oak, Scots pine, birch and beech. This northern half is known on the nineteenth-century OS map as Wiggonholt Common and therefore was not part of the park. Woods and shelterbelts of smaller size occupy parts of the southern half of the park, i.e. the deer park, but here are also areas of open parkland with many oaks and other parkland trees, either solitary or in small groups. Two large ponds have been created in the western part of the deer park. The deer park hosts a herd of c. 350 fallow deer especially bred for their dark brown colour (see Figure 7-1).

There are many sections of parkland within the deer park that have unimproved grassland; examples can be found on Windmill Hill and on the slopes below West Plain to the west and southwest of the house, where bracken is common. Low-lying unimproved open grassland areas have water-logged soils with hard rush (*Juncus inflexus*). West Plain (with a cricket pitch) and East Plain are virtually treeless expanses of improved or mown grassland. In the north of the deer park, around the Fangrove Hill enclosure, is a section of pasture woodland with some large oaks, but the pressure of the many deer on shrubs and small trees is evident here. In more open parts on dry slopes there are large colonies of rabbits that help the deer to prevent the natural regeneration of oaks and other trees, although many tree seedlings were seen on 27th March 2014. The deer park has a good number of ancient and veteran oaks, the largest of which is 7.52 m in circumference; 67 were recorded in the 5.00–7.50 m circumference range. On Fangrove Hill, there is a large oak which appears to be multi-stemmed and possibly grew up from an earlier coppiced oak. Other recorded (native) parkland trees include ash, beech, hawthorn and field maple; there are also a few large sweet chestnut trees. Apart from some groups of Scots pine, a few cedars, some horse chestnuts and a small avenue of lime trees, there are few non-native trees in the deer park. The deer park is notified as an SSSI on account (in part) of the ancient oaks and associated biodiversity, in particular lichens and rare beetles. Parham Park suffered much damage during the Great Storm in October 1987 and a programme of replanting, especially of native oaks in the parkland, is ongoing. Hazel and hawthorn saplings are grown in a few small enclosures, presumably for planting when beyond deer-browsing size. This park would, from the perspective of the ancient trees and their habitat, benefit from a reduction in the numbers of deer (and rabbits!) so as to allow natural regeneration to succeed.

Richmond Park in Surrey

The present Richmond Park is a large (*c.* 1,000 ha), partly landscaped deer park surrounded by a brick wall and was imparked by Charles I in 1634–37. It is a pseudo-medieval deer park, imparked following medieval concepts and traditions at a time when elsewhere such parks were being transformed into pleasure grounds for other purposes (Brown, 1985). This park, which became known as 'New Park', is not to be confused with the Sheen Park of 1292 or with Richmond Old Deer Park, which has also been documented as 'Richmond Park'. The extent of each of these parks is not exactly known, but Sheen/Richmond Park probably occupied the level land from Richmond Palace (of which a gate survives) along the east side of the Thames as far as Kew Gardens. A map of 1637 drawn by Nicholas Lane (Figure 9-23) shows the outline of the brick wall of the present Richmond Park, then just completed, which imparked a complex of ancient semi-agricultural

Figure 9-22. Veteran oaks in Parham Park, West Sussex

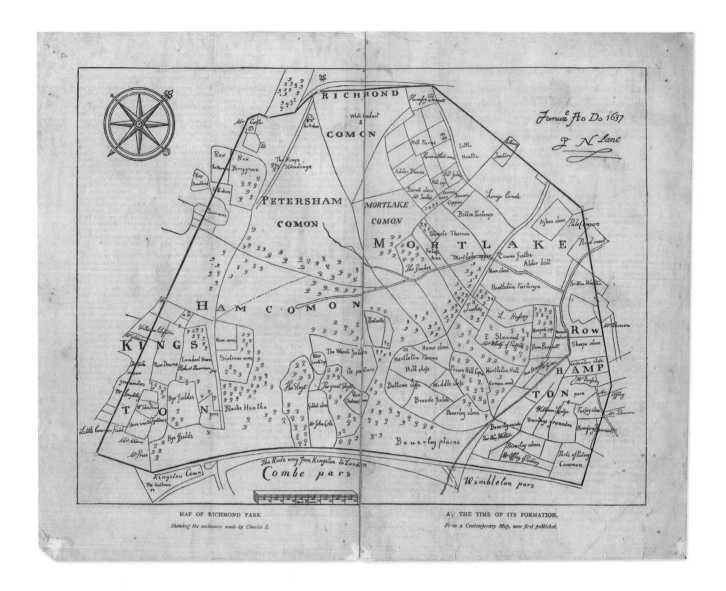

Figure 9-23. Map of Richmond Park showing the completion of the imparking wall in 1637 and the situation within the wall before any landscaping had been done. Reproduced with permission from the Richmond Local Studies Collection, Old Town Hall, Richmond, Surrey

and essentially medieval commons. This wall enclosed land that in the Middle Ages had belonged to six royal manors, which later became the parishes of Sheen, East Sheen, Mortlake, Roehampton, Kingston and Ham. Also imparked were parts of Putney and Mortlake pertaining to the manor of Wimbledon, which was not a royal manor.

The park lies on the higher ground (50 m) above Richmond and the Thames Valley. The soils are thin and impoverished, overlying gravelly sand and loam of glacial outwash and ancient river beds, underlain in turn by London Clay which is exposed in a few marshy places. This imparked land included pasture woodland commons as well as 'waste' (heath with gorse and rough grassland with thorn scrub) known as Great and Little Heath Commons, enclosed fields, some coppice woods, meadows along Beverley Brook, and medieval farmland,

as attested by ridge and furrow. In the sixteenth century, this area was known as Shene Chase, implying occasional hunting parties. Landscape 'improvements' took place mainly in the eighteenth century, with the Queen's Ride oriented on the White Lodge (built 1730) cut through existing oak woodland in 1736 and the Pen Ponds (divided by a causeway) created in 1746. Several smaller ponds (some ancient) are scattered throughout the park; most were created in the nineteenth century in connection with drainage works to dry out swampy areas (Rabbitts, 2014).

From the early nineteenth century through to the late twentieth-century, new plantations of trees were enclosed against the deer. These included non-native trees and rhododendrons; the best known of these enclosures is the Isabella Plantation. Smaller enclosures contain various dwellings; a few go back to farms which were established prior to the park. Petersham Park, on the slope below King Henry's Mound, which had been alienated earlier, was reincorporated into the park in 1833–34 along with a small part of Sudbrook Park, the remainder of which is now a golf course.

Topographically, Richmond Park has changed little from the situation depicted on the nineteenth-century OS map: several unpaved sand paths or rides are now tarmac roads with busy commuter traffic; the road gates close after sunset. The non-SSSI section east of Beverley Brook is now a golf course. Ploughing of more or less level plains for agriculture and other war-time activities during the twentieth century, as well as air pollution, have all left their marks on the ancient landscape and vegetation. There are large herds of red and fallow deer and extensive areas of acid grassland (with much encroachment of bracken) and unimproved parkland (with thousands of nest hills of the yellow meadow ant (*Lasius flavus*)), enclosed patches of gorse, and many old hawthorns but very little heather.

The open parkland has numerous ancient and veteran pollard oaks as well as younger planted maiden oaks. A concentration of >100 ancient and veteran oaks southwest of Isabella Plantation is marked as 'Sixteene Acres' on the 1637 map and is now known as High Wood. Elsewhere, some ancient oaks stand in rows, marking former hedges along fields, as near Holly Lodge. My survey of the ancient and veteran oaks of Richmond Park recorded by the ATI up to February 2014, found 131 ancient and veteran oaks with a circumference of ≥5.00 m. Most of these trees (106) are between 5.00 m and 6.00 m in girth, 22 oaks are between 6.00 m and 7.00 m in girth and only three are between 7.00 m and 8.00 m in girth. The largest measured oak at 7.96 m in girth is just under the 8.00 m threshold. No truly big oaks (>9.00 m) exist in Richmond Park. The oldest oaks in the park, growing as they do on sandy (but not poor) localities, could date from the late fifteenth to early sixteenth century and could be around 550 years old. These trees were, as elsewhere, not planted but remnants of ancient pasture woodland in which the trees were wild-growing. Most of the ancient oaks are pollards, but some with the characteristics of ancient trees (see Chapter 1) are maidens, and many of the recorded veteran oaks are also maidens. Many of the latter could date from the time when the park was created and so would be around

380 years old. The ancient pollards are mostly older. In a response to a proposal by George III (reigned 1714–27) to raise £2,000 by selling timber, the Surveyor General commented that there were "Several thousand pollard trees fit for nothing but fuel…" Although pollarding continued into the eighteenth century and perhaps, here and there, up to the beginning of the nineteenth century, the evidence suggests that most of the oaks planted since Charles I and his appointed Ranger, the Earl of Portland, took charge were not cut as pollards. There was, according to Horace Walpole and others, "great decay of the woods", meaning that trees were felled and not replaced and perhaps that pollards were unmaintained.

Many maiden oaks were planted in the eighteenth and nineteenth centuries, often in groves, sometimes in enclosures. Today, there is an active programme of surveying ancient and veteran trees in the park. All veteran trees have individual management plans. Nevertheless, the attrition rate is high and the exigencies of weather frequently necessitate unscheduled interventions on public safety grounds. Oak planting was accelerated after the great storms of 1987 and 1990. Most of the planted trees were and are temporarily protected until large enough to survive attention from deer. In the open park, oak seedlings occur after a good mast year but stand no chance of becoming a tree because of high deer densities. Planting of oaks therefore continues to provide the replacement for losses. A programme to make new pollards has also been initiated, and by early 2014, around 100 young oaks had been cut as pollards (pers. comm., Adam Curtis). There is currently a trial in which pollarding of 'second generation' oaks is being tried. Richmond Park is an NNR, a SAC and a SSSI. (The golf course is excluded from these designations.) Richmond Park is the principal British locality where the cardinal click beetle (*Ampedus cardinalis*), which is associated with ancient oaks, is clearly well-established. It is managed for the Crown by The Royal Parks, an executive agency of the Department of Culture, Media and Sports (DCMS).

Savernake Forest in Wiltshire

Savernake Forest is mentioned (as 'Safernoc') in an Anglo-Saxon charter from King Æthelstan in 934 CE. It was declared a Royal Forest in the twelfth century and is mentioned in the 1130 Pipe Roll as 'the Forest of Marlborough'. Savernake Forest once legally encompassed land to the extent of 39,000 ha, extending as far south as Harewood Forest near Andover in Hampshire (Walwin, 1976). The 'physical forest', i.e. that part which was uncultivated land and in this part of England would have been more or less wooded, was always much smaller

Figure 9-24.
Richmond Park,
Surrey in spring (May
2013, right), summer
(July 2013, below),
autumn (November
2009, opposite top)
and winter (February
2009, opposite
bottom)

(Rackham, 1976). As was the case with most other Royal Forests, disafforestation gradually reduced this vast area to a much smaller jurisdiction. Yet in the eighteenth century, Savernake Forest was still nine times its present size of 1,820 ha and by this time all cultivated land would have been excluded. The area occupied by Savernake Forest on the nineteenth-century OS map is almost identical to that of the present forest. Savernake Forest is situated on a Cretaceous Upper Chalk plateau; its soil is in most parts clay with flints, but some areas have acid sands and gravels. This heavy type of soil was difficult to plough in the Middle Ages and therefore was often left to 'waste' suitable for the royal hunt. The topography is formed by gently rolling hills and level plateaus, and by some dry valleys at altitudes between 150 m and 185 m a.s.l.; in some areas, drainage is poor due to impermeable clay.

Shortly after the Battle of Hastings (1066), the Anglo-Saxon forest at Savernake was given in care to Richard Esturmy as the first Warden of the Forest. Hereditary forest wardens continued to keep the Royal Forest for the king until 1547, when the Seymour family who had succeeded the Esturmy family in 1427 took over its ownership outright; it is now the only privately owned Royal Forest in England (Oliver & Davies, 2001). There are remains of an old park pale, indicating the imparkment of a section of the Forest as a deer park; apparently there were various later attempts to impark further sections in the sixteenth and seventeenth centuries, but none lasted very long. The outline of one of these parks can be traced on both the nineteenth-century OS map and on the present OS map. On a map *Wiltonia sive Comitatus Wiltoniensis* (Wiltshire) dated 1660 and held in the Map Room of the Royal Geographical Society in London, 'Savernake Forest' is shown as a park. In the sixteenth century, Henry VIII came here regularly to hunt deer (and to court Jane Seymour).

Both the remains of a Roman road and a section of the late eighteenth-century Grand Avenue, which follow a converging course, cut through the length of the forest. Six other straight drives were laid out in a radial orientation from a central wooded hub, Eight Walks, on a point along the Grand Avenue. The Grand Avenue was planted with beech and was originally 6.3 km (3.9 miles) long and, together with its subsidiary star avenues, was designed by Lancelot ('Capability') Brown. In the seventeenth and eighteenth centuries, fields were created and cottages built; these assarts were farming enclosures in the forest, now mostly managed as grassland for grazing. Coppices were also enclosed; these include Cobham Frith, Little Frith, Chisbury Wood, Haw Wood, Birch Copse and Horseleaze Wood. Beech and oak were planted for timber trees in other

parts of Savernake Forest. In the late nineteenth century, these hardwood plantations were followed by the first conifer plantations, such as in Ashlade Firs and adjacent Savernake Wood, which were accompanied by straight rides at nearly right angles. Most conifer plantations were established in the coppice woods; at present, very few coppice woods remain due to this conversion. Nevertheless, tradition as well as other interests in the ancient forest prevented the owners from wholesale conversion to plantation forestry. In 1870, most of Savernake Forest was imparked by a 25 km fence and in use as a deer park. This fence no longer exists, but four species of deer still occur in the forest. Areas that were clear felled during the World Wars were often replanted with conifers. Since 1939, the Forestry Commission has held a 999-year lease on the forest's timber and trees, but under restrictions imposed by the trustees of the Brudenell-Bruce family (the hereditary wardens) that limit commercialisation, i.e. replanting solely with conifers. Recreational values were set high on the management agenda, opening the forest to the public for all but one day of the year. The planting of conifers continued into the 1980s, even on sites with ancient oaks. Many ancient oaks were killed (shaded out) by the growing conifers (and others by earlier beech plantations) but a programme to rescue survivors by haloing them has recently begun.

A small section of the forest around the valley of Red Vein Bottom has retained its open character as pasture woodland, and since 2003 has again been grazed by cattle; a similar area is present along the southern edge of the forest. Although now managed mainly to retain its ecological values, most of the core of Savernake Forest has become closed-canopy native deciduous forest, especially in the southern part that is dominated by beech but with oaks the second most abundant trees. Much secondary young deciduous woodland, often with a scattering of planted larches, has filled in the formerly open glades. Two busy roads traverse the forest, the A4 cuts through the northern section and the A346 skirts the western boundary, just lying within.

There are still high numbers of ancient and veteran trees, mostly beech and fewer oak, in Savernake Forest. The best known oak is the Big Belly Oak (11.18 m in girth), which unfortunately stands right on the verge of the A346 road. Two other oaks that were still alive in September 2014 are >9.00 m in circumference: King of Limbs (10.25 m) and Cathedral Oak (10.15 m) (Figure 9-25). The Duke's Vaunt Oak (once recorded as *c*. 10.00 m in girth) is now a mere wreck, with just one live branch and at most 6.00 m in circumference. In addition, three ancient oaks are between 8.00 m and 9.00 m in girth, and six are between 7.00 m and 8.00 m. Most of these large oaks are pollards and they must date from medieval times, i.e. prior to 1485. Analysis of the records of the ATI shows *c.* 110 ancient oaks >5.00 m, of which 96 have girths between 5.00 m and 7.00 m. Many of the oaks in this list are not obviously pollards, and if these are younger than most of the pollards, they may indicate that a change in management took place at some time, perhaps in the late sixteenth to seventeenth century when hunting had declined as a primary activity and oak timber became more important. Oak pollards are particularly rare in the most southern part of Savernake, which is mostly tall beech forest with scattered conifers and secondary woodland with birch and oak, the oak both planted and from natural seedlings (Oliver & Davies, 2001). Many hundreds of oaks with girths below 5.00 m are recorded as veteran trees, together with several thousand beeches. Most of these trees date from eighteenth and early nineteenth-century plantations. Wych elm, field maple, birch, rowan and hawthorn are also common, together with the broad-crowned ancient oaks that are indicative of a more open structure of the woodland in the past. Holly is often an under-storey tree in beech-dominated stands, while remnants of hazel stools indicate former coppice woods.

Most of the forest where not converted to conifer plantations can be classified as beech woodland (*Melico-Fagetum*) on neutral soil, tending to oak-beech woodland on some more acidic areas on sand. The oaks, a mixture of pedunculate oak and sessile oak and their putative hybrids according to Oliver & Davies (2001), will be gradually outcompeted in the presence of beech but have been favoured by centuries of woodland management. Pasture woodland should be extended to those areas that have concentrations of ancient and veteran oaks, where young secondary woodland and conifer plantings should be cut and grazing with cattle restored. Savernake Forest is an SSSI for its native deciduous woodland parts and unimproved grassland glades (905 ha) and on the basis of its high diversity of lichens and fungi, as well as of invertebrates associated with abundant dead wood. Natural England considers the SSSI to be in an "unfavourable recovering condition" (Marren, 2014) because even if no further forestry felling and planting were to occur, natural succession would increase tall forest and phase out the remains of pasture woodland with its ancient oaks and pollard beeches.

Staverton Park in Suffolk

Staverton has been listed as a deer park since 1306 according to Cantor (1983), but an earlier mention dates

from the 1260s, when it was seized by the Crown from Hugo le Bygot, Earl of Norfolk (Rackham, 2003b). It is situated on the Suffolk Sandlings near the village of Butley along the road to Orford. The medieval deer park had a more or less rectangular shape with rounded corners. It was bounded in the south by the Woodbridge to Orford road and in the north by the Butley River. The eastern park pale coincided with a parish boundary, and in the west, the limits of the park ran where the present conifer plantations begin. This shape is still visible from the air as well as on the OS maps (both nineteenth-century and current) and indicates a park of *c.* 150 ha.

Woodland was already scarce on the Sandlings in Anglo-Saxon times and it is likely that the park was established in the only wooded part of the manor of Stauertuna, which could support 30 swine as recorded in Domesday Book in 1086 (Peterken, 1969). It was therefore already pasture woodland at the time of imparkment for deer. A map by John Norden of 1600–01 shows open areas surrounded by parkland trees where there are now arable fields; given the type of soil, these areas were presumably heathland when this map was made. This may have been the situation long before the seventeenth century; in Domesday Book, there were 80 sheep recorded for the manor. An unpublished study of maps and GIS data by Gary Battell has demonstrated that, apart from a core area, tree cover has fluctuated considerably over time since 1600. By 1820, when the first OS survey was made, the open areas had attained their present extent and only some of these were marked as heathland. Heath reclamation of the open areas was nearly complete by 1846, but later maps show a return to heath in several fields, until by the end of the nineteenth century there was no cultivated land in the park (Peterken, 1969). By 1846, several areas that had been open on Norden's map were covered in so-called 'ancient' oaks, the implication being that these trees at least are not older than 400 years. Three sections of the ancient park woodland remain: 'The Park', which is the central core of oak pasture woodland (55 ha); 'The Thicks', a denser oak-holly woodland section to the south of 'The Park' (26.5 ha); and 'Little Staverton', west of 'The Thicks' (4 ha). The remainder of the medieval park is now again arable fields. Nevertheless, most of the perimeter is still recognizable on the ground as an eroded bank and ditch; along the northern boundary this separates the park from marshy meadows along the Butley River. Staverton Park lies on Pleistocene sands, with the first 25–30 cm windblown material and sand with gravel deeper down. No podzol profile has been found under the core wooded area, indicating that there was never heath here and that the tree cover has been permanent from 'wildwood' times. The soil of the arable fields is disturbed, but podzol was present in a remnant of heath just south of the road (Peterken, 1969). This poor acid soil and the lowest annual precipitation in Britain (550–600 mm) have determined the vegetation as oak-birch woodland.

In 'The Park', free-standing pedunculate oak pollards dominate, accompanied by birch, holly, rowan, hawthorn and scattered elder, often under the largest oaks. The ground layer is dominated by bracken and here and there brambles, but grassy glades and rides are also present. Both bilberry and heather, typical for oak-birch woodland and/or its degradation to heath (Van der Werf, 1991) are absent. Instead, the presence of grasses such as creeping soft-grass (*Holcus mollis*) and cock's foot (*Dactylis glomerata*) in the glades and rides indicate the input of nutrients, possibly from the surrounding fields (see below). In 'The Thicks', the tree layer is similarly composed, but with a nearly closed canopy, and here some areas are dominated by large hollies. The increase of holly to local dominance appears to have begun in the eighteenth century, perhaps when its use for winter fodder declined. Some of the hollies are considered to be the largest in England.

Staverton Park has perhaps the greatest concentration of ancient oak pollards in Europe. The ATI had recorded only 42 ancient and veteran oaks with girths of between 3.00 m and 7.00 m by January 2014, and at that time, these records were evidently very incomplete. Oaks may grow very slowly here, given the ecological conditions, so unlike elsewhere, a 4.50 m oak could be 'ancient' in age as well as appearance. Estimates of the densities of the 'mature oaks' cited in Peterken (1969) give 49 ± 29 oaks per ha in 'The Park' and 66 ± 27 oaks per ha in 'The Thicks'. This would give a total of *c.* 4,300 mature oaks in Staverton Park, which seems a highly implausible figure for such a small area. An attempt to count tree crowns on the aerial image in Google Earth gave 1,465 for 'The Park' (27 trees per ha), 942 for 'The Thicks' (36 trees per ha) and 206 for 'Little Staverton'. This includes holly, birch and rowan, with holly probably accounting for half of the trees in 'The Thicks' and birches most common in 'The Park' (35% of all trees). An 'ancient tree survey' of Staverton Park in the mid 1990s (Sibbett, 1999 and Excel spreadsheet of this survey) counted 2,854 living oaks in 'The Park' and 'Little Staverton' ('The Thicks' was not surveyed), but this survey included all oaks whatever their size (lowest size class <20 cm diameter). In order to give an estimate of the number of ancient oaks, it is necessary to exclude all oaks below a certain girth as improbably ancient (ancient here meaning >400 years old; see also Chapter 2). I have taken a minimal size of 4.50 m girth for an ancient oak in

Figure 9-25. The Cathedral Oak in Savernake Forest

Staverton (see Table 9-2). When using this limit, the number of truly ancient oaks in Staverton Park according to the survey is 258, excluding those in 'The Thicks'. Some of these are not true pollards but maidens or natural pollards. No oaks were measured in the ATI with a circumference greater than 7.00 m, but the English Nature survey gives four oaks of between 7.30 m and 8.20 m. This inconsistency is probably due to the fact that the English Nature measurements were made at 1.20 m, 30 cm lower than is recommended by the ATI. Some age estimates based on ring counts were carried out by Peterken (1969), but as he acknowledged, all pollards are hollow and the estimated ages (up to 420 years in 1969) were obtained from smaller than average oaks. Peterken concluded that the uneven ages of the trees are evidence of natural generation, not planting. This question is no longer in dispute: here as in other medieval pasture woodland, the ancient oaks are all wild-growing oaks. Planting of oaks in Staverton Park began in 1949 to replace the natural death of the old oaks and is still ongoing. Deer are still held within a moveable fence, but there are also free-roaming deer that come in from the surrounding forestry plantations. More recently, other grazing animals have been introduced, in particular cattle and some sheep.

Staverton Park is one of the most important sites for ancient oaks in England; it contains an SSSI and an SAC of 81.5 ha for this reason. On the arable fields in Staverton Park, pig farming has been undertaken in recent years, with massive input of animal feed to virtually immobile pigs. Crops are raised on other fields. The perpetually bare ground on the pig farm fields means that animal waste is not taken up by crops or vegetation; it either percolates to groundwater or, in dry weather, the

resulting nutrients (phosphates and nitrates) are blown off with dust and in part end up in the pasture woodland of the park. This may be the reason why we find the grasses mentioned above instead of heather and bilberry. The park and the surrounding conifer plantations and farmland are privately owned by a single landowner. The pasture woodland parts of the park are managed on behalf of the owner by Natural England. It is highly desirable that farming, and in particular pig farming, be relocated and

the fields within the medieval park restored to extensive pasture, where grassland and/or heathland will offer opportunities for the spontaneous regeneration of oak woodland, which is apparently now virtually absent in Staverton Park (Peterken, 1969; Rackham, 2003b). It is also notable that, while Staverton Park was right in the path of the hurricane of 16th October 1987, few ancient trees were blown over or even heavily damaged. Not very much regeneration is required to replace these ancient, squat pollards, just a few hundred oaks to survive in a few hundred years.

Windsor Great Park in Berkshire and Surrey

Windsor Great Park was first mentioned as a royal deer park in 1132 and in medieval time was the largest deer park in England. Large sections of the medieval park pale can still be traced on the ground, but its southern boundary has been lost in the landscaping of Virginia Water and surrounding plantations. In other parts, for example around Flemish Farm, there are various traces indicating changes made to the boundary in medieval times. The present deer park, surrounded by a wire fence, occupies the northern part of the medieval park from Queen Anne's Gate to around Snow Hill, but excludes sections west of the A332, and is some 500 ha in extent. In this site description, I include the current deer park, plus adjacent areas which have retained a parkland landscape without major alterations. This site boundary follows the ancient park pale as indicated on the current OS map from Queen Anne's Gate in a south-easterly direction to Cookes Hill (where it is lost), thence to the northern boundary of the Savill Gardens and westward to Cumberland Lodge and the Hollybush Cottages, following the road to Sandpit Gate near where the park pale remains are met again leading north to Ranger's Lodge. Here the park pale divides, but we leave Flemish Farm out (there are no ancient oaks there) and include the Cavalry Exercise Ground; the boundary is now marked by an ancient moat near Spital and thence comes back to Queen Anne's Gate. Also excluded from the area described in this account, as it does not historically belong to Windsor Great Park, is the area around Cranbourne Tower; it is described above as Cranbourne Chase or Park.

Figure 9-26. Ancient and veteran oaks in Staverton Park, Suffolk

With few exceptions, this area contains all of the ancient and most of the veteran oaks of the medieval park and falls within the ancient park boundary. I also exclude the farms of Windsor Great Park, the forestry plantations and the landscaped gardens surrounding Virginia Water in the southern half of Windsor Great Park, also known as The Royal Landscape. How far the medieval park extended southward is difficult to establish, but evidence of a lodge ('Manor Lodge'), which existed *c.* 1240, has been found at a place called Manor Hill just north of Virginia Water (Roberts, 1997). A few ancient oaks and several old stumps remain in these parts, but almost all that may have existed have been lost in the changes made to the landscape since the eighteenth century.

The history of Windsor Great Park before the reign of Elizabeth I (1558–1603) is shrouded in legend, as there are no surviving written records (Menzies, 1864). In 1240, Henry III defined the borders of 'The Park' at Windsor, but it is unclear if this coincided with the building of the park pale as presently traced. During his reign, the royal residence and hunting lodge was the manor house (or palace) at Old Windsor. Nearby in the park at Bear's Rails is the moat of another house, Wychamere, residence of William of Wykeham who extended the fortress of Windsor Castle for Edward III in the 1350s. Another ruined moat is situated in the north end of the park near Spital to the west of Queen Anne's Gate. It is probable that there was effectively no defined border between 'The Park', Cranbourne Chase and Windsor Forest until the former was surrounded by a closed park pale; all three were used for the royal chase. The traces of two parallel park pales from Queen Anne's Gate to Ranger's Lodge, and what seem to be the remaining trees of an "avenue of oaks dating from *c.* 1450 in the reign of Henry VI" (Smith, 2004), may in fact mark a broad ride from Windsor Castle cut through the wooded parts of Windsor Park until it reached the open heath on the sands of Cranbourne Chase, effectively dividing the park into two parts. Even after imparkment, Windsor Great Park with more or less its present extent was large enough for the chase and thus very different from an average medieval deer park.

The landscape is gently undulating at altitudes between 25 m in the north and 80 m in the south. Its large open deer lawns are apparent on the earliest detailed map of the park made by John Norden in 1607 (reproduced in Menzies, 1864). These spaces were either acid grassland or heath on the higher and drier hills. Areas in which solitary trees were depicted were probably pasture woodland with ancient and veteran oaks. Small woods and coverts are also evident, but the present coverts show no evidence of coppice woods.

Cumberland Lodge (1652) and Royal Lodge (present by 1662) became the residences of the Ranger and Deputy Ranger of Windsor Great Park in the mid-seventeenth century. Within the area here described, the most prominent landscaping feature is the Long Walk, a 4.26 km long avenue originally planted with a double row of elms in the 1680s that stretches from Windsor Castle to Snow Hill, 70 m a.s.l. Another avenue, but not paved, is Queen Anne's Ride, which leads from Queen Anne's Gate to the Queen's Statue (on the border of our site) and beyond; it is longer than The Long Walk. The Village was built in the 1930s in order to house Crown Estate staff.

Throughout this area of the park are coverts of variable shapes and sizes, which nearly all date from the seventeenth century or later, except the older enclosure of the moat near Bear's Rails. A stream, Battle Bourne, runs through the deer park and is dammed to create ponds near and in The Village. Other ponds, such as Ox Pond, Prince of Wales Pond, Rush Pond and Bear's Rails Pond, are older and except for the latter do not result from the damming of streams. Although some parts of the parkland have unimproved acid grassland (as in an area east of Snow Hill) or even heather (as in the patch near Cumberland Lodge), much of the grassland has been 'semi-improved' or at least is regularly mowed, as can be seen from aerial Google Earth images. As a consequence, only a few areas have nest hills of the yellow meadow ant, a sure sign of undisturbed acid grassland on sandy soil.

The ancient and veteran oaks occur throughout the site, but with concentrations in some areas. Of particular interest as ancient pasture woodland is the wooded area now called Bear's Rails; here are many ancient and veteran oaks (most are maidens) with in-filled secondary woodland of mainly birch. The ditch and bank of the medieval park pale run through this area, so part of it was originally outside the park. A programme of removing exotic trees and rhododendrons is ongoing in this section of the deer park. The largest living oak in Windsor Great Park (as here defined) is known as the Signing Oak (9.72 m girth) and stands near The Gallop. Two ancient oaks have girths of 8.95 m and 8.84 m (data from the ATI), 12 oaks are between 7.00 m and 8.00 m in girth and 42 oaks are between 6.00 m and 7.00 m in circumference. A total of 87 oaks that I have selected as ancient or veteran have been measured with girths of between 5.00 and 6.00 m, giving a total of 144 ancient and veteran oaks ≥5.00 m in girth in the northern half of Windsor Great Park as here circumscribed. This site covers about half of the 2,000 ha of Windsor Great Park and is therefore comparable to the 1,000 ha of Richmond Park. The ancient and veteran oak numbers of

Girth classes	Richmond Park	Windsor Great Park
9.00–10.00 m	0	1
8.00–9.00 m	0	2
7.00–8.00 m	3	12
6.00–7.00 m	22	42
5.00–6.00 m	106	87
Total numbers of oaks ≥5.00 m	131	144

Table 9-1. Comparison of ancient and veteran oaks by size class in two royal parks.

the two parks then compare as shown in Table 9-1.

Although the total numbers of ancient and veteran oaks in the two parks are similar, these trees tend to be larger in Windsor Great Park, where a greater proportion of the oaks exceed 6.00 m and 7.00 m in circumference; the two highest girth classes are not represented in Richmond Park. In both parks, most oaks ≥5.00 m are pollards, but in Windsor Great Park, more oaks in the smallest class are maidens than in Richmond.

The soils of Windsor Great Park are variable from predominantly London Clay and silt deposits in the northern parts to Thames River terrace gravel and sands and in central and southern parts with cappings of poor Bagshot Sand and local podzolisation. The clay areas may give better growing conditions than the poor sands, but given the great individual variation in tree growth that occurs, it is difficult to correlate this to the trunk diameters of oaks, especially as we do not know their ages. The ancient and veteran oaks are subject to documentation (with >7,000 'significant trees' tagged across the whole estate, Alexander & Green, 2013), but their monitoring and active management is restricted to addressing health and safety issues in localities accessed by the public. In recent years, special measures to protect and preserve outstanding specimens have increased, while other areas of former pasture woodland, such as Bear's Rails, are being cleared of conifer plantations and other exotic trees. More details on current management are given in Alexander & Green (2013). Windsor Great Park is an NNR, a SAC and an SSSI, the latter in combination with parts of Windsor Forest. The importance of especially the large ancient oak pollards for biodiversity is acknowledged (Alexander & Green, 2013). The park has one of the richest invertebrate faunas associated with ancient trees in England, as well as several nationally rare species of fungi (see Chapter 10). Windsor Great Park is owned by the State and managed by The Crown Estate.

Woodend (Shute) Park in Devon

Shute Deer Park, now known as Woodend Park, is a partly wooded park west of Shute. A manor house in the form of a simple hall house existed here around 1380 and was much enlarged in the second half of the sixteenth century ('Old Shute House' or 'Shute Barton'). Cantor's (1983) gazetteer of medieval parks does not mention a park here, but according to the Ancient Tree Forum cited at shute.co.uk/shute_places.html, the deer park "dates back to the twelfth century". In Domesday Book (1086), the hundred of Colyton ("Culitone") was listed as "King's land", and the park is reputedly a medieval royal deer park. This could explain its absence in the registers, as a royal park would have needed no license to impark. The park could also be late medieval. When the property fell to the Grey family (Marquesses of Dorset) after the Wars of the Roses (ending 1487 with the Battle of Stoke), they may have started a deer park here. They also had owned Bradgate Park in Leicestershire since 1445.

On the nineteenth-century OS map, Shute Deer Park is partly parkland, partly farm fields and woods. Parts of its original boundary can be recognised from aerial photography as wood banks or wooded lanes. The park boundary at the western end probably followed a spur track from Pacehayne Lane around the hill leading to Chandler's Hall. Along the park's southern boundary, on the hillside above Pennyhayes Farm, are the well-preserved remains of a medieval park pale: a (now wooded) earth and stones bank with a ditch on the inside, stretching for several hundred metres (see Figure 4-1). The old pale is also visible along the western edge and in part of the north. The park may well have extended from the old manor house (Shute Barton) to the hill in the west that has 'Druid's Circle' on or near its summit. The parkland on the hill is known on the nineteenth-century OS map as Shute Deer Park. To the north and east of the hill, the former park boundary is obscured by the creation of farm fields with their hedges. Most of the park is situated on Triassic mudstones, which form a clay soil, but on the hill are deposits of Lower Cretaceous Greensand. The parkland on the hill seems to have been compartmented a long time ago as many of the ancient and veteran oaks stand in ancient and fragmentary hedgerows, but some areas of pasture woodland remain unaltered. A separate section of parkland, known as Shute Lawns, lies between 'Old Shute House' and 'New Shute House' and reaches as far east as Haddon Corner. Here occur five ancient oaks with girths of between 5.70 m and 7.02 m.

In the entire area of what I think could have been the medieval deer park there are 42 ancient and veteran oaks with girths ≥5.00 m, of which seven exceed 7.00 m in circumference. A majority of these are typical broadly spreading parkland oaks, a few are probably pollards. The largest ancient oak, with a circumference of 10.04 m, is situated on the hill and known as King John's Oak. This name is part of typical folklore: if this tree had existed during the reign of King John (1199–1216), which is not impossible (it would now be at least *c.* 800 years old), it would then have been at most an unremarkable immature tree. No medieval kings, nor anyone else at that time, planted native oaks; an activity that was not undertaken before the reign of Elizabeth I (1558–1603). As oaks were present already when medieval parks were created and not planted for the occasion, the King John's Oak would likely

Figure 9-29. Ancient oaks in Woodend Park (Shute Park), Devon

have been one of these. It is now almost dead. There are also veteran trees of common lime (*Tilia* x *europaea*), sweet chestnut, beech, ash and field maple, but oaks dominate. Hawthorns are common in some parts on the hill, for example in hedgerows, but they are not ancient. In the lower part of the park, oaks have been planted, these are now mature trees. A few old Scots pines have been planted on the top of the hill, but apart from this, there is little or no evidence of a landscaping effort. Farming (Woodend Farm) has created the largest changes to the ancient park landscape, with both arable and grassland fields separating the two park sections. Much of the grassland on clay in the lower sections of the present park is improved and some fields are ploughed for other crops and then reseeded with English rye grass (*Lolium perenne*). Unimproved grassland, partly invaded by bracken, is found mostly on the western

slopes of the hill, now partly belonging to Pennyhayes Farm. It has abundant bluebells flowering in spring and is rich in fungi, among which are wax caps (*Hygrocybe*). No deer have been kept in the park since *c.* 1947 and grazing is by farm animals. This park was described by Harding (1980) as important pasture woodland for the conservation of wildlife. 'Old Shute House' is owned by the National Trust, but 'New Shute House' and its surroundings as well as Woodend Park are in private ownership. The park is designated as a Devon County Wildlife Site.

Analysis

There are 23 sites in England which meet (most of) the criteria mentioned in the introduction of this chapter. They are listed and compared in Table 9-2.

Deer parks account for 20 of the 23 sites in this analysis and are thus the most important type of landscape and land use associated with ancient and veteran oaks, a conclusion that is also arrived at from different analyses

Site	Oaks ≥6.00 m in girth	Oaks ≥7.00 m in girth	Medieval deer park	Later deer park	Royal forest or chase	Modified in the eighteenth/ nineteenth century	Unimproved grassland present	Area in ha	SSSI/ NNR/ local reserve	Plan
Ashton Court Park	14	9	√	√		√	√	340	SSSI	√
Ashtead Common	>60*	1					√	200	SSSI/NNR	√
Aspal Close	28*	1	√				√	12	Local	√
Brampton Bryan Park	24	11	√	√		√	√	191	SSSI	
Birklands (Sherwood Forest)	40	10			√		√	228	SSSI/NNR	√
Bradgate Park	±40*	2	√	√		√	√	332	SSSI	√
Calke Park	20	4	√	√		√	√	243	SSSI/NNR	√
Chatsworth Old Park	63	17	√	√		√	√	72.5	SSSI	√
Cranbourne Chase (Windsor Great Park)	19	7		√	√		√	±40	SSSI	√
Grimsthorpe Park	16	4		√			√	110	SSSI	√
Hatfield Park	±15	6	√	√		√		±250		√
Hamstead Marshall Park	21	8	√	√			√	220		
High Park (Blenheim)	±100	29	√				√	125	SSSI	√
Ickworth Park	33	12	√			√		370	Local	√
Kimberley Park	20	10	√	√		√	√	±40	Local	
Moccas Park	54	22	√	√		√	√	137	SSSI/NNR	√
New Forest	31	8			√	√	√	14,600	SSSI/NNR	√
Parham Park	21	8		√		√	√	120	SSSI	√
Richmond Park	31	4		√		√	√	1,000	SSSI/NNR	√
Savernake Forest	84	14	√		√	√	√	1,820	SSSI	√
Woodend Park (Shute)	15	7	√			√	√	42.4	Local	
Staverton Park	258*	4	√	√		√	√	85.5	SSSI	√
Windsor Great Park	71	19	√	√		√	√	1,000	SSSI/NNR	√
Total	**1,060**	**217**	**16**	**15**	**4**	**16**	**21**	**21,400**	**21**	**19**

in Chapters 4 and 5. Sixteen parks are certainly or most likely medieval in origin, and nine of these are still (in part or whole) deer parks. Four deer parks are of a later date, of which two, Cranbourne and Grimsthorpe Park, are no longer deer parks. Royal Forests and chases account for four sites and just one, Ashtead Common was neither a deer park nor a Royal Forest or chase. Only eight sites have mostly 'escaped' landscaping or other major modifications: Ashtead Common, Aspal Close, Birklands, Chatsworth Old Park, Cranbourne Chase, Grimsthorpe Park, Hamstead Marshall Park and Blenheim High Park. Staverton Park is often thought of as essentially unmodified, but its more open areas with heath and scattered trees have been converted to arable farmland, with abrupt boundaries to the remaining pasture woodland. The modifications vary in intensity and extent; if they are extensive, the hectares given in Table 9-2 refer to the less modified parts only. Sometimes the modification is diffuse, with grassland improvement alternating with unimproved grass or bracken in which

Table 9-2. The most important oak sites in England.
*The pollard oaks in Ashtead Common and Aspal Close are smaller than elsewhere and here include trees ≥4.00 m in girth; the oaks in Staverton Park include those ≥4.50 m in girth; and the oaks in Bradgate Park include those ≥5.00 m in girth. The actual number of recorded oaks ≥6.00 m in girth in the 23 sites per 21st April 2016 was 695.

the parkland trees occur, or with enclosed or unenclosed plantations distributed randomly in the parkland. In such cases the entire area of the park is given.

Of the three Royal Forest sites, only Birklands has almost escaped plantation forestry and is here listed as not modified. Cranbourne is listed here as a chase, although the site has been a Royal Forest and a deer park in the past. A medieval deer park was also present in Savernake Forest. The part of the New Forest included in Table 9-2 includes only open pasture woodland; this is an area of 14,600 ha, much larger than any of the sites discussed in this

chapter. By contrast, Aspal Close is a mere 12 ha, and size discrepancies such as this make direct comparisons of the numbers of ancient and veteran oaks between the 23 sites less meaningful. Of the eight sites with ≥10 ancient oaks ≥7.00 m in girth, two are large with ≥1,000 ha; it is notable that the very large New Forest has fewer than 10 great oaks. The other six sites are 40–370 ha, and five of these were deer parks at least during parts of their history. Although the sample is relatively small, it seems that greater densities of very large ancient oaks have been retained in deer parks than in other forms of pasture woodland, even where the ancient landscape remained relatively unaltered. This could of course be just a function of the statistics already given above: there are more deer parks with ancient oaks than other kinds of sites. Alternatively, it could indicate that oaks in deer parks had a higher long-term survival rate, growing to become very large more often. These questions lead us into how and why ancient oaks in England have survived in such numbers, the subject of Chapter 7.

In identifying the most important sites for ancient oaks in England, a list of sites was first compiled on the basis of estimated numbers of oaks or numbers of oaks recorded by the ATI of the Woodland Trust. No tree size limits were set initially, but oaks that were recorded as ancient and veteran were all included when compiling the list. As a second step, the sites on the list were viewed on the OS maps (nineteenth-century and current) and Google Earth aerial imagery to investigate the probable nature of the sites and their boundaries. All of the sites, a total of 35, were visited at least once and several more than once.

Using the criteria outlined at the beginning of this chapter, 12 sites were considered and rejected as they fell short of the standards that I set for a most important site for ancient oaks. Several of these have ancient oak numbers similar to those at the 23 sites that were accepted, which are described in detail above and summarised in Table 9-2. This is not a contradiction because I considered criteria other than the presence of ancient and veteran oaks, as explained above. In any other country of Europe, these sites might qualify as 'most important' for ancient oaks, but in England, there are many sites with substantial numbers of such oaks and to list them all as 'most important' would devalue the list. These sites were rejected because their ancient landscape has been altered substantially, often to the detriment of the ancient and veteran oaks. I am interested in the conservation of an ancient landscape and

ecosystem: pasture woodland dominated by ancient and veteran oaks; the presence of the oaks alone is insufficient in this context. The sites listed below were considered, visited and rejected, chiefly for the reasons given in brackets. All are briefly described in Chapters 4 or 5; some again in Chapter 11.

- Attingham Park, Shropshire (agriculture)
- Brocket Park, Hertfordshire (golf course)
- Cornbury Park, Oxfordshire (landscaping)
- Croft Castle Park (agriculture and forestry)
- Cowdray Park, Sussex (golf course)
- Duncombe Park, North Yorkshire (agriculture, forestry)
- Forthampton (Corse Chase), Gloucestershire (agriculture)
- Kedleston Park, Derbyshire (golf course and landscaping)
- Lullingstone Park, Kent (golf course)
- Oakly Park, Shropshire (agriculture and landscaping)
- Packington Park, Warwickshire (golf course)
- Stoneleigh Park, Warwickshire (golf course)

It appears that the construction of an 18-hole golf course (or two) among the oaks was a reason for rejection of half of these 12 sites; four were once medieval parks and two Tudor deer parks. In all cases, ancient and veteran oaks are still present. It is difficult to know how many (if any) oaks were destroyed to lay out the fairways and tees, although very few occur outside the roughs on the actual fairways, for obvious reasons. On some sites, I found the distribution and density of oaks to be such that it is difficult to envisage a natural distribution that would have avoided the intensively managed course. There is at least anecdotal evidence of the removal of oaks from courses that were established earlier, before awareness of their importance for nature conservation became more general. Unless the site is protected, these ancient oaks have no legal conservation status and SSSI sites exclude golf courses. Although situations vary somewhat, much or most of the natural vegetation is removed or altered when a golf course is built; the trees may stand in the roughs but these often retain little of the unimproved acid grassland that was once present. Glades and launds in the former deer park would have been incorporated into the fairways and tees where possible. Dead wood is removed almost entirely from most golf courses, again for obvious reasons. This leaves just the dead wood inside hollowing ancient oaks where these trees are not considered a health and safety hazard. Animal grazing, except by rabbits and the occasional feral deer, is excluded, with one exception where the park deer have access. Drainage is often extensive to keep golfers and

their sport dry; but how this may affect the oaks remains to be seen, as in most cases the golf course is quite new and its effect on the oaks in a series of dry summers such as that which occurred in the mid 1970s is still awaited. My exclusion of golf courses from the list of sites that are most important for ancient oaks, it should be stressed, does not make the ancient oaks on golf courses less important and every reasonable effort should be made to preserve them.

Agriculture and modern forestry (mostly a form of agriculture with long crop rotations) are seen as the main causes for disqualification in five sites. These activities have definitely led to the removal, deliberately or as collateral damage, of ancient and veteran oaks. Where ancient oaks are still present in numbers, they mostly stand in intensive farmland, which in modern agriculture may switch from grassland to arable and back within a few years. In some cases, the ancient park is now all or mostly farmland. As we have seen in Chapter 4, such conversion has usually led to the disappearance of the ancient oaks, but, oddly, they remain in a few sites. One such site is Forthampton (now Home Farm), where many oaks are standing dead or dying. Plantation forestry with conifers has often decimated the old oaks by shading them, if they were not cut beforehand to make space for the young trees to be planted. The Forestry Commission is now attempting to rescue such ancient oaks by haloing them, but only one of the sites here rejected is managed by that organisation. In most cases, haloing comes almost too late, as in Croft Castle Park, where the National Trust owns the land but the conifer plantations are under a long-term lease to the Forestry Commission (Brian Muelaner, pers. comm. 2013). For sites where much farming and/or forestry is present, I had to decide not only how many living oaks remain on a site, but also how much, if any, of the ancient pasture woodland remains or can be restored on the site.

The third reason for rejection of a site is 'landscaping'. Almost all ancient parks and even Royal Forests or chases have been touched by the 'improvements' of the seventeenth, eighteenth and nineteenth centuries. The question is how much has this activity altered the landscape and vegetation of the ancient pasture woodland and have the ancient oaks been removed or retained. I have also considered whether a return to the old situation is possible, for instance by removal of planted exotic trees or by undoing the improvement of the grassland. In cases where it is obvious that the landscaping was extensive and has cultural and historical value that prohibits a return to an 'unkempt' deer park with only native local trees, the site does not qualify even where ancient oaks have been retained. We cannot now destroy Lancelot ('Capability')

Brown's or Humphrey Repton's creations and if ancient oaks are present in such landscapes (as they are for example at Cornbury Park and Kedleston Park), they are important and should be protected, as most of them are. But very much the same factors that affect golf courses are in place at these landscaped sites, making them less valuable for ancient oaks and their historic context and for ecology.

Also excluded from the 23 most important sites is a site in Suffolk, near the village of Risby, which has some 112 veteran pollard oaks and is known as Oak Pin on the OS map. The largest oak there was measured by me on 24th

April 2014 with a girth of 5.44 m at 1.30 m above the ground. These oaks are reminiscent of those in Aspal Close and Ashtead Common but the setting is very different. They stand in a meadow of limited size on Breckland Sands, now grazed by sheep. It is unclear how old they are, but none are ancient and it seems likely that they were planted a few centuries ago. Rackham (1976) presented a photograph of this site (plate XIX) and commented that such farmland oak fields were a "speciality of the Breckland edge". A brief description of this site is given in Chapter 5 under 'Commons'.

10

The biodiversity of ancient oaks

Introduction

The significance for biodiversity of the association of many different organisms with ancient oaks, or with ancient trees in general, has been commonly acknowledged only recently, arguably with the launch of the Veteran Tree Initiative in March 2000 by several organisations involved with nature conservation (Green, 2001; but see the Swedish experience, Chapter 7). This despite the fact that inventories of fungi, lichens and invertebrates had given accumulative evidence of this important habitat for at least a century before this date. It is now undisputed that ancient trees, and particularly oaks, collectively support greater biodiversity than any other type of habitat in Europe. In this chapter, the significance of ancient and veteran oaks for the biodiversity of fungi, lichens and invertebrates is explained by three experts in these fields, Martyn Ainsworth (fungi), Pat Wolseley (lichens) and Keith Alexander (invertebrates).

Non-lichenised fungi

A. Martyn Ainsworth

Fungal lifestyles and recording

Of the three groups of oak associates considered in this chapter, the non-lichenised fungi, those without photosynthetic partners and hereafter referred to as 'fungi' in this section, are probably the hardest to detect and assess in terms of population and community structure. An absolute requirement for sunlight ensures that their lichenised counterparts (lichens) are relatively conspicuous components of biodiversity. Although they may not always be identifiable until examined 'back at the lab.', they are nevertheless quite amenable to field survey and lend themselves to studies of distribution and population trends. Some of the oak-associated invertebrates may demand more effort to locate within woody tissues and similarly require specialist study for their identification, but nevertheless a similarly informative picture of their populations and distribution is emerging. The challenges involved in knowing exactly where fungal species are in their natural habitats, how long they persist and what they are doing in response to environmental change have undoubtedly slowed progress. Unfortunately, much fungal recording is still carried out on a casual basis and there are fewer dedicated site surveys relating fungi to habitats, ecological continuity and levels of pollution than there are for invertebrates and lichens.

Although once regarded as simple or 'lower' plants, fungi are now generally accepted as members of an independent kingdom (or kingdoms), and modern DNA studies support their closer evolutionary links to animals rather than to plants. Oak-associated invertebrates and fungi have to obtain their carbon from living or dead parts of the tree, but differ in their method of feeding. The cobwebby bodies (mycelia) of most fungi consist of masses of woven microscopic tubular filaments (hyphae), which extend at their tips, branch and fuse together into a living network. These networks spread within and around carbon sources and extend into the surrounding environment to explore and forage. They can enter and spread through woody tissues, digesting complex molecules such as cellulose and lignin and absorbing the breakdown

products to fuel further territorial expansion. They provide a hidden transport system that shuttles nutrients between sites of capture, storage and use. Mycelia are fundamental to the absorptive feeding behaviour of fungi, to their establishment, to their ability to control territory, to their mating interactions and, not least, to the restructuring, relocation and recycling of nutrients in natural ecosystems.

Mycelia are rarely conspicuous despite their ubiquity and potential longevity (sometimes living for centuries or more) as they are almost always concealed within soil or plant tissues such as living leaves and dead wood. Even when exposed and visible, very few mycelia can be identified to species unless culturing or DNA sequence analysis is available because they are just so similar to each other. Almost all the key identification characters used in contemporary identification guidebooks are to be found in the reproductive structures: the fruit bodies of larger species (macrofungi) and the microscopic sporulating structures of microfungi. To be identified therefore, a fungus usually has to be found when in reproductive mode. For the fungus, this necessitates acquisition of sufficient resources, receiving and responding to poorly understood environmental cues, and exiting from a food source to produce spores. For many larger fungi, the process also involves recognising, fusing and mating with a compatible partner. Fruiting is an infrequent, irregular, unpredictable and, for the fungus hunter, often a highly frustrating event. For most of their lives, fungi exist as hyphal networks and yet this stage in their lifecycle is the most difficult to detect, identify and quantify.

Fungal collection, identification and recording have flourished in Britain since Victorian times and it is now routine to identify, whenever possible, not only the target fungus but also the associated organism. In most cases this is an attempt to specify what the fungus is feeding on. This can be straightforward, as in the case of recording the species of tree whose attached leaves or fallen fruits are being blemished or whose trunk is being internally decayed and sculpted into an interconnected cave-system of microhabitats. On the other hand, it might be merely an inference that a mushroom is connected by underground hyphae to microscopic fungal 'gloves' (mycorrhizal fungi), which envelope living root tips where nutrients are traded between fungus and tree. Clearly such inferences are likely to be relatively sound when recording in monocultures but less so in mixed woodlands where the nearest tree, often recorded as the associated organism by default, is not necessarily a mycorrhizal partner of the target fungus. Although such inferences are likely to remain an inherent part of nationwide fungal recording, below-ground investigations that are based on the identification of fungal DNA extracted from individual mycorrhizal root tips are radically changing our perceptions of fungal community composition. These limitations and the unique constraints and characters of fungal lifestyle diversity should always be borne in mind when interpreting historic fungal records, distributions and habitat preferences.

Oak-associated fungi

Oaks are generally regarded as having more associated species of wildlife than any other native tree species in England. Field mycologists would probably concur that a similar claim could be made for the fungal component of oak-associated biodiversity. Fortunately, over two million accumulated records are now in national fungal recording databases and can be interrogated. Searching the Fungus Conservation Trust's online CATE2 database, for example, yields some figures with which to test this claim. The records therein do indeed bear out the view that, in Britain as a whole, there is not a tree genus with more fungal associates than Quercus, and almost all of the records are on the native species, Q. robur and Q. petraea. Native beech, however, shares the lead. Quercus and Fagus are each associated with a total of c. 2,600 fungal species in Britain, a figure that does however include some lichens and slime moulds. This constitutes around one sixth of all fungi that occur nationally. Of course the figures are necessarily estimates; a few fungi are likely to be misidentified, as are some of the fragments of decaying wood supporting them or their associated mycorrhizal trees, and some of the species might only have been recorded on introduced oaks. Nevertheless, these represent the most readily-accessible data currently available but, in view of the fact that only fungi in a visibly reproductive state are being recorded, they must be regarded as underestimates and interpreted with caution. A selection of fungi typically associated with oak is listed in Spooner & Roberts (2005).

Oak-associated mycorrhizal fungi

Oak roots, like almost all tree roots in temperate and boreal countries, have root tips that are not in physical contact with the soil but interact with it through fungal intermediaries – they have mycorrhizas (fungus roots). Based on fruit-body data, oaks are particularly rich in mycorrhizal species, upon which they depend for supplies of water, nitrogen and phosphorus, and a few genera are particularly well represented. For example the CATE2 data indicate that the three mycorrhizal mushroom genera *Russula*, *Cortinarius* and *Inocybe* have, respectively, 129, 105 and 72 oak-associated species in Britain.

DNA analysis of oak mycorrhizal root tips from sites across Europe has shown that these fungal communities are highly diverse, but that their species richness is sensitive to nitrogen pollution from man-made sources (Suz *et al.*, 2014), recalling a trend familiar to lichenologists. This technique was then used to analyse and compare the below-ground mycorrhizal species diversity (sampled over a five-month period) in oak-dominated old-growth semi-natural stands in the New Forest (which had been in continuous existence for more than 1,000 years) with that in similar nearby (less than 1 km distant) areas that had been clear-felled and replanted (with oak dominant) between 160–204 years ago (Spake *et al.*, 2016). Similar mycorrhizal species richness and community composition were found in both woodland types. This offers the hope that conservation of mycorrhizal fungi dependent on dwindling old-growth oak-dominated habitat could be assisted by setting aside oak plantations, at least those planted *c.* 180 years ago on former old-growth sites in the near vicinity. Although this study involved old-growth oak woodland, it was not concerned with the fungi associating with old trees *per se*. It would be very interesting to compare the baseline results obtained by Spake *et al.* (2016) with those obtained from deliberate sampling from the root tips of ancient oak trees. Personal observations over 20 years in Windsor Great Park have revealed that the oldest oaks were not associated with the greatest mycorrhizal species diversity, based on fruiting records alone, nor did there seem to be a suite of fruiting mycorrhizal species which were characteristic of the oldest trees. Furthermore, oak-associated mycorrhizal fungal assemblages that are of current conservation concern, such as warmth-loving boletes and tooth fungi, were similarly not found fruiting near the oldest oaks at Windsor. The bolete hotspots were observed among trees approaching or in early maturity, whereas mycorrhizal tooth fungi, which are found fruiting in association with oak in the New Forest, are mysteriously only seen with sweet chestnut on the Windsor Estate.

Oak-associated saprotrophic fungi

Although there are many fungal saprotrophs (feeding on dead tissues) associated with oak, those found fruiting on leaves and acorns are, like oak's fungal pathogens, not particularly characteristic of the oldest trees. Fruiting communities of many wood-inhabiting saprotrophs also start to develop when the tree is quite young, for example on the trunks of dead saplings, dead attached twigs and lower branches of living trees, and on fallen wood. Unlike most mycorrhizal species, wood-inhabiting fungi are amenable to routine laboratory cultivation from spores, fruit body tissue or colonised plant tissues.

Figure 10-1. *Laetiporus sulphureus* hollowing oak heartwood in Grimsthorpe Park, Lincolnshire

Figure 10-2.
Fistulina hepatica
fruiting within a
hollow oak trunk on
Ashtead Common,
Surrey. Photograph
by A. M. Ainsworth,
RBG Kew

Matching mycelia from these various sources by allowing them to recognise self and non-self when confronted in the laboratory afforded new insights into the ecology of fungi living inside trees. Indeed, some of the first studies of woodland fungal communities involved mapping the mycelial territories in shade-suppressed lower branches and young trees of oak (Boddy & Rayner 1983a; Boddy & Thompson 1983). Such communities typically comprise common decay pioneers with thin waxy or leathery fruit bodies, such as cinnamon porecrust (*Fuscoporia* (*Phellinus*) *ferrea*), *Peniophora quercina*, *Phlebia rufa*, bleeding oak crust (*Stereum gausapatum*) and waxy crust (*Vuilleminia comedens*). This work, led by decay pioneers Alan Rayner and Lynne Boddy, led to a more general reappraisal of wood decay concepts. Almost all of the oak-associated species they studied cause a white rot (involving lignin and cellulose decomposition) of sapwood. They noted that this tissue was very decay-resistant when alive and actively conducting sap, but was rapidly decayed after death. They hypothesised that a failure of water-conduction (by aeration and disruption of water columns in water-conducting tissues), was a pre-requisite for saprotrophic mycelial development and that such mycelia were not only generated by fungi arriving as spores on broken, cut or damaged wood surfaces, but also developed following the conversion of already present fungal propagules from a relatively inactive or latent state into an active exploratory mycelial form (Boddy & Rayner 1983b; Cooke & Rayner 1984). Hence the question posed by Parfitt *et al.* (2010), "Do all trees carry the seeds of their own destruction?"

Fewer oak-associated saprotrophs are specialised for life in the relatively durable and chemically hostile (rich in phenolic compounds) environment of the central cylinder of dead heartwood. Nevertheless, some are fairly common and are frequently seen fruiting on sawn or fallen larger branches; examples include oak curtain crust (*Hymenochaete rubiginosa*), forming wavy tiers of thin dark brown brackets, and clustered bonnet (*Mycena inclinata*), a white-spored mushroom with a distinctive oily and soapy smell. Trunk heartwood also supports some commonly recorded and highly conspicuous bracket fungi, such as the bright yellow or orange tiers of chicken of the woods (*Laetiporus sulphureus*) (Figure 10-1), also seen on a range of other trees including yew, and the meaty red tongues of beefsteak fungus (*Fistulina hepatica*) (Figure 10-2), which is more selective, although sometimes seen fruiting on sweet chestnut. Both of these cause a brown rot (cellulose is decomposed but brown lignin is retained) in which the wood is characteristically split along and across the grain into cubes. Since ancient oaks have massive volumes of dead heartwood available for mycelial occupation and as an increasing proportion of this heartwood becomes exposed to the air with age, both inside and outside their hollow centres, these trees also support good fruiting populations of specialist heart-rotting fungi. It is among these species that most of the fungi characteristic of sites with good populations of ancient oaks are likely to be found (see below).

Some oak saprotrophs are root-inhabitants; examples include zoned rosette (*Podoscypha multizonata*), which fruits in soil near to the trunk, and spindle toughshank (*Gymnopus* (*Collybia*) *fusipes*) fruiting where the trunk meets the soil. The latter is a common white-spored mushroom, whereas the former is a European rarity, producing arrays of pinkish concentric fans as large as a cauliflower. It has a stronghold in southern English open-grown oak sites, such as the former and current pasture woodlands of Epping Forest, the New Forest and Windsor, occasionally also fruiting on beech and other broadleaved trees. It is officially recognised as one of 61 English fungi of principal conservation importance, a so-called 'Section 41 species', and is one of the few fungi that have been mapped across Europe with around 120 known sites, of which *c.* 80% are in Britain and France (Fraiture & Otto, 2015). However, the below-ground lives of both zoned rosette and spindle toughshank are still poorly understood and it would be particularly interesting to discover whether they are restricted to dead and dysfunctional roots and how far along the saprotrophic–pathogenic spectrum they can operate.

Characteristic saprotrophs at the most important ancient oak sites

Ainsworth (2004, 2005) devised an indicator list of beech woodland saprotrophic fungi as a tool for ranking English sites against a European reference list. An attempt was made during the current project to devise something similar for fungi characteristic of ancient oaks. However, there seem to be fewer suitable indicator species for oak than were identified for beech, a greater proportion of very common species, such as those mentioned above; and fewer

site-specific survey results available, so any emerging list may be of rather limited utility.

The rarities (white-rotters) include: robust bracket (*Fomitiporia* (*Phellinus*) *robusta*), unofficially red-listed as Vulnerable (Evans *et al.*, 2006) with English strongholds at Windsor and in the New Forest, usually fruiting high on trunks and sometimes just below the point of decapitation; tufted bracket (*Fuscoporia* (*Phellinus*) *torulosa*), a southern European species fruiting low on the trunk and listed as Near Threatened (Evans *et al.*, 2006) but without any English stronghold sites; *Fuscoporia* (*Phellinus*) *wahlbergii* (Ainsworth 2008a), a tropical/subtropical species macroscopically resembling *F. torulosa* but discovered in England too recently to be assessed for conservation purposes; and *Ceriporia metamorphosa*, which forms white, skin-like poroid fruit bodies often accompanied by a profusion of bright orange, powdery asexual spores so distinct that two different scientific names were given to the same fungus.

Lying between the commonest and rarest species are the following bracket-formers: oak mazegill (*Daedalea quercina*), named after Daedalus and having a highly distinctive labyrinthine lower surface to its beige and woody fruit body, which is usually recorded on stumps and fallen wood but can also fruit much higher up on dead parts of living trunks and main branches; hen of the woods (*Grifola frondosa*), producing a short-lived, leathery profusion of grey-brown brackets at the base of the trunk with a notable smell of mouse urine; two reddish *Ganoderma* species, *G. resinaceum* and the rarer lacquered bracket *G. lucidum*, both of which produce short-lived, resin-coated fruit bodies near to the ground, those of the latter with a shiny resinous stem and those of the former with or without such a stem (Ainsworth, 2001); oak bracket (*Pseudoinonotus* (*Inonotus*) *dryadeus*), producing an initially sponge-like golden bracket notable for its copious droplet exudation, maturing into a more sombre brown corky bracket located low down on the trunk; and oak polypore (*Buglossoporus* (*Piptoporus*) *quercinus*), which is treated in more detail below.

Table 10-1 shows the 23 most important sites in England for ancient oaks, as identified in Chapter 9, scored for the fruiting presence on native oak of 16 fungi selected as being characteristic of ancient oak-dominated sites and recorded over the past 50 years in the British Mycological Society's online FRDBI database. These data were supplemented by personal records, those in Overall (2008) and those obtained directly from field recorders and surveyors (see Acknowledgements below). The fungi on the top row are arranged left to right in decreasing order of the numbers of CATE2 records on *Quercus*. Hence, if a site has fewer

	Fistulina hepatica	*Hymenochaete rubiginosa*	*Gymnopus (Collybia) fusipes*	*Mycena inclinata*	*Laetiporus sulphureus*	*Daedalea quercina*	*Pseudoinonotus (Inonotus) dryadeus*	*Buglossoporus (Piptoporus) quercinus*	*Grifola frondosa*	*Ganoderma resinaceum*	*Podoscypha multizonata*	*Ganoderma lucidum*	*Fomitiporia (Phellinus) robusta*	*Ceriporia metamorphosa*	*Fuscoporia (Phellinus) torulosa*	*Fuscoporia (Phellinus) wahlbergii*	**Species total**
The New Forest, Hampshire	Y	Y	Y	Y	Y	Y	Y	Y	Y	Y	Y	Y	Y	Y		Y	**15**
Windsor Great Park, Berkshire/Surrey	Y	Y	Y	Y	Y	Y	Y	Y	Y	Y	Y	Y	Y	Y			**14**
Epping Forest, Essex	Y		Y	Y	Y	Y	Y	Y	Y	Y	Y	Y		Y			**12**
Ashtead Common, Surrey	Y	Y	Y	Y	Y	Y	Y	Y	Y		Y			Y			**11**
Langley Park, Buckinghamshire	Y	Y	Y	Y	Y	Y		Y			Y			Y		Y	**11**
Moccas Park, Herefordshire	Y	Y	Y	Y	Y		Y	Y	Y	Y	Y	Y					**11**
Cranbourne Chase or Park, Berkshire	Y	Y	Y	Y	Y	Y		Y			Y	CW	Y	Y	CW		**10–12**
Richmond Park, Surrey	Y		Y	Y	Y	Y	Y	Y		Y	Y	Y		Y			**10**
Savernake Forest, Wiltshire	Y	Y	Y	Y	Y	Y	Y		Y	Y							**9**
Staverton Park, Suffolk	Y	Y	Y	Y	Y		Y	Y							Y	Y	**9**
Bradgate Park, Leicestershire	Y		Y	Y	Y	Y	Y	Y				Y					**8**
Grimsthorpe Park, Lincolnshire	Y		Y	Y	Y	Y	Y	Y									**7 (IRA)**
Birklands (Sherwood Forest), Nottinghamshire	Y		Y	Y	Y			Y									**5 (IRA)**
Aspal Close, Suffolk	Y		Y		Y			Y									**IRA**
Ashton Court Park (near Bristol)							Y	Y									**IRA**
Brampton Bryan Park, Herefordshire			Y				Y										**IRA**
Calke Park, Derbyshire	Y		Y		Y			Y									**IRA**
Chatsworth Old Park, Derbyshire			Y						Y	Y							**IRA**
Hamstead Marshall, Berkshire			Y		Y		Y										**IRA**
Hatfield Park, Hertfordshire	Y		Y	Y													**IRA**
High Park (Blenheim Park), Oxfordshire	Y				Y			Y									**IRA**
Ickworth Park, Suffolk	Y						Y										**IRA**
Kimberley Park, Norfolk																	**IRA**
Parham Park, Sussex	Y				Y		Y		Y	Y							**IRA**
Woodend (Shute) Park, Devon	Y				Y		Y	Y								Y	**IRA**

Table 10-1. The 23 most important sites in England for ancient oaks (see Chapter 9) scored for the fruiting presence (Y) on native oak of 16 characteristic saprotrophic fungi recorded over the past 50 years in the British Mycological Society's online FRDBI database (with some additions, see text). 'Cranbourne Wood' (CW) indicates that these species were recorded in the woodland immediately adjacent to, but not within, Cranbourne Chase or Park. Epping Forest and Langley Park have been added to the list as these sites have had some oak-associated fungal survey work. Well-surveyed sites are ordered by decreasing fungal species diversity; sites that are less well surveyed are listed alphabetically commencing with Aspal Close. IRA, insufficient records available.

than four of the five fungi listed at the far left of the top row, it can be assumed that it has only received occasional, if any, mycological study, and requires further survey. The rows are arranged in descending order of the fungal species totals per site. Sites that currently have fewer than four of the commonest species (indicating that insufficient records are available) are listed in alphabetical order towards the bottom of the table.

The higher-scoring sites are generally mycologically well-recorded areas such as those on the Crown Estate at Windsor (Crawley, 2005), the New Forest (Dickson & Leonard 1996) and Ashtead Common, which has had commissioned oak-associated fungal survey work (Ainsworth 2008b). In view of this, two further sites with known relevant survey reports, Epping Forest, Essex, and Langley Park, Buckinghamshire, have been added to the table (Ainsworth 2006a,b; 2008c,d,e; 2010). As might be expected, Windsor and the New Forest (both with SSSI status) top the lists when scored either for their ancient oak saprotrophic fungi or for their saproxylic fauna, whether the latter is assessed by the Index of Ecological Continuity (IEC) or Saproxylic Quality Index (SQI) methods (see also sections on lichens and invertebrates in this chapter). More surprising is the importance of Langley Park, Buckinghamshire, which was ranked among the top five sites based on fungal and faunal (SQI only) assessments, although not considered to be among the most important ancient oak sites (Chapter 9) and currently not a notified SSSI. This former medieval deer park is mentioned in Chapter 4 and has only two recorded oaks with ≥6.00 m girth.

Oak polypore, the flagship fungus of ancient oak

Oak polypore (Buglossoporus (Piptoporus) quercinus) is another brown-rotting heartwood inhabitant fruiting on both native species of oak and, as Buglossoporus pulvinus, was afforded legal protection against picking by Schedule 8 of the Wildlife and Countryside Act in 1998. Like zoned

rosette, it is a 'Section 41 species' (see above) and is also listed as a conservation priority species in Scotland and Wales. There are only 275–300 oaks currently thought to be occupied by this species in England, estimated to represent fewer than 900 individual mycelia, and yet despite being proposed twice for official British red-listing as Vulnerable (Ainsworth 2008f; Smith et al., 2015), at the time of writing, it is not officially recognised as a threatened species.

The British, European and world stronghold of oak polypore is, as far as is known, the Windsor Crown Estate. There are at least 100 occupied oaks on the Windsor Estate, including the Conqueror's Oak and the Signing Oak; these trees are in the Great Park, Cranbourne Park, Swinley Park and other remnants of Windsor Forest, but most are concentrated in OS grid square SU97. The area around Cranbourne Tower shown on Figure 9-11 (Chapter 9) is one of the hotspots and there are 26 known occupied trees, a quarter of the Windsor total, within the mapped area (Ainsworth, unpubl.).

The spongy brackets of oak polypore are pure white when young and fleetingly bruise purple before turning rusty brown where damaged. With maturity, they become yellow then speckled with brown on the upper surface, strongly resembling the surface of crème brûlée, before darkening and drying in situ (see Figure 10-3). In sheltered locations, such as within hollow trunks, hardly recognisable mummified brackets can persist for a year or more, although most specimens are soon consumed by other organisms such as bacteria, other fungi, invertebrates, rodents and deer.

Oak polypore starts fruiting in June at the earliest, usually appearing a few weeks after the first brackets of chicken of the woods and at the same time as those of the earliest beefsteak fungus. July is the best month to see mature fruit bodies and to search for the rarer Ceriporia metamorphosa (see "Characteristic saprotrophs", above) in similar habitat. Occasionally, fruit bodies of chicken of the woods, beefsteak and oak polypore are found on the same tree, but all three are expected to be slow-growing, long-lived stress-tolerant species with one, or at most a few, genetically distinct individuals per tree (Rayner & Boddy, 1988); this assumption is in need of further supporting evidence. Preliminary results show that the English population of oak polypore is lacking genetic diversity (Crockatt et al., 2010), a finding with implications for conservation that needs to be probed further with molecular analytical tools.

Oak polypore fruits on the central deadwood of living or dead standing trees, on their exteriors or hollowed interior surfaces, even if slightly charred, and on fallen branches

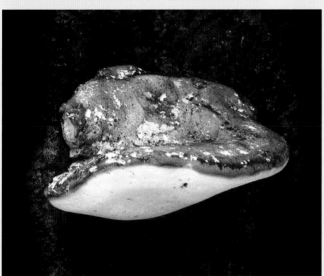

Figure 10-3. *Buglossoporus*
(*Piptoporus*) *quercinus* showing
fruit body developmental stages.
An old fruit body (left) is shown
at ground level fruiting next to
Fistulina hepatica (right) in the
final frame. Photographs by A. M.
Ainsworth, RBG Kew

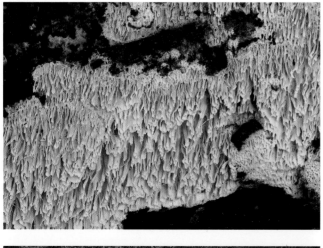

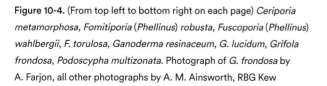

Figure 10-4. (From top left to bottom right on each page) *Ceriporia metamorphosa*, *Fomitiporia* (*Phellinus*) *robusta*, *Fuscoporia* (*Phellinus*) *wahlbergii*, *F. torulosa*, *Ganoderma resinaceum*, *G. lucidum*, *Grifola frondosa*, *Podoscypha multizonata*. Photograph of *G. frondosa* by A. Farjon, all other photographs by A. M. Ainsworth, RBG Kew

and collapsed hulks, even if translocated from formal areas to nearby woodland. It can fruit 10 m up on oak trunks and branches, a situation in which long-distance spore dispersal is likely to be favoured, but most brackets develop closer to ground level. Many are found on fallen branches in bracken, a microhabitat whose humidity might promote fruiting but also attracts grazing molluscs and is likely to favour much shorter-range spore dispersal. Fruiting can cease on such fallen wood if it is accidentally fragmented by habitat management activities, such as mechanical bracken control, or if it is buried beneath the piled aftermath of conifer removal, haloing and/or pollard canopy reduction. On the other hand, oak polypore seems to respond well if

understorey vegetation is carefully cleared and removed, as witnessed around Epping Forest pollards at Lord's Bushes between 2005 and 2008 (Ainsworth 2006a,b; 2008c,e).

A staggering 96% of oak polypore records, commencing in the 1840s, held in the online FRDBI database were made since 1998. It fruits before the autumn peak of field mycological recording activity, which has undoubtedly led to some historical under-recording. Nevertheless, several field recorders have recently started to see this species on old oaks for the first time in woodlands that they have regularly patrolled between June and August for decades, so other factors are likely to have contributed to the recent significant upsurge in records. It is suspected that this is a long-lived fungus that establishes in young trees but only fruits when heartwood is exposed in ancient or veteran trees; the youngest known trees supporting fruit bodies are *c*. 100–150 years old and have extensively damaged trunks. One collapsed oak hulk at Windsor burst into life in 1998 and over the following eight years supported seven further annual rounds of oak polypore fruiting activity.

Unless there has been a recent upsurge in the colonisation of old trees, it seems plausible that our lengthening spring seasons, exacerbated by climate change, are allowing old mycelia to gather sufficient resources from heartwood much more frequently than hitherto, and hence fruit more often. It is to be hoped that the supply of young oaks and the rates of entry and establishment of oak polypore within them will continue to be sufficient for this heartwood inhabitant to remain a flagship species of ancient oak sites across England.

Acknowledgements

Thanks to Liz Holden (2003) for recommended English names and to Keith Alexander, Neil Mahler and Kerry Robinson for contributing extra fungal records.

The sub-fossil data from this species alone provides strong evidence that open pasture-woodland conditions existed across much of lowland England. This is supported by the evidence for the other beetle species too (Alexander, 2007, 2012, 2013).

Lichens

Pat Wolseley

Lichens are essentially fungi that have adopted a symbiotic lifestyle quite different to the parasitic and saprotrophic fungi, in which the fungus provides a body in which a photobiont (a green alga or cyanobacterium) thrives and provides the nutrient that the fungus (mycobiont) lives on. The photobiont needs light as well as moisture, hence the association of lichens with trees, which provide a substrate for the lichen to grow on. The majority of lichenised fungi that live on trees are ascomycetes or cup fungi, which reproduce sexually from ascospores shed from an apothecium, while the photobiont reproduces by cell division in the lichen thallus. However, this means that a spore, on germinating, must find a photobiont in order to develop into a lichen. This may be a difficult step in the life of a lichen and as a result, many lichens reproduce vegetatively from propagules which contain a mixture of fungal hyphae and algal cells. Lichens that rely on sexual reproduction may be restricted to a tree species or a microhabitat on a tree where conditions are suitable for establishment and where they can obtain a photobiont from another lichen.

Habitats provided by an ancient oak

Oak trees provide a great many habitats for lichens, from new grown twigs to ancient bark on the trunk and branches, and to the lignin provided by long-dead heart wood. These habitats continue to be present throughout the life of the tree into senescence, when the structure and nutrient content of the bark changes with increasing age. This together with the different aspects occurring around a single tree — the degree of sun or shade reaching a micro-site, the variation in exposure and moisture content due to drainage channels from the crown to the base of the tree — provide a huge range of niches that can potentially support lichens (James et al., 1977; Rose & Wolseley, 1986). This is especially true of veteran trees that have been on the site for several hundred years and may have become hollow with extensive areas of exposed dead wood. There is yet another phase when the tree falls as the hard heart wood of oak

takes another century or so to rot, providing a substrate for a specialist group of lichens.

Lichens and oak trees

The relationship between oak trees and lichens goes back to prehistoric times when oak trees were a dominant feature of the British landscape, Q. petraea forming extensive woodlands in the north and west and Q. robur in the middle and east of England. Vera (2000) postulates that much of the 'wildwood' would have been open in structure because of the large numbers of grazing animals and that dead and dying wood, especially ancient oak trees that continued to survive for centuries after maturity, would have been an important component of the landscape. The longevity of these ancient trees suits lichens as many of them are dependent on environmental stability and continuity, so that with increasing age of the site and the substrate, the chance of the right habitat niche becoming available increases with time.

The concept of ecological continuity, where a site continues to provide a habitat that supports a specialist community of lichens over millennia, was largely researched in British and European deciduous woodlands, of which oak woodlands with a long history were a major source of data. Francis Rose (in Morris & Perring, eds., 1974) listed 303 lichens associated with oaks, but few of these were restricted to oak and even fewer to veteran trees. Further work on the lichen communities of sites where ecological continuity could be established led to the distinction of indicator species and to the use of a Revised Index of Ecological Continuity (RIEC) to evaluate ancient woodland sites. The index was based on a list of 30 lichens that are associated with sites of long ecological continuity and a scoring system of 1–100 (Rose, 1976). This was widely applied in both woodlands and in parklands of wood pasture origin, often based on surveys of veteran oaks in England. Extensive sites, such as the ancient oak-beech forests in the New Forest (a Royal Forest since 1079), have a score of 100 due to the presence

Figure 10-5. The different habitats for lichens (and mosses) provided by a veteran oak in Dunsland Park, Devon, with marked differences in niches on the dry (left) and wet side of the trunk. Photograph by N. A. Sanderson

veteran trees, these could support relict lichen species, but where sites are reduced in size or converted to agriculture, the specialist lichen communities may be lost or restricted to a single tree in its final stages, creating a time lag or extinction debt for the rarer species. Perhaps the greatest environmental effect on the lichen communities of veteran trees was caused by industrial and domestic pollution from sulphur dioxide during the nineteenth and twentieth centuries. Produced from the burning of fossil fuels, this gas was deposited across much of central and northern England as acid rain, creating a lichen desert that is still obvious today on veteran trees and sites that were affected in the past. Hence, many of the parks in the central part of England included in this book have low RIEC scores resulting from former pollution events. The change in management from wood pasture to arable agriculture is another factor that affects lichen communities of veteran trees, so that even where veteran trees remain in hedgerows and field margins, the increasing deposition of nitrogen and dust causes a change in the lichen community and their replacement by nitrogen-tolerant species such as the bright yellow *Xanthoria parietina*.

of over-mature ancient trees in a mosaic of conditions that extended over more than 3,000 hectares (Tubbs, 2001; Sanderson, 2010).

Many of the lichens associated with ancient oaks are slow-growing and dependent on the stability and continuity of environmental conditions for their survival, suggesting that specialists have developed in a niche associated with the same micro-habitat over several hundred years. In places where the enclosure of 'wildwood' in medieval deer parks allowed the survival of ancient and

Distribution of lichen communities on veteran and ancient oak trees

The bright green (when wet) leafy lichen Lobaria pulmonaria, also called lungwort due to its resemblance to lungs, gives its name to the main macro-lichen community associated with the well-lit moister aspect of veteran trees in ancient forests and parklands — the Lobarion pulmonariae. This large lichen is a light and moisture lover and was formerly frequent on the trunks and branches of open-grown trees in pasture woodland and parklands across Britain. However, it is highly sensitive to the air pollution that has resulted from the increasing use of fossil fuels and has been lost from large areas of central and northern England that have been particularly affected by industrial and domestic pollution. It is also sensitive to changes in land management and to agricultural intensification, so that even if veteran trees are still present, this lichen and its associated species are rarely present. Hence the scatter of locations for veteran trees in England does not always coincide with the areas that are rich in lichen indicators of ecological continuity. In sites where the population of Lobaria has fallen very low there is an additional problem, as the production of fruiting bodies is only possible if different mating types are present in the site. If there is only one mating type remaining in a site, only vegetative reproduction is possible, which restricts the recombination of alleles (Singh et al., 2012). The combination of all of these factors means that, among the sites that are included in this book, Lobaria is now only frequent on veteran oaks in the south west of England, where there are ancient woodlands in areas that have not been affected by atmospheric or agricultural pollution.

Other species of Lobaria occur in the Lobarion pulmonariae community, together with species of Leptogium, Nephroma, Pannaria, Parmeliella and Sticta, all of which contain cyanobacteria as photobionts. Species of these genera are also mainly associated with the west of Britain and are rare in central and eastern England, so yet again, it is hard to use these species as indicators of ecological continuity in areas where lichens have been eliminated from veteran trees due to anthropogenic factors.

The drier side of a veteran oak tree supports a very different lichen community that is characteristic of rough dry bark. This community is dominated by closely appressed crustose species with a different photobiont — a green alga that appears yellowish when scratched. The community — Lecanactidetum premnae — is, as a whole, less sensitive to air pollution than the Lobarion pulmonariae community and has survived on veteran trees across the south of England but is less common in the north. Cresponea premnea is a characteristic species of this community. This species forms pale sheets on the dry bark with abundant dark-coloured disc-like apothecia that are distinguished by the yellow pruina on the surface of young fruiting bodies. Other genera that are also included

in this community with a trentepohlioid photobiont that are also indicators of ecological continuity include species of Schismatomma, Enterographa, Lecanactis and Lecanographa. In sites where leafy Lobarion species have long-since disappeared, characteristic species of this dry bark community may continue to occur as in Windsor Great Park where crustose species of ancient oaks survive, such as Lecanographa lyncea and Buellia hyperbolica.

Veteran oak trees are widely distributed across Britain in a range of climatic conditions, from the wetter warmer west to the drier cooler east, and from the more base-rich lowland south to the base-poor upland regions in the north. This range of conditions supports a large proportion of our epiphytic lichen 'flora' and in order to accommodate this diversity, a New Index of Ecological Continuity (NIEC) (Coppins & Coppins, 2002) expanded the number of species to 70 and added a number of bonus species with a more limited distribution but which had a high conservation value and could be used to identify sites of high conservation status. The NIEC was further expanded to include indices for the west and east of Britain and for eu-oceanic calcifuge upland areas, in order to cover the high diversity of species across a wide range of climatic and environmental conditions.

More specific lichen communities including rare species with a high conservation value are associated with particular microhabitats on veteran and ancient trees, and many of these are included in the 40+ species listed as bonus species for the NIEC. When oak bark becomes old and rough, the bark pH also increases becoming more base-rich, and a special community rich in species of Gyalideopsis, Pachyphiale, Phyllopsora and Porina may become established, the latter genus including three species that are found in the New Forest but that are rare elsewhere. Another specialist lichen community found on dead wood exposed inside the hollow trunks of ancient oaks contains shade-tolerant and slow-growing endemic species, such as Ramonia dictyospora and R. nigra, the latter being restricted to ancient oak and ash in a few scattered sites in England (such as the New Forest and Savernake Forest).

One of the characteristic lichen communities of the 'wildwood' is associated with the hulks of fallen trees, which in our civilized environment are rarely left where they fall. An exception to this is a 72.5-ha site of open pasture woodland in Chatsworth Old Park, where huge fallen ancient oak trees support a number of nationally rare lichens including Biatora veteranorum. This is the only site in England for Hertelidea botryosa, which occurs on many of the fallen oaks (Price, 2010). Although some

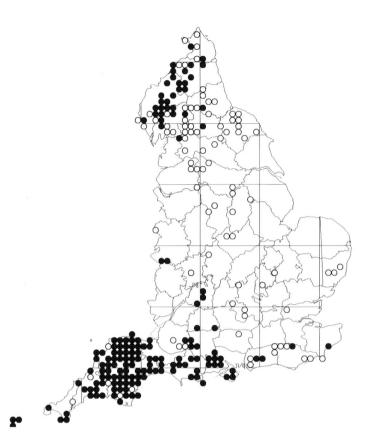

Figure 10-6 (above and above right). The distribution of the lichen *Lobaria pulmonaria* in England. Black dots represent recent recordings that are still extant and white dots past recordings now extinct. Map by the British Lichen Society, photograph by M. W. Storey

of these inconspicuous lichen species may have been overlooked at other sites, there are also species that are now very rare or considered extinct. This is the case for several of the 'pin head' lichens associated with the dry lignum or bark of ancient oak trees. *Calicium adspersum*, now considered critically endangered, was known in the nineteenth century on ancient oaks in Blenheim Park and in Westmoreland and Ingleby in Yorkshire. Another pin head, *Calicium quercinum*, is now considered extinct although evidence from herbarium specimens shows that it was formerly widely distributed in the lowlands of southern England (Church *et al.*, 1996). Other crustose species known on ancient oak but not recorded in England since the nineteenth century include *Chaenotheca phaeocephala*, *Cliostomum corrugatum*, *Sclerophora farinacea* and *Trapeliopsis viridescens*. In cases where the species was only known from a few trees, the loss of a host tree is a major threat to the lichen species. However, many of the veteran

tree sites identified in this book have not been revisited recently, so a targeted search of these sites might relocate some of these species.

The special nature of pasture woodland

Vera (2000) regards 'wood pasture' as having ecological continuity with the prehistoric 'wildwood', where grazing animals were an integral part of an extensive woodland habitat. These animals maintained open glades in a mosaic of conditions, from dense woodland to open heath, that supported many different species (see also Chapter 8). With increasing management of woodlands and creation of plantations, much of our ancient pasture woodland is now restricted to small remnants, particularly in areas which are well populated, which is where the deer parks and hunting enclosures that survive contribute to supporting veteran trees. In these situations, open-grown trees survive into maturity and senescence, maintaining conditions that were part of the 'wildwood'. However the chief difference is in the extensive nature of the grazing that occurred over a mosaic of conditions — including wetlands, woodlands and open heathland — all with a high proportion of mature and senescent trees that would decay *in situ*, compared with a deer park where animals are confined within an enclosure.

Across much of the New Forest, wood pasture management has been maintained by commoners, providing a continuity since it was declared a Royal Forest for hunting by William the Conqueror in 1079 (Tubbs, 2001). This classic site supports a great many lichen species on ancient trees of oak and beech that are rare outside the New Forest. Other sites with extensive areas of old-growth forest that support rich lichen communities occur in the valleys of the west country as at Holne Chase.

Environmental and atmospheric change

Although imparked sites in central England are historically ancient and may continue to support veteran trees there have been big changes both in the surrounding land management and in atmospheric quality, which have drastically affected their lichen communities. This has meant that in central and southern England there are few sites where the more sensitive lichen communities associated with veteran trees have survived, as illustrated by the loss of

Figure 10-7. Six species of lichen commonly found on ancient oaks: (from top left to bottom right) *Pachyphiale carneola* (photograph by J. Seawright), *Pannaria conoplea* (photograph by P. Cannon, RBG Kew), *Lecanographa lyncea* (photograph by P. Cannon, RBG Kew), *Cresponea premnea* (photograph by P. Cannon, RBG Kew), *Calicium adspersum* (photograph by P. Cannon, RBG Kew) and *Buellia hyperbolica* (photograph by N. A. Sanderson)

Lobaria pulmonaria in these areas (see Figure 10-6). These macro species are still frequent in sites toward the west of the country in places where good air quality is combined with environmental continuity and the continuing presence of ancient trees, but there are many crustose species associated with the microhabitats of veteran oaks that are now rare and only occur in widely scattered sites. Even here a major threat is still the loss of the ancient trees and the conversion to agricultural land where trees become dominated by nitrogen-loving lichens (see Chapter 11).

Veteran and ancient oaks ≥6.00 m in circumference are of an age that exceeds most historic records of lichens which become sparse in pre-industrial Britain. The British Lichen Society started a mapping scheme in 1963 and included earlier pre-1960 records of species where these existed so that we can trace the decline of species associated with veteran trees in England, such as *Lobaria pulmonaria* and *Cresponea premnea*, over time since 1963. Many of these data were abstracted from herbarium specimens as well as published records but this inventory is far from complete and at present we have little information on lichen communities from before the Industrial Revolution. A recent study of lichens found on the bark of roof timbers of vernacular buildings in Exmoor National Park from the eighteenth century or earlier showed that species loss from this date included near-threatened species occurring on ancient trees of elm, including *Wadeana dendrographa* that is also known from ancient oak lignin

(Yahr *et al.*, 2014). The major loss of veteran trees took place in the eighteenth and nineteenth centuries when clear-felling replaced the felling of individual trees for a particular use (Rose & Wolseley, 1984), but most of our ancient trees predate this period. Further investigation of archaeological timbers could help to build information on the presence of lichens in pre-industrial Britain, and to distinguish species that are associated with veteran oaks in earlier periods from species that are associated with ongoing changes in environmental conditions.

Management issues

Veteran oak trees may have dead and/or decaying parts and living branches on the same tree. These trees not only provide a continuity of habitats for epiphytic species but a continuous source of propagules that are carried in a variety of ways to other trees and locations. Spores and propagules may pass through the gut of invertebrates such as snails or be carried in the feathers of birds as well as by wind. Where there is public access to a site and there are health and safety concerns about the risks of rotten trees and branches causing damage to people or property, manageing veteran trees may require specialist attention, both to make the trees safe and to identify the organisms that are associated with it.

Many lichens that are associated with pasture woodland require an exposed well-lit trunk. It is important, therefore, to maintain the conditions around the trunk and to avoid fencing that may allow undergrowth to shade the trunk or excludes grazing animals that keep the vegetation down.

Where trees have fallen due to old age or fungal colonisation, the trunks still support many organisms associated with the decaying bark and lignin for several decades if the trunk is left whole and lying *in situ*. This represents the last phase of succession of lichen communities containing a great diversity of species as well as a long-term source of propagules.

Invertebrates

Keith Alexander

Ancient oak trees have the potential to support a rich assemblage of invertebrates, and especially species which are of special conservation interest through being rare and threatened across their British and European ranges. Any species which requires a host that has taken many hundreds of years to become suitable — and which also requires landscapes full of such trees as well as younger trees which will provide suitable hosts in the future — is inevitably rare and threatened in modern Britain and Europe. Concentrations of ancient oaks have become increasingly scarce over recent centuries as the human population has increased and has increasingly made impossible demands on the land for the resources it requires. Today, one must seek out the few remaining treasuries of ancient oak, such as Windsor Great Park and Forest (Berkshire/Surrey), Sherwood Forest (Nottinghamshire), and the much smaller Duncombe Park and Castle Hill (North Yorkshire), Richmond Park (London), Moccas Park (Herefordshire), Calke Park (Derbyshire), Blenheim High Park (Oxfordshire), Ashtead Common (Surrey) and others. While individual ancient oaks are important in their own right — culturally and biologically — it is the larger concentrations of ancient oaks that are needed to maintain viable populations of the specialist invertebrates.

As an oak tree ages, so it begins to accumulate areas of dead woody tissues. The rarest invertebrate habitat provided is the dead heartwood, which accumulates within the core of living oak trees as they lay down additional annual rings year after year. This dead woody tissue is colonised by specialist wood-decay fungi, which may break down just the cellulose and hemicellulose content of the wood, leaving the lignin behind as red- or brown-rotted wood, or both the cellulose and lignin may be broken down, forming a white-rot. This partially decayed material is physically broken down by insect activity and gradually builds up in the base of the hollowed trunk or limbs, where it composts and provides habitat for a further range of invertebrates. The succession from undecayed wood to composted wood mould is followed by a parallel succession of colonisation by invertebrates. Decay may also be occurring in the wood core of the living branches and roots of the tree. Thus, as a single individual tree grows old and ancient, it attracts successive waves of invertebrate species which colonise, breed and ultimately die out as conditions change. The duration of each wave increases as the tree ages, and ancient trees may be regarded as supporting relatively stable assemblages of invertebrates which require the particular conditions provided by the tree. As an example, the first species to colonise freshly dead or dying wood only breed for one year or maybe two, exploiting the moist nutrient-rich cambial layer which is soon depleted and dried out. By contrast, a species developing in large accumulations of wood mould in the base of large hollow ancient oaks may go through very many generations before the tree eventually becomes unsuitable. Indeed, it may not actually be necessary for these invertebrates to even leave the tree for decades, as there will be plenty of food and mates available to them, year after year.

Brown-rot is probably the key habitat for invertebrates provided by older oaks. This fauna is dominated by nationally scarce (NS) and red list (RL) beetles. Fresh brown-rot is colonised by the hairy fungus beetle, *Mycetophagus piceus*, which feeds on the mycelial mats which develop through the dead heartwood, while the woodworm beetles *Anitys rubens*, *Dorcatoma chrysomelina* and *D. flavicornis* burrow through the blocks of brown-rotten heartwood. This is probably the basic fauna, but the larger concentrations of ancient oaks are able to support populations of the much rarer associates. Oak click-beetle, *Lacon querceus*, which is Endangered in Britain (Hyman, 1992) and Vulnerable across the EU (Nieto & Alexander, 2010), develops exclusively in red-rotten heartwood of oak and the larvae are thought to feed on the larvae of the hairy fungus beetle. In Britain, oak click-beetle has only ever been found in Windsor Great Park and Forest (Alexander, 2002a; Alexander & Green, 2013). The beautiful red and black cardinal click-beetle, *Ampedus cardinalis* (Figure 10-8), which is Vulnerable in Britain and

Figure 10-8. The cardinal click-beetle (*Ampedus cardinalis*) is a predator of other beetles that live in ancient oaks. Photograph by R. S. Key

Figure 10-9. This handsome beetle, *Pseudocistela ceramboides*, has mealworm-like larvae that live amongst the dry debris in hollow trees. Photograph by S. Krejcik

Near Threatened across the EU, has larvae which feed on the larvae of the three woodworm beetles. It is somewhat more widespread than the oak click-beetle and is best known from Windsor Great Park and Forest and from Sherwood Forest (Alexander, 2011), but it also has smaller populations at another 10–15 sites including Donington, Moccas, Parham and Richmond Parks, as well as Kensington Gardens in the centre of London, where it relies on the old sweet chestnuts as a supply of suitable brown-rot. The ecology of other rare brown-rot invertebrate species is less well understood, but they include the click beetles *Ampedus nigerrimus* and *Brachygonus ruficeps*, both with Endangered status in Britain — and Near Threatened across the EU — and confined to the Windsor oaks. The small yellow darkling beetle, *Pentaphyllus testaceus*, which has Vulnerable status in Britain (Alexander *et al.*, 2014), also burrows in the blocks of brown-rot at Windsor and turns the rot into a very fine orangey-red dust (Hammond, 2007a). The minute beetle *Micridium halidaii* is also thought to be associated with the mycelium of chicken-of-the-woods, *Laetiporus sulphureus*, and has only been found in association with the oaks at Windsor, Sherwood, Richmond Park and Calke Park (Johnson, 2008).

As the remaining brown-rot begins to fragment, it is colonised by species such as the false darkling beetle (*Hypulus quercinus*), which is Near Threatened in Britain and appears to favour drier rot, and two much rarer beetles, the variable chafer (*Gnorimus variabilis*) and the Windsor weevil (*Dryophthorus corticalis*) (Hammond, 2007b), which require moister conditions, presumably arising from percolation of rainwater into the decaying trunks. The Windsor weevil has long been known as a Windsor speciality in Britain, but variable chafer was formerly also known across the old wood pasture commons of what is now south London, with records from Tooting Common in the 1840s for example (Welch, 1987). For many years, the last surviving population was thought to be the large colony at Windsor but a smaller colony was discovered at Parham Park in West Sussex (Alexander, 2004a). The false darkling

beetle has a somewhat wider distribution, with about 10 to 15 localities currently known, including Castle Hill (Duncombe Park Estate), Grimsthorpe and Croft Castle Parks, and Leigh Woods NNR, Bristol. Another Vulnerable beetle, *Plectophloeus nitidus*, is associated with brown-rot in the ancient oak hulks of Windsor Great Park, High Park (Blenheim), Moccas, Bradgate and Donington Parks, and Sherwood Forest.

Moccas beetle (*Hypebaeus flavipes*), which has Vulnerable status in Britain (Alexander, 2014), is something of an oddity in requiring ancient oaks but only being known in Britain from Moccas Park (Harding & Wall, eds., 2000) where it has been found on seven of the ancient oaks (J. Cooter, pers. comm.). The larvae are thought to live in the galleries made by woodborers in the brown-rotten heartwood of ancient oaks growing in open sunny situations. It has once been reared (in Germany) from brown-rotten wood mould removed from a hole in a dead standing oak. The other beetle species present in this rotten wood were the woodworm *Dorcatoma chrysomelina* (Anobiidae) and the yellow darkling beetle (*Pentaphyllus testaceus*, Tenebrionidae); the former is known from Moccas but not the latter so their presence may have been incidental rather than any species association. Moccas beetle has also been reported from a large rotten beech stump in Sweden, where the anobiid *Ptilinus pectinicornis* was present, as were other wood-boring beetles. In Germany, it is more regularly found by beating hornbeam. Brown-rot can also occur in beech and hornbeam, although white-rot (caused by different species of bracket fungi) tends to be much more frequent in these tree species. The continental sites where Moccas beetle is found all tend to be close to open water, and so the Lawn Pool or the Wye floodplain may be significant for this beetle in Moccas Park. Its presence there but not at Windsor or Sherwood is really inexplicable.

As the decayed heartwood begins to fragment and become more powdery, later to mulch down into a black wood mould which begins to accumulate in the base of the hollow trunks and branches, so the brown-rot species disappear and a very different range of invertebrates — sand-worm type organisms rather than wood-worm types — begin to colonise the decaying tree. The accumulations of powdery brown rot are the home of the forest stiletto-fly, *Pandivirilia melaleuca* (Therevidae), which has Endangered status in Britain and which has long narrow larvae that are superbly adapted to a life amongst loose 'dust' — its closest relatives live amongst loose sand (Stubbs & Drake, 2001). All stiletto-fly larvae are active predators and the forest species specialise in feeding on the larvae of darkling beetles, such as *Prionychus ater* and *Pseudocistela ceramboides*, which have mealworm-like larvae living amongst the dry debris in hollow trees. The forest stiletto-fly has long been known as a Windsor oak speciality but has recently been found at a few other sites. Adults are needed in order to identify the species with certainty but are very rarely seen, and the larvae — once found — can be very difficult to rear successfully away from the host trees.

The final stage in the heartwood-decay process is the formation of a fine black wood mould, similar in consistency to a fine garden compost. One of the most extraordinary finds to have been made in Windsor Great Park in recent years is the false scorpion *Larca lata*, known as the oak-tree chelifer. It is a well-known inhabitant of wood mould in ancient oaks on the continent but was assumed to be absent from Britain. One was eventually found; interestingly, it had been overlooked amongst invertebrate samples taken from nest material found in a rot-hole in a fallen decaying oak tree in the Cranbourne Chase area of Windsor in 1982, but it was not recognised and reported until 1998. None have been found since. This species remains something of a 'Holy Grail' amongst English invertebrate ecologists.

Once the ancient oak has passed through the main hollowing phase, much of the ancient oak fauna described above may be able to maintain itself in the freshly decaying wood of the main branches or in the outer, relatively

Index of Ecological Continuity	Saproxylic Quality Index
Windsor Great Park and Forest	New Forest
New Forest	Windsor Great Park and Forest
Richmond Park	Langley Park
Bushy Park	Richmond Park
Moccas Park	Bushy Park
Sherwood Forest	Sherwood Forest
Calke Park	Parham Park
Langley Park	Moccas Park
Duncombe Park Estate	Croome Park
Grimsthorpe Park	Ashtead Common
Ashtead Common	Ickworth Park

Table 10-2. Top 11 sites in England for the saproxylic fauna of ancient oaks.

Species	Site	Period	
Ampedus nigerrimus	Runnymede	Late Bronze Age	Table 10-3. Sub-fossil
Dorcatoma chrysomelina	Hatfield Moors	Bronze Age	records for the
	Thorne Moors		ancient oak beetle
Dorcatoma flavicornis	Langford, Nottinghamshire – log jam in the River Trent	c. 4000 BP	fauna. (Data from
Dryophthorus corticalis	Hampstead	Mesolithic	Buckland & Buckland,
	Hatfield Moors, Yorkshire	c. 3618–3418 BC	2006.)
	Runnymede	Mesolithic to Late Bronze Age	
	Shustoke, Warwickshire	Late Holocene	
	Sweet Track, Somerset	Early Neolithic	
	Thorne Moors	From c. 5000 BP	
	Worlds End, Shropshire	From c. 5000 BP	
Gnorimus variabilis	Brigg, Lincolnshire	Bronze Age, c. 2600 BP	
Hypulus quercinus	Thorne Moors	Bronze Age	
Mycetophagus piceus	Thorne Moors	Bronze Age	
Plectophloeus nitidus	Thorne Moors	Bronze Age	
Trinodes hirtus	Thorne Moors	Bronze Age	

freshly developed, heartwood of the trunk or alternatively recolonise the tree once sufficient decay has formed. The main large cavity in the trunk tends to support more generalist species of hollow trees — those not requiring brown-rot specifically, and therefore species common to both white and brown-rots and therefore common to a wide range of broad-leaved tree species.

Moving away from the very special brown-rot fauna, ancient oaks also support a few other unusual invertebrates. A striking feature of large old oaks is the accumulation of old thick bark which remains attached to the trunk as the trunk gradually expands and new bark forms below. The cavities behind this bark provide attractive places for spiders to spread their webs so as to catch the rain of invertebrates which breed in the trees. Old webs accumulate and the dried husks of the invertebrates fed upon by the spiders also accumulate. This resource is exploited by a small number of beetles which are closely related to the 'woolly-bear' pests found in our homes and warehouses. The larvae are clothed in stiff bristles so that spiders are not able to get their jaws close to the more vulnerable softer tissues, and the larvae have the freedom to roam around the cavities feeding on the remains left from the spiders' meals. The common cobweb beetle, *Ctesias serra*, is almost ubiquitous in these situations and is clearly highly mobile as an adult beetle. There are, however, rarer species which may also be present. The two rarest are *Globicornis rufitarsis* and *Trinodes hirtus*. The former has been regarded as a Windsor speciality for many years, but has relatively recently been discovered at High Park (Blenheim) (Alexander, 2003) and in the Croome Park area (Worcestershire) (Foster, 1996; Lott *et al.*, 1999).

Trinodes hirtus is also known from very few sites across Britain, almost invariably where there are large numbers of ancient oaks – Windsor of course, but also Croome, Richmond and Shrubland Parks, and in the ancient oaks at Forthampton Home Farm, near Tewkesbury (Alexander, 1992 & 2002b).

While the development of large volumes of heartwood decay and accumulations of old trunk bark are key features of ancient oaks, one additional invertebrate which is characteristic of such trees appears to have other requirements from them. This is the oak pot-beetle (*Cryptocephalus querceti*), which has Vulnerable status in Britain. It is a member of a group of leaf beetles (Chrysomelidae) which preferentially feed on the dead leaves of their hosts and which construct a larval case from sections of leaves as a protection from predators (Piper & Hodge, 2002). The larvae are thought to feed on debris in cavities within the trunks of ancient oaks; oak leaf litter appears to be the preferred food. The adult beetles feed on the living foliage of the same ancient oaks, especially tender young leaves. The species favours open parkland conditions rather than closed canopy woodland. It has primarily been found on the ancient oaks of Windsor and Sherwood, reports from other areas have not been substantiated.

It does quickly become clear that this ancient oak invertebrate fauna is most species-rich at the main concentrations of ancient oaks across England. Two parallel systems have been developed for site conservation assessment for this fauna. The first was based on the model developed by Francis Rose for lichens — an Index of Ecological Continuity — using the presence or absence of species with strong associations with a long-

Figure 10-10 (opposite). Six beetle species associated with the decaying wood of ancient oaks: (from top left to bottom right) *Dryophthorus corticalis* (photograph by R. S. Key), *Gnorimus variabilis* (photograph from natur1 wikipedia), *Hypulus quercinus* (photograph by S. Krejcik), *Lacon querceus* (photograph by R. S. Key), *Mycetophagus piceus* (photograph by A. Haselböck), *Trinodes hirtus* larva (photograph by S. Krejcik)

documented site history, and grading the species scores according to the apparent strength of the association. It was first proposed in 1988 but has been revised more recently (Alexander, 2004b). The other system is a more complex statistical approach, whereby each saproxylic species is scored according to their national conservation status, the total summed and then divided by the total number of qualifying species – the Saproxylic Quality Index (SQI) (Fowles *et al.*, 1999). While one score is based on presumed ecological continuity, the other is based on rarity. The top sites however are rated much the same, just ordered slightly differently. Table 10-2 shows the top sites for the oak fauna using the two schemes and presented in order of importance.

The differences also reflect the extent of survey work that has been carried out, with the IEC approach more affected by this variable. This is clearly seen with a site such as Parham Park, which has been subject to only limited surveys to date. Windsor has a considerably larger old oak population than the New Forest, something that the IEC approach detects better than the SQI.

The very detailed habitat requirements of these invertebrate species make them potentially very valuable to palaeoecologists studying the structure of past landscapes. Unfortunately, the extent of research into palaeo-entomological records from the present Interglacial has been very limited. Nevertheless, a few of the ancient oak beetles have been found in the sub-fossil records (Table 10-3). While the palaeontological record of the native beetle fauna of ancient oaks is far from complete, a good proportion of the species have been demonstrated to have been present in the earlier periods of the present Interglacial. The most striking feature is the Windsor weevil, which occurred very widely across lowland England. This does require large volumes of brown-rotten heartwood and therefore ancient trees – it is known to use large old pines (*Phaeolus schweinitzii* is the main brown-rot fungus in Scots pine) as well as oaks on the continent and this seems also to have been the case in Britain in the past (Alexander, 2005). The important feature is the wide range of this weevil in the past in comparison to its Windsor restriction in modern times. Clearly, large old trees with suitably large volumes of brown-rotten heartwood were commonplace in the past, something which could not have happened under the closed-canopy forest hypothesis. The sub-fossil data from this species alone provides strong evidence that open pasture-woodland conditions existed across much of lowland England. This is supported by the evidence for the other beetle species too (Alexander, 2007, 2012, 2013).

11

Conservation of ancient oaks

Decline and its causes

Probably the most comprehensive list of large ancient oaks in England (and beyond) of a considerable time ago is the enumeration by Loudon (1838, Vol. III). It is notable that although some of these trees also feature in the list of Table 5-4 in Chapter 5, many have gone the way of all life and are no more. The seven largest oaks in England in about 1830 were the Cowthorpe Oak in Yorkshire (with 14.50 m girth the largest English oak ever measured), the Merton Oak and the Winfarthing Oak in Norfolk, the Hempstead Oak in Essex, the Grindstone Oak in Surrey, the Newland Oak in Gloucestershire (Forest of Dean) and the Salcey Oak in Northamptonshire. None of these now exist. A later inventory of some kind was given by Charles Hurst (1911) accompanied by a map. The 68 numbers on the map include a few groups of oaks and not all were ancient trees, also some listed oaks were already dead by 1910. Of those presumably alive in 1910, say around 55, 27 have not been recorded by the Ancient Tree Inventory (ATI) in recent years and have probably died in the century since 1910. Elwes & Henry (1906-13) mentioned other great oaks that no longer exist. This result suggests a fairly high rate of death of ancient oaks; indeed, some that were reported alive in the ATI database only a few years ago, such as the Chaceley Oak in Gloucestershire, I found to be dead in 2013. In November 2014, the King John's Oak in Woodend Park, Devon was seen to be nearly dead and a month later I observed that the last living branch on the Sidney's Oak at Penshurst Place in Kent is now dying back (Figures 11-1 and 11-2).

Of the well-known pair Gog and Magog at Wick in Somerset (there were other oaks so named, see Harris et al., 2003: 178), Gog is dead and the other moribund. If only 1.4% of the ancient oaks die each year, it takes just 50 years to reduce the population by half. Of course, other oaks now mature will become veteran and these will become ancient and thus there may not be a decline in real terms at all. We will return to this later in the chapter.

In order to come to an understanding of a possible rate of decline in ancient and veteran oak numbers overall, we have to separate and, where possible, quantify a number of variables. First, there are both natural causes and unnatural causes of death. Natural causes would be senescence finally resulting in death, but also catastrophic death by lightning, storms, or a sudden defoliation by insects, mildew and other pests and diseases of an already weakened tree. The more vital and intact an oak tree is when these events occur, the greater the likelihood of its survival and *vice versa*. Unnatural causes are due to human interference, either by deliberate destruction or as an unforeseen or unintended consequence of interference with the tree or its habitat. Natural causes may be assumed to produce a steady rate of death over time, but this assumption is only valid if we assume other factors, such as climate, to be (or to have been) constant. And if climate has changed in the past and/or is changing at present, we need to know how this may affect or have affected rates of death. Some diseases may have been introduced by human interference and would then be semi-natural in this country. Human interference has certainly not been a constant factor. Then

death, are unlikely to have had such impact on overall numbers. The causes are, insofar as is possible, discussed in order of importance, beginning with the most severe. In making this assessment, we will primarily consider causes that have been at work more recently and/or are still relevant. The history of a century or more ago and its results are mentioned but treated as the *de facto* situation. The changes in land use due to agriculture, although undoubtedly having had the greatest impact on the abundance of ancient oaks, mostly fall into this category. We must examine if and how these adverse forces affect ancient and veteran oaks at present, either directly or by inhibiting their replacement by succeeding generations.

Competition for light with other trees

The planting of conifers on sites which have or had ancient oaks has undoubtedly resulted in a decline of ancient and veteran oaks on these sites. Examples can be seen in Alice Holt, Forest of Dean, Salcey Forest, Savernake Forest, Windsor Great Park and Forest and Croft Castle Park, all occurring during the late nineteenth century and most of the twentieth century. Likely victims of plantation forestry include the Conquerer's Oak or King Oak in Cranbourne Chase (Windsor Forest), now a rotting stump but in 1838 a hollow tree with a girth of 8.00 m and in *c.* 1876 pictured as a large hollow tree with two great limbs standing in what seems to be a glade in the oak-dominated woodland (image reproduced in Hight, 2011: 72). It is situated on the boundary of what was the White Deer Enclosure, of which remnants of an iron fence remain. Since when it has been dead is unclear, but Wilks (1972) still reported some living branches in the early 1970s. (Pakenham (2003) identified another oak as the Conquerer's Oak.) The surrounding woodland is now dominated by sycamores, sprung up after the felling of conifers, of which a few remain. Other large stumps of oaks occur in plantations in nearby Quelmans Head and in the southwest corner of Windsor Great Park.

This situation is repeated elsewhere in the country, especially on sites managed by the Forestry Commission. About 50% of the *c.* 70 ancient and veteran oaks recorded in the conifer plantations of Croft Castle Park, owned by the National Trust but leased to the Forestry Commission, are dead. It has been reported that the Forestry Commission had ring barked the veteran oaks in order to kill them so they could be replaced by "beggarly spruces" (Massingham,

we need to estimate the rate at which veteran oaks become ancient, which requires an understanding of the life of an oak (see Chapter 1). In the following paragraphs, we will examine some of the causes of death, while noting that the distinction between natural and unnatural can be blurred in some situations. We will assess which of these causes of death are likely to have contributed to a decline in the numbers of ancient oaks and which, although causing

1952). Very few of the ancient oaks in Salcey Forest outside the central meadows of Salcey Lawn are still alive, even though in 2013 they were still mentioned in a nature walk leaflet distributed on site by the Forestry Commission. Nothing but a dead stump on a lawn remains of the ancient and veteran oaks in Alice Holt Forest, but here we have no records of how many there were before conifer plantation took off. Historical records suggest that The Forest of Dean was certainly once rich in veteran oaks (British History Online (BHO): Forest of Dean), but heavy timber exploitation as well as mining greatly reduced the oaks as well as the beeches, and enclosures for tree planting began in the seventeenth century. Today, just two ancient oaks (of 6.90 m and 7.20 m, the latter known as the Verderer's Oak) and 10–12 veteran oaks (all with <6.00 m girth) remain, which must at least in part be attributed to the plantations.

More recently, however, an understanding and valuation of ancient trees has not only stopped this destruction in most cases, but has actually led to attempts to rescue the surviving oaks by haloing, that is, by removing the planted conifers (or other trees) surrounding them. This will mean that the decline caused by plantation forestry will be significantly reduced in the future. Yet, some further losses will occur as an aftermath of this kind of interference because a number of ancient oaks have only barely survived

and may be beyond recovery. An example is the Duke's Vaunt Oak in Savernake Forest, Wiltshire. In 1802 it was a 9.00 m (30 ft.) circumference hollow maiden oak with a full secondary crown of reiterated branches. The same oak was recorded by the ATI in November 2005 as a 10.00 m ancient tree, although this was presumably an estimate of how large it had been rather than a measurement of the actual bole. Miles (2013: 142–3) republished an 1802 illustration in *The Gentleman's Magazine* together with a photograph of "the last remaining shard". On my visit on 15th September 2014, a true measurement of its rotting bole remnant was found to be impossible, so I gave it the benefit of an estimate of 6.00 m. Just one branch with a little foliage showed this oak to be still alive but probably not for long.

Another threat to ancient oaks, not dissimilar to that presented by plantations, comes from the fact that much former pasture woodland with ancient oaks, with or without conifer plantings, is now left unmanaged and unused. This leads to spontaneous growth of trees of which particularly beech is a menace to oak conservation; as a shade-tolerant tree with a sympodial growth pattern, beech is soon able to overtop any oaks that are present. German foresters had good reason to call beech the Eichenmörder (oak murderer). If beech or other shade-tolerant trees

Figure 11-3. The Duke's Vaunt Oak in Savernake Forest was nearly dead in September 2014

are present, the long-term survival of oak is problematic without some form of management or use which suppresses the competition (Alexander *et al.*, 1996; Vera, 2000). Ancient oaks, which often have a low stature and a reduced crown, are especially vulnerable to competition for light. Examples of pasture woodland which had ancient and veteran oaks now almost gone due to competition from spontaneous beech, holly and also tree planting (but not of conifers) are Ebernoe Common and The Mens in Sussex. These wooded commons, briefly described in Chapter 5, have become tall closed-canopy forest in which the present living oaks are mostly planted timber trees.

Even where management has begun to reverse the dominance of spontaneous trees by some cutting (including haloing of old oaks) and/or grazing with livestock, this work usually happens on a small scale and unless greatly expanded will make little amends for the losses of ancient oaks already suffered. I do not see The Mens (160 ha) returning to pasture woodland anytime soon; the present management appears to be essentially hands-off, letting

Figure 11-4 (left). A beech killing an oak in Ebernoe Common, Sussex

Figure 11-5 (right). Oak and beech pollards entwined in Epping Forest, Essex/Greater London

it succeed into semi-natural forest dominated by beech and holly. Another example can be seen in Epping Forest. Here tall beeches, many of which are large but lapsed pollards, dominate in many areas to the disadvantage of the remaining oaks, many of which are senescent even while rather small. This presents a serious problem: unless managed and used in the traditional way most pasture woodland will become closed-canopy forest in which oaks are at a disadvantage. Although beech is the predominant competitor, it is not the only threat and hornbeam can be important as a replacer of oak locally. In the north and west especially, sycamore can also be a succeeding tree. Neglect of parks can have similar effects. An example I have seen is Aldermaston Park in Berkshire, once a medieval deer park. In what is left of the pasture woodland, neglect (no cutting, no grazing) has caused the remaining ancient oaks to be overgrown with birch, willow, sycamore (here feral) and other trees. Nine out of the 14 remaining ancient oaks were senescent or nearly dead in summer 2015. This first and possibly most serious cause of decline in numbers of ancient oaks can thus be summarised as competition for light with other trees, most of which will eventually exclude the oaks from the forest.

Eighteenth-century landscaping of parks

The alteration of parks, which began on a large scale in the eighteenth century with the advent of the English Landscape Movement led by park designers such as William Kent, Charles Bridgeman, Lancelot ('Capability') Brown and Humphrey Repton, has undoubtedly contributed to the loss of ancient oaks. They could be preserved and incorporated, as for example in parts of Blenheim Park, Oxfordshire, but were probably just as easily replaced by new trees, often planted in groups deliberately positioned to enhance views out into the wider landscape as well as back to the mansion. It is often difficult to establish whether and where such tree replacement has actually happened, but it seems likely to have occurred at places such as Petworth Park in Sussex, where just four large ancient oaks survive amidst many hundreds of planted parkland trees in a large deer park that goes back to medieval times. In Sherborne Park, Dorset, the medieval deer park of the Bishops of Salisbury, Brown was called in to create the lake between the old castle ruins and Sir Walter Raleigh's new castle in 1753. A vast sweep of open treeless lawn and fields now stretches to the hills in the distance, and the trees on the slopes of these hills, including native oaks, date back to the eighteenth century at most. Just one ancient oak with 7.47 m girth, now in woodland, survives from the ancient deer park.

By the eighteenth century, the deer had become irrelevant and notions about aesthetic views and sights determined the landscape design and tree planting. The main thrust of the English Landscape Movement was to expand the park and merge it visually with the countryside (Fleming & Gore, 1982). The Enclosure Acts made this possible by privatising common land. These acts by Parliament deprived commoners of their traditional rights such as grazing and collection of firewood, and enabled landowners not only to 'rationalise' agriculture but also to redesign the landscape on a much larger scale than before. Alexander Pope, an influential writer on this subject (and much else), wanted the park to look natural, but his "Unerring Nature, still divinely bright" did not mean in the eighteenth century what it might mean now. 'Nature' was very much seen through the eyes of the landscape painter, which is to say

that the visual aspect was predominant. Ecology was an unknown concept, as was biodiversity; man was the God-given master of nature and should manipulate it to satisfy his own ideas and wishes. A move into romanticism in the nineteenth century undoubtedly helped the preservation of particularly old, large or picturesque specimen oaks, but this was also the time when park owners discovered ornamental trees from abroad. Conifers, especially such as Lebanon cedars and later giant sequoias, grew faster and became more impressive than most of the oaks in a landscape design context. In Chapter 4, I analysed the alterations and losses to medieval deer parks and found them to be substantial. However, it is difficult to demonstrate, let alone quantify, the losses of ancient oaks resulting from the landscaping of parks. Unlike plantation forestry, where in several cases dead stumps remain, the parks were altered long ago and nothing remains. We can only make some rough inferences, in some cases by looking at old maps of the park, as I have done above.

Golf courses

A more recent development is the transformation of parks with ancient and veteran oaks into golf courses. There are numerous examples, several of which are mentioned and briefly described in Chapter 4, that still have ancient and veteran oaks. There is usually no direct information about the numbers of ancient oaks present before the creation of the golf course. We can, however, make some inferences from comparisons with the nineteenth-century OS maps, which show individual large trees where they stand in open fields and hedgerows. Some examples of former medieval and Tudor deer parks where ancient oaks still occur may illustrate the possible losses of ancient and veteran oaks due to the creation of golf courses.

Brocket Park in Hertfordshire

This large park is probably medieval (see Chapter 4) but much of it is now a golf course. There are at least 10 ancient oaks recorded here, of which eight are on the golf course; in addition, there are around 30 veteran oaks but most of these have a girth of less than 5.00 m and are not old. In the nineteenth century, this park included open parkland with few trees, parkland with many trees, and wooded areas. The golf courses, laid out in the mid-1990s, occupy most of these elements of the former landscaped park. Losses of ancient and veteran oaks are difficult to assess, because according to the recent survey, the site has a variety of veteran trees, including planted exotics which may have been among the trees indicated as solitary on the nineteenth-century OS map. Another factor against the inference of losses is

Figure 11-6. Oaks in Georgian parkland at Hatchlands Park, Surrey

the relatively recent construction of the fairways, at a time when awareness of the value of ancient native trees was well established. On a visit in February 2015, I noticed metal number tags on several oaks and a *c.* 8.95 m ancient oak near the formal garden was fixed with a lightning conductor, the first I have seen on an ancient oak. It remains to be seen whether the golf courses may have negative effects on the ancient oaks that were preserved that may cause losses in future. Plantings of ornamental conifers among the ancient oaks between the mansion and the formal garden may in future cause problems related to competition for light and water with the much slower-growing oaks.

Cowdray Park in Sussex

The golf course in this park covers a large part of the medieval deer park, which was situated north of the road from Easebourne to Tillington (now the A272). Fields along The Race, now arable, were still part of the park in the nineteenth century. A section around Broomhill Plantation is still ancient parkland and here stands the Queen Elizabeth I Oak, with a girth of 12.67 one of the largest ancient oaks in England. A number of ancient oaks survive on the golf course, but the nineteenth-century OS map shows that there were more solitary large trees (presumably mostly oaks) in some areas. This was also true for the section around Broomhill Plantation, but here there still is a mixture of ancient and veteran oaks and sweet chestnuts, so many of these trees may have been chestnuts. Other parts of

the golf course already had few trees by the late nineteenth century, so it seems that losses of oaks resulting from the creation of the golf course were moderate; eight ancient and 15 veteran oaks still occur in the golf course area.

Easthampstead Park in Berkshire

This park was already partly converted to farmland in the nineteenth century and other parts were built up in the twentieth century, including a large school. Most of what remained of the park became a golf course. Just one large ancient oak (with 9.05 m girth), the only remaining oak of the medieval deer park, stands on the edge of a fairway. Yet the nineteenth-century OS map shows many solitary trees in this area, many in rows that suggest hedges but with no indication of the hedge lines as is usual when there was a hedge. It seems evident that here too ancient and veteran trees were lost when the last section of the park was urbanised with housing, a school and a golf course. The survival of the last big ancient oak seems almost a coincidence.

Kedleston Park in Derbyshire

Kedleston Park has a golf course on the north side of the string of artificial lakes between Kedleston Hall and Kedleston Road. This golf course was created in 1947, well before the National Trust's acquisition of Kedleston in 1987, and is managed by the Kedleston Park Golf Club. It is therefore not mentioned in NT brochures and websites,

and in *Wikipedia*, the park is claimed to "remain mostly unaltered" since the late eighteenth century. This area was more or less open, unimproved parkland in the late nineteenth century, when the OS map showed scattered (groups of) solitary parkland trees and a concentration of these in the northernmost corner, marked by the 'Lion Oak' (now gone). Most of that corner is beyond the golf course and still has a number of mostly veteran oaks. On the golf course, 14 veteran and three ancient oaks have been recorded, which seems to amount to a decline in the number of ancient and veteran oaks as most of the trees in this area marked on the nineteenth-century OS map would have been large oaks.

Lullingstone Park in Kent

The fairways of the two golf courses at Lullingstone Park are mainly laid out in areas where there were no or few solitary trees on the nineteenth-century OS map. The roughs in these areas have unimproved chalk grassland, and presumably this was the vegetation over a larger area before the fairways were made in the mid-1960s. Before this, a decoy airfield was situated here briefly in 1944. In some areas where ancient and veteran oaks now occur on the golf course, there appear to have been more large trees in the nineteenth century and some of these were probably ancient or veteran oaks. Several oaks were destroyed to make space for the golf course in the vicinity of the club house. Others were probably cut down much earlier to make space for plantations in Lower Beechen Wood (Pitt, 1984). Pittman (1983) reports losses of some ancient oaks due to the airfield and the golf courses; the losses during World War II went unrecorded and no more than 10 large oaks were destroyed for the golf courses. However, the park was only lightly touched by the landscaping fashions of the eighteenth and nineteenth centuries and lost its medieval character as late as the 1960s largely due to the golf courses. Tree planting in the 1940s and 1950s in Upper Beechen Wood is probably responsible for an additional number of dead ancient oaks there, and this may also have been the case in the Workmen's Enclosure. As seen on a visit in June 2015, the shading of ancient pollard oaks in the woodland sections of the park is still a threat to some of them.

Moor Park in Hertfordshire

The losses here could have been substantial, especially in the northern and western sections of the park where the nineteenth-century OS map shows many solitary large trees but where no ancient or veteran oaks are left. Avenues extant in the nineteenth century (possibly of sweet chestnuts, of which a few trees survive) were also taken out to create the golf courses, which opened in 1923. Most of the remaining ancient and veteran oaks stand in the roughs, many of which are now wooded. On a visit in December 2014, I observed several of the *c.* 40 recorded oaks to be dead or dying.

Packington Park in Warwickshire

The ancient and veteran oaks are still quite numerous on the golf course in this park, and stand mainly in the roughs which are overgrown with bracken. However, on the nineteenth-century map, there are many more solitary trees that are more or less evenly distributed, with no visible correlation to the layout of the fairways and with many situated where there are now fairways. In this location, almost all of the trees present in the nineteenth century would have been native oaks, many ancient or veteran. A few old sawn stumps of oaks on the edges of fairways were seen on a visit in August 2014. This means that substantial losses were incurred, probably as a consequence of the creation of the golf course, which opened sometime in the 1980s.

Stoneleigh Park in Warwickshire

The present ancient and veteran oaks on the golf course stand scattered in the roughs as well as on or closer to the fairways and are still quite numerous. Nevertheless, quite substantial losses can be inferred from the nineteenth-century OS map, especially in the central parts of the park along the River Avon. When I visited in May 2013, I observed mostly planted oaks and other trees in these parts that were much younger than 100 years, many probably dating from the time when the golf course was laid out. Earlier replacement of 'dodders' with newly planted oaks in these parts remains another possible cause for losses of ancient oaks. A large ancient oak on the bank of the River Avon near the two-arch stone bridge pictured in a late nineteenth-century photograph does not exist any more. This tree is unlikely to have been removed for a fairway.

Impact of golf courses on ancient oak numbers

Six of the parks converted to golf courses discussed here were considered for inclusion in the list of most important sites for ancient and veteran oaks in England (Chapter 9). Some would most likely have passed the criteria but failed because of the

Figure 11-8. An ancient pollard oak and a natural pollard oak on a golf course in Moor Park, Hertfordshire

Figure 11-9. An ancient pollard oak in Eastwood Park, Gloucestershire

creation of the golf course and its effects on the ancient landscape and probably on the oaks. There are medieval and Tudor deer parks converted to golf courses where no ancient or veteran oaks remain. This does not necessarily mean that they disappeared as a result of the creation of a golf course. And certainly, in the present context of a wider understanding of the importance of these ancient trees, they are not likely to be cut to make space for fairways in future. But, it seems likely that the creation of golf courses, like the earlier landscaping of parks, has resulted in the destruction of ancient oaks, either by deliberate removal or by slow death resulting from alterations in the drainage of the land, damage to roots under fairways and soil compaction.

Perhaps a lesser but not unrelated recent cause of the destruction of ancient oaks has been the tendency to 'tidy-up' public parks and golf courses (Rackham, 2004). Linked to this are ongoing health and safety concerns where ancient oaks have sometimes become victims of the modern litigation culture. In parks that are open to the public, owners may prefer to remove a tree rather than risk injury to a member of the public from a falling branch. Fortunately, sensible management that can mitigate such risks is possible and can actually stimulate the maintenance of pollards. In Richmond Park, Surrey, a heavily visited large deer park with many ancient oaks, continuous vigilance ensures that hazardous dead branches are removed but the sawn pieces are left on the ground under the tree for fungi and invertebrates. Despite this precautionary action, accidents do sometimes happen.

Agriculture

Agriculture was identified in Chapter 4 as the main destroyer of medieval deer parks. The undoubtedly negative impact

Figure 11-10. This ancient oak with a girth of 9.90 m at Easthampton, Herefordshire may have stood in the medieval park of Shobdon

that farming has had on the presence of ancient and veteran oaks is very difficult to quantify. That such destruction of parks occurred until relatively recently is testified by comparison of the OS maps. It may of course have had a long history. Two examples of this are given here.

Sheriff Hutton Park in Yorkshire

Ralph de Neville, Lord of Raby, obtained a licence to impark Sheriff Hutton in 1334, probably expanding an existing smaller park (Dennison, ed. 2005). It was a large park mostly created from carr. It also included some areas of drier land, one around the present site of Sheriff Hutton Hall and another on the site of the deserted village of East Lilling. In 1382, a large house was built in the park; the present Sheriff Hutton Hall was built on the same site in the seventeenth century. The park and its surrounding land once belonged to the Forest of Galtres (www.british-history.ac.uk: Sheriff Hutton Parish) and in 1471 the park's ownership

reverted to the Crown. Royal hunts in the Forest and park continued into the seventeenth century; the park was sold by James I in 1622. A survey by Robert Norton carried out in 1624 recorded "near 4,000 decayed and decaying oaks, the most of them [stag]headed" for Sheriff Hutton Park (Muir, 2005: 172). As Muir points out, the majority of these trees, which were part of the Royal Forest, must have existed before imparkment. On the nineteenth-century OS map, some parkland remains around and to the north of the house, but the greater part of the park is farm fields separated by hedgerows in which trees are indicated. To the east of the house lies Sawtry Plantation, a newly created wood. On the current OS map, the park is further reduced to a small area around the house and the farm fields are larger with fewer hedgerows in which there are only a few mature trees. Just one ancient oak, a maiden with a girth of 8.15 m, survives in a bit of parkland close to the farm buildings near the hall. The park has become a farm.

Oakly Park in Shropshire

Oakly Park, situated south of Bromfield between Priors Halton, Lady Halton and the River Teme, was an early Tudor deer park dating from *c.* 1490 (called 'Ockley Park' on John Speed's map of Shropshire dated 1611). In the Middle Ages, it was part of the chase attached to Ludlow Castle. It was still a large park on the nineteenth-century OS map, with farm fields only at its southern end near the hamlets of Lady Halton and Hill Halton. These and other fields near Priors Halton were included in the park after 1733 and so they only show hedgerow trees on the map. Oakly Park house, gardens and 'pleasure grounds' are near the river, downstream from Bromfield, and here a few more enclosed fields can be seen on the nineteenth-century map. For the most part, the park appears as open parkland with numerous parkland trees, some pheasant coverts and coppices, and two avenues (the southernmost one called Duchess Walk), while woods (such as Ham Bank) border the river in the west. Some large ancient and veteran oaks remain in the Ham Bank area, now called High Trees, indicating that this was also parkland before the planting of the wood there in the 1780s. The park had been landscaped in the late eighteenth century with more avenues, but most are now gone. What remains of the late nineteenth-century park landscape, apart from the Duchess Walk, is now situated between Halton Lane, which was planted as an avenue between Priors Halton and Bromfield, and the river. The rest of the park is farmland, mostly arable or temporary grassland. Nearly all of the ancient and veteran oaks that would have occurred here in abundance are gone, and the few remaining living oaks are all younger. Those ancient and veteran oaks that remain in Oakly Park, apart from some in the High Trees area, are in the remaining parkland between Halton Lane and the River Teme. Some of the largest of these (of 7.00 m to 8.05 m in girth) may date back to the time of the Tudor deer park, which no doubt incorporated the oaks that stood on the chase of Ludlow. The destruction of the oaks to make space for farming occurred during the twentieth century. The whole site is still called Oakly Park on the current OS map and on estate websites, but a park that has been converted to farmland (or a golf course) is no longer a park in our context.

Current farming practice and its impact on oaks

Although examples can easily be found, there is to my knowledge no published record of damage to or the death of ancient and veteran trees (oaks) caused or probably caused by agriculture. Apart from the deliberate removal of trees, two main causes of damage or death are suggested by field observations. The first is damage to the roots and trunk base by farm animals. Horses with shod hooves can scrape off bark and cambium, and if this damage becomes extensive, the tree is effectively ring-barked. This is often the cause of death of mature oaks with girths <6.00 m, but where bark is present on only a part of the stem and root base of an ancient oak, only that bark needs to be damaged to finish the tree off. Animals' rubbing against a tree, if sustained over years, can have similar effects. The behaviour undoubtedly varies depending on the kind of animals doing the damage, and if the tree is given a good break, it can often restore damage. I observed an example of this kind of damage to ancient and veteran oaks on 14th January 2015 in Bulstrode Park, Buckinghamshire, where large numbers of horses have free access to the trees. Sheep and cattle often congregate for shelter under a solitary big oak in a field. If sustained over long periods, this may lead to soil compaction over the root system, a problem now widely recognised in public parks where compaction is caused by people's feet (Kirkham, 2009). In addition, chemical alterations to the soil resulting from heavy input of fertilisers in the form of animal dung and urine are likely to have a negative impact on the mycorrhizal associations in the tree roots, thereby damageing the health of the tree.

Corse Chase, a medieval chase in Gloucestershire, has been transformed to agriculture. On the nineteenth-century OS map, two fields near Forthampton Farm (now Home Farm) remain as pasture woodland with many solitary large trees, presumably all oaks. These fields are well above the flood plain of the Severn and separated from the river by a tree-lined bank and hedge; the large trees in this hedge have disappeared. There is some evidence of ridge and furrow in the northern part of the fields and perhaps a ditch of the park pale of Bushley Park in Worcestershire. Another hedge with large trees in the field nearest the farm has been destroyed and most of the solitary trees present in the nineteenth century have also gone. Of 40 ancient and veteran oaks observed in these fields during a visit with Brian Jones on 25th July 2015, just 17 were still alive, some of these barely. The 23 dead oaks include ruined stumps. One dead tree had a girth of 8.50 m (Figure 11-11), and one that was alive and healthy had a girth of 8.27 m. What has caused the (recent) death of such a high percentage of ancient and veteran oaks in these fields? The fields are now further compartmented by barbed wire fences and they are heavily grazed by dairy cattle from the farm. Many ancient oaks appear to stand in a small depression and have damaged root bases. In some

Figure 11-11. A dead ancient oak with a girth of 8.50 m near Home Farm, Forthampton, Gloucestershire

cases, callus growth has formed new bark, restoring the connection between foliage and roots and these trees had survived, but in most cases, the damage was too severe. There may have been horses in this field before the cattle replaced them. The grassland is, in most parts, improved rye grassland and seems to have been ploughed in the past. The drought of 1976 may have exacerbated the decline of the oaks, but the evidence points at ongoing damage begun well before that date. This field with ancient oaks on a farm is an example of management that is detrimental to the preservation of the oaks.

A second cause of death of ancient and veteran oaks on farms is ploughing too close to the tree. Many ancient oaks stand in the middle of farm fields. In many cases, these are arable fields, and in modern farming there is no permanent separation of grassland and arable on many sites. Conversion of ancient unimproved grassland in parks to improved grassland or to a rotation of arable and seeded grassland is perfectly legal unless the park is a protected nature reserve or an SSSI. The root system of a large oak may spread well beyond the drip line of the canopy, and most of the fine roots and especially ectomycorrhizal fungi connected to these that take up water are in the upper 30–40 cm of the soil. "Any deep ploughing less than 5 m beyond the limit of the canopy is likely to cause damage and even ploughing beyond this zone can be destructive." (Muir, 2005: 110).

The Chaceley Oak in Gloucestershire, an enormous low pollard measured with a girth of 12.51 m, was dead when I visited it on 19th June 2013. It had a full crown when it died, apparently quite suddenly and at least as early as 2008, as a photograph taken on 28th July 2008 posted on the ATI website of the Woodland Trust shows no leaves in the crown. Its death has been attributed to flooding in July 2007, when the field in which it stands was under some 10 ft (3 m) of water. However, on my visit, I observed that the grassland is not permanent but re-sown after ploughing, with only a small circle of a few metres radius around the bolling left untouched. It seems unlikely that flooding alone

could have killed it. This ancient oak, on the floodplain of the River Severn, must have experienced numerous floods. Owing to periodic water-logging, the roots of trees on a flood plain are shallower than those on well-drained soils. If traumatised by damage to much of its root system, the 2007 flood may have given this ancient oak the 'coup de grâce'.

A maiden oak with a girth of 9.86 m stands on farmland near Easton Lodge, in the former Easton Park, Essex. Deep-ploughed arable land and concrete pavement leading from farm sheds into the field are close to the tree on two sides, and a trench has been dug between the concrete and the oak. The upper trunk and branches were dead (more than seemed normal in an oak with mostly intact bark on the bole), but the tree appeared to have formed a secondary crown when I visited in October 2013. This growth pattern is likely to have been caused by partial severing of the root system, but at least in one direction, where there are some shrubs and grass, the roots must have remained intact, allowing the ancient oak to recover with a smaller crown. It would probably have been beyond recovery had the land been ploughed there too. On a revisit in October 2015, no changes to the crown were seen, but we cannot yet conclude that no permanent damage has been done.

Other causes of destruction

Vandalism of ancient oaks is a problem, especially in or near urban centres. Many of the ancient and

Ancient oaks in the English landscape

Figure 11-12. The Chaceley Oak in Gloucestershire died in 2008 or possibly 2007

veteran oaks that I have visited in the Greater London Area have charred black insides of hollowing trunks, showing them to have been burned. This is no accident from a lightning strike but is caused by someone deliberately lighting a fire inside. In more remote countryside, ancient oaks rarely show signs of having been burnt. In several cases, the oak is now dead, probably killed by the fire. These fires are sometimes accompanied by graffiti (spray painting) which seems to rule out the kind of accident we might have heard of in the past, such as rural people making a small fire to boil water for a picnic using the shelter of the hollow tree. In West Wickham Common and in Burnham Beeches, the City of London Corporation has tried to ring fence or even board up hollow oaks to prevent damage of this kind, and where bramble or holly thickets have grown as a result, this action may have some positive effect. An obviously hollow oak in Hatfield Park, Hertfordshire has been protected by a split chestnut slat and wire fence, which is impossible to climb (Figure 11-13). A more indirect cause of fire damage to ancient or veteran oaks is carelessness. In exceptionally dry conditions, fires that are (accidentally or deliberately) started in dead bracken or gorse may transfer to the oaks. In Ashtead Common, Surrey, a large number of veteran oaks have been killed or damaged in this way (see Chapter 9).

Neglect of ancient pollard oaks can be the cause of destruction, or hastened death. The activity of pollarding ceased in many cases almost 200 years ago (Chapter 8). The last crop of branches to regrow was allowed to increase in size and weight despite the fact that the bolling was deteriorating with brown rot and becoming too hollow. The remaining shell, especially when gaps in it are widening, is no longer able to support the increasing weight. These forces will tear what remains of the bolling apart, especially in high winds when leaves are still on the tree. There are several ancient pollards that are now in a ruinous state due to this neglect. Props to support the limbs may postpone breakage but will not stop it. The Major Oak in Sherwood Forest is a famous example of a tree that is supported in this way. Storms can blow off limbs as a natural form of pollarding, and if an ancient tree is strong enough, it may be able to regrow a smaller set of branches that are, at least for a time, more in balance with their natural support — the Queen Elizabeth I Oak in Cowdray Park is a good example. Although there have been some moderately successful attempts at re-pollarding besides these natural ones, it is considered a risky business as the trauma could also kill

Figure 11-13. An ancient oak protected from vandalism by a fence at Hatfield House, Hertfordshire

the tree. This poses a dilemma, to which perhaps the only solution is assisted retrenchment of the crown (Read, 1999) to ease the weight-pull exerted on the hollow bolling.

Diseases and pests

Are diseases and pests that affect the two native oak species serious threats to ancient and veteran oaks? We must first distinguish between acute or direct threats — the diseases and pests that are killing ancient and veteran oaks — and potential causes of decline affecting populations of younger oaks, from which veteran and then ancient oaks are emerging in due course. There are numerous diseases and pests affecting oaks, among which are some that have only recently been introduced into or observed in the UK. Some of these are thought to (potentially) impact oaks at

the population level, but most either kill just a few trees or inhibit growth. Ancient and veteran oaks may die from a number of causes or their combined effects, and we can only be concerned if it is probable that a cause increases the rate at which they die. Among the direct threats, we can dismiss as irrelevant all 'pests' — mainly insect damage to foliage but also bark stripping of young oaks by the grey squirrel — as rarely causing the death of a healthy oak. Unless such damage occurs repeatedly, preventing a second flushing of leaves, and is followed by pathogen invasion it is probable that the tree will recover. The exception is if the pest attack defoliates the last remaining foliage on an already very senescent oak. By and large, however, these damageing agents are merely growth inhibitors (Tyler, 2008).

Inhibitors of oak regeneration could be more serious if they turn out to be widespread and sustained. Oak mildew (*Erysiphe alphitoides*), now known as a fungus of tropical origin (and possibly conspecific with *Oidium mangiferae* found on mango trees), has been found in continental Europe since 1907 and in the UK since 1908. Rackham (2003b, p. 297) theorised that this fungus has affected oak seedlings by making them less shade tolerant and implicated it in the apparent reduced success of oak regeneration in woods (but see above for competition with other trees in 'neglected' woods). This disease is commonly found on seedlings and saplings, but also on coppice and lammas shoots and on young parkland trees (as seen, for example, in Chatsworth Old Park in summer 2014) and, less conspicuously, on mature oaks. It can also affect new growth on pollards, compromising the renewal of this kind of tree management. Several factors mitigate the pathogenic effects of this disease on native oaks, including the late spring development of the fungus when most leaves have grown and its relatively mild effects on photosynthesis (Hajji *et al.*, 2009). Nevertheless, oak mildew is likely to have a negative effect on regeneration, especially under certain forestry conditions, and could become more serious with climate change (Marçais & Desprez-Loustau, 2014). Theoretically, the negative effects of oak mildew on photosynthesis could be (partially?) offset by increased CO_2 in the atmosphere. On the other hand, oak mildew impairs the regulation of moisture loss and thus aggravates drought stress in hot summers (Lonsdale, 2015).

Another group of 'pests' with effects similar to those of oak mildew, inhibition of regeneration, are browsing mammals. Native deer, and especially several species of introduced deer, are all increasing their (feral) populations and have a marked effect on tree regeneration. Foresters are obliged to protect new plantations of virtually every broadleaf tree in plastic tubes, or success beyond the

sapling stage will be severely reduced. However, oaks and other trees do grow up through thickets of brambles and thorns (Vera, 2000 and my own observations). Experiments indicate that if there is enough light, oak seedlings and saplings tend to recover from browsing (Tyler, 2008). Owing to the very long time span involved — an oak that can live 500 years needs to produce just one successful tree to replace it during that time — the long-term effects of growth inhibitors on regeneration are difficult to assess. Further studies, perhaps including simulative modelling, are required to look at this issue, but most such studies tend to focus on forestry issues (i.e. on shorter-term forestry cycles and growth inhibition affecting timber quantities). Such a study should instead focus on the long time-span of an oak's potential life.

Some diseases could turn out to be more serious. One of these is known as Acute Oak Decline (AOD) and, in the UK, affects mostly native oaks (pedunculate and sessile); it was first observed in England around 1980. AOD is a bacterial disease, most probably involving *Brenneria goodwinii* and *Gibbsiella quercinecans*, both newly described species of unknown origin. Their implicated vector is the two-spotted oak bupestrid beetle (*Agrilus biguttatus*), a native species; the adult produces characteristic D-shaped exit holes in the bark. However, although there is a correlation between the occurrence of the disease and the presence of exit holes made by the beetle, there is no evidence that the beetle is actually the

vector (Alexander, 2015). It is assumed, but still under study (April 2015), that these bacteria produce necrogenic enzymes that cause the breakdown of cambium cells. Affected trees develop numerous lesions oozing black exudate from small splits in the bark. Mature oaks 'aged over 50 years' are mostly affected, but it has also been found on smaller trees. The spread of AOD is, as yet, limited to the southern half of England and eastern Wales, with greatest disease frequencies in South East England, the Midlands and the Welsh borders (Forest Research Information, 2015). During extensive field work all over England, I have observed the symptoms of AOD on mature oaks with a girth below 4.00 m but very rarely on veteran oaks; one example being a 6.95 m girth pollard oak in Bulstrode Park, Buckinghamshire, seen on 14th January 2015. Unlike ancient oaks, these much younger oaks (up to 215 years old, see White, 1998) are almost always planted. Could it be that tree nursery provenance with limited genetic variation plays a role in susceptibility to this disease? The disease may have been observed in continental Europe since 1918, spreading westwards and southwards since then. (See Denman *et al.*, 2015 for more details on AOD research).

Another condition recognised by Forest Research as a disease with complex causes is known as Chronic Oak Dieback (COD). This affects mostly pedunculate oak and has been observed in many areas throughout England for at least a century. The condition is also widespread on the continent (Tyler, 2008) and appears to occur with fluctuating frequencies. The most obvious symptoms are progressive dieback in the crown, leading to reduced density of foliage and death of main branches causing stag-horn crowns. This should not be confused with the process of crown retrenchment in veteran oaks (see Chapter 1), which normally does not lead to death. When mature oaks that do not show other signs of ageing, such as a hollowing trunk, develop significantly thinned foliage and eventually stag-horn crowns, COD may be involved. The causes can include environmentally induced stress, such as from droughts, weakening the tree's resistance to defoliation by insects, and attack from pathogenic fungi (but not dead-wood decaying fungi). The process of dieback is 'chronic' in the sense that it can take many years for the oak to die. The most serious recent episode of chronic dieback occurred in 1989–1994, when drought damage weakened trees which were then exploited by the two-spotted oak buprestid beetle, frequently leading to tree death (Forest Research data). More recent thinking about this form of dieback implicates 'compromised root health' involving root-attacking pathogens in combination with drought as the primary causes (Denman & Webber, 2009). While progress of dieback in mature oaks may be diagnosed as COD, this diagnosis is much more difficult in ancient or even veteran oaks. However, when uncommonly large numbers of ancient and veteran oaks are dying over a time span of decades, as seems to be the case on several sites, we may suspect that COD is happening. Examples I have observed where I believe this to be the case are at Birklands in Sherwood Forest (Chapter 9), at Crab Hill Iron Age Fort in Gerrards Cross, Buckinghamshire, and in the High Wood section of Grimsthorpe Park, Lincolnshire. For the oaks affected at these locations, environmental factors besides climate may lie at the core of the problem. In these examples, there is a suspicion that changes caused by what can be summarised as 'urbanisation', including nearby mining or extraction of ground water, may have impacted on the health of the trees. In none of these three sites is there a direct influence from farming or use as a golf course. I am therefore inclined to search for the primary environmental causes of COD suspected in ancient and veteran oaks, and consider the biotic factors as a corollary to be expected from these.

I mention the pathogen *Phytophthora ramorum*, which causes Sudden Oak Death (SOD), here only in passing for the sake of completeness. It kills species of oak native to North America and is also known to attack beech; disease on beech was first observed in Britain in 2003, but native oaks have not been affected (Forest Research data; Tyler, 2008).

An uncertain factor not yet discussed is 'climate change', a shorthand term for the long-term global warming trends observed since the mid-nineteenth century and especially since the beginning of the twentieth century. The global increase in the temperature of the atmosphere has so far been modest: about 1°C since 1860 and 0.7°C since 1900 (see numerous sources, including some doubts as to the accuracy of these figures, e.g. Plimer, 2009). Similar, if not greater, warming has already been experienced by some of the oldest oaks during the Medieval Warming Period (900–1300 CE), and this was followed by marked cooling during the Little Ice Age (1300–1850 CE). We have no data on how the native oaks responded to these changes. Ecological relationships are highly complex, and average annual temperatures are only the roughest indicators for changes in these. In a warmer climate (if sustained for a few centuries or more), periods of dormancy become shorter for the trees as well as for other organisms associated with them, and distribution may expand upwards and northwards (Kremer, 2013). Reductions in dormancy periods have been observed for many trees in northern Europe since the early nineteenth century (Firsov & Fadeyeva, 2012). If precipitation levels change or shift seasonally, this could also affect the oaks, but they already occur over a wide range of precipitation patterns. It would appear that the native oaks will be limited more by cooling trends than by warming trends, given where they now occur both in Britain and in continental Europe. 'Predictions' based on computer models indicate much higher rises in global temperatures for the next century. This is not the place to discuss the veracity of these model-based climate prognoses, but the inherent uncertainties with models of any kind in my opinion render a discussion of what their predictions may mean for ancient and veteran oaks in England highly speculative. If continued, warming of the atmosphere would result in regular droughts and this, combined with other stressors, could become the most significant impact of climate change on the ancient oaks. If such droughts are likely to impact beech more severely than oak, however, the oaks would benefit in woodlands where these two species occur together.

Rates of decline

At Windsor Great Park and Forest, the Crown Estate has begun a desk study using information from the First Edition of the OS maps to identify which of the marked trees still survive (Alexander & Green, 2013). This could provide more data that can be used to estimate rates of decline under various management regimes, ranging from pasture woodland to conifer plantations. But even if this study yields results, it cannot tell us whether these results could apply to the whole country or even across the south of England. Is it possible to infer rates of death and/or decline from analysis of the data on ancient and veteran oaks in the ATI database? Of the 3,609 records ≥6.00 m in girth oaks in England available in the database on 2nd February 2016, 148 or 4.1% were reported dead. We do not know if there is a bias in recording; for example, recorders might prefer to collect data for living trees, so that we are underestimating the proportion of dead oaks. If the recording bias is negligible, we could assume a figure of around 4% of ancient and veteran oaks in England to be dead; we may take this figure as a working hypothesis. In southern Sweden, the database of 'Tredportalen' (www.tradportalen. se/), accessed on 29th July 2015, contained 1,334 native oaks ≥6.00 m in girth, of which 75 or 5.6% were dead, a slightly higher figure. Of the 147 oaks ≥6.00 m in Blenheim Park in Oxfordshire recorded for the Blenheim Estate in 2002, seven or 4.8% were dead. A survey of Staverton Park in the mid 1990s reported c. 5.6% of all oaks dead (Sibbett, 1999, and Excel spreadsheet of this survey).

What might the proportion of dead oaks be under natural circumstances, i.e. if none of the negative factors caused by human interference applied? Can we apply theories of population dynamics to these figures, or to related data we have gathered? One potential difficulty in interpreting the data for oaks is that dead standing oaks tend to remain in the landscape for a long time, and would only fail to be recorded when reduced to a mere short stump of rotten mould and wood. Dead oak trees do disappear eventually, but does the rate of decay of dead oaks keep pace with the rate of death among still-living oaks? The rate at which dead oaks 'disappear' is an unknown variable, but to let our working hypothesis function, we could make a second assumption and keep this rate equal and constant. There is a theoretical 'minimum mortality', a constant for a population, which represents the loss of trees under ideal or non-limiting conditions. The rate of death

in a population, which is not a constant, depends on the actual longevity achieved by all individuals and is always greater than the theoretical minimum mortality (Odum, 1971). We want to know whether the observed mortality range of 4.1–5.6% is normal ('ecological') or abnormal (influenced by human interference). Each of the size classes 1–8 in Figure 11-15 includes a range 0.50 m in girth larger than the preceding class; class 9 covers a range 1.00 m larger than that covered by class 8; and class 10 includes all trees ≥11.00 m in girth. According to White (1998), we may assume a fairly constant thickness of annual rings and therefore a similar number of rings to be added in each of the first eight size classes. Oaks above 10.00 m girth often hardly increase in girth as they usually have bark only around sections of their bolling or trunk. I have divided these largest oaks into two classes of undeterminable age.

If size can stand as a proxy for age, Figure 11-15 shows that the number of oaks decreases exponentially with age. We are looking back in time in this graph, perhaps as much as 500 years from trees of 6.00 m girth to those of 13.00 m girth. A declining population would be expected to have a relatively large proportion of old individuals (Odum, 1971), indicating that the oldest trees are not being replaced in the population by younger trees as they die. Does the distribution of size/age classes in Figure 11-15 show a small, normal or large proportion of old individuals? The percentage of dead oaks was 4.1% of the total number of oaks ≥6.00 m in girth recorded by February 2016. Of the 3,197 oaks with girths of 6.00–8.99 m (size classes 1–6), 126 or 4% were dead. Of 143 oaks with girths ≥9.00 m recorded

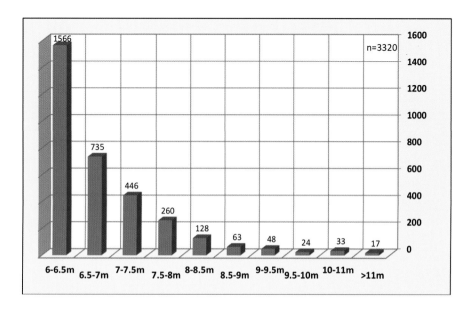

Figure 11-15. Distribution within 10 size classes, starting at 6.00 m girth, of 3,320 living ancient and veteran oaks in England as recorded by 1st April 2016. Calculations and graphics by Torsten Moller

by February 2016, 11 or 7.7% were dead. We can expect to see this increased mortality among the largest oaks even if mortality rates overall are low. The distribution of the 10 size classes shows the shape typical of an expanding population, with greatest numbers in the smaller (i.e. younger) classes. Our 'population' is, of course, limited to ancient and veteran oaks with ≥6.00 m girths and does not include smaller (younger) oaks, but these are vastly more numerous and it is very likely that their inclusion would only exaggerate the shape of the size/age distribution.

In the beginning of this chapter, I stated that if only 1.4% of the ancient oaks die each year, it takes just 50 years to reduce the population by half. This statement assumes that there is no replacement of dead individuals by living trees. We found that, at present, 4.1–5.6% of recorded oaks with girths ≥6.00 m are dead. This is not equivalent to an annual death rate because dead oaks remain standing for many years. Dead oaks disappear at one end of this time line and appear at the other, but evidently at a much lower rate than 5% of the population of ancient and veteran oaks. If dead trees stand to be recorded for an average of 20 years, the annual death rate would actually be 0.25%, provided that the proportion of dead oaks remains constant. The distribution of the size classes in our dataset indicates that the numbers of living ancient and veteran oaks are sufficient to replace the dead ones. Nearly half the number (46.8%) of all oaks ≥6.00 m in girth fall in size class 1 (6.00–6.49 m) alone. Among these, dead trees were 3.7% of the total, only slightly less than the 4.1% calculated for all size classes combined. It appears that the proportion of dead oaks remains fairly constant among the size classes 1–7 and only increases among the largest oaks, which on average

Figure 11-16 (top). A veteran oak with a girth of 7.34 m near Graft's Farm, Whorlton, Durham

Figure 11-17 (above). A veteran oak with a girth of 6.15 m in Sarnesfield Park, Herefordshire

would also be the oldest. These data do not indicate an increase in death rates above that which would occur naturally. Combined with the evidence of the regression curve obtained from the distribution of size classes, these percentages of standing dead oaks indicate that no decline of ancient and veteran oaks is imminent in England, unless the negative factors discussed in this chapter were to increase dramatically.

Analysis of datasets of ancient and veteran oaks recorded at National Trust properties and at Blenheim's High Park and Staverton Park (both SSSIs), all of which are protected sites in respect of the veteran and ancient trees, show size/age distributions similar to those discussed above for large oaks in the ATI dataset. Furthermore, a local study of the oak population at High Park in Blenheim Park, Oxfordshire using a different statistical representation seems to support the prognosis discussed above. In this study, all 'veteran' oaks, dead and alive, were compared by size (dbh) and it was found that dead oaks follow the bell curves of live oaks and all oaks. "There does not appear to be a particular age at which the proportion of dead to live trees increases substantially. This suggests that trees die for a variety of reasons at all ages, furthermore there is not a particular point at which the oak population will collapse and the majority die in a short period of time" (Blenheim High Park Management Plan, 2014 at www.wychwoodproject.org). Obviously, High Park is in a relatively good condition; there are sites where the outcome of a similar study could be different.

Data from a survey of ancient and veteran oaks in Richmond Park, Surrey carried out since 1993 were analysed in 2001–02. Trees were categorised into one of seven different states – dead fallen, dead collapsed, dead standing and alive with 1/4, 2/4, 3/4 and 4/4 of live bark present around the base of the tree. Outside some fire damaged areas, there appeared to be a "very low rate of attrition" (Adam Curtis by e-mail, 11th February 2014). Richmond Park is heavily used by the public and suffers from air pollution as it is located under the Heathrow Airport flight path and with busy car traffic through the park.

The question of recruitment over the very long time span that is relevant to ancient oaks is complicated. The results of analyses presented here are indicative only and there may be factors influencing this that have not been considered. Further research is needed to resolve this question more conclusively. Even if my analysis proves to be correct, it should not be concluded that threats are of no concern or not real. They do cause the premature death of ancient oaks, every one of which has a high value historically and in terms of biodiversity habitat and must be preserved when possible. As Oliver Rackham has said, a thousand 100-year-old oaks are not a substitute for one oak of 800 years. At the other side of the argument, we will have to accept that no life form is eternal; ancient oaks will die but they will be replaced by succeeding oaks as they have been in the past.

Conservation action

We can summarise the threats identified in this chapter as follows:

1. Overshadowing from competing trees
   a. plantations of conifers
   b. spontaneous forest regeneration
2. Landscape design or alteration of parks
   a. picturesque designs and planting schemes
   b. utilitarian designs
3. Agricultural malpractice
   a. damage by farm animals
   b. ploughing near trees
4. Vandalism
5. Lapsed pollards
6. Diseases and pests
   a. inhibition of regeneration
   b. AOD and other pathogens
   c. COD linked to environmental change

All of these threats are currently of concern and need to be addressed. Some have peaked in the past and their impacts are abating (1a, 2a), whereas others are still having a negative impact (2b, 3a, 4, 6a) or may be increasing (1b, 3b, 5, 6b, 6c).

What is to be done about these various threats to ensure that ancient oaks will remain numerous in the English landscape far into the future? Usually in nature conservation, the first necessary condition is an adequate legal framework offering protection. For ancient trees, legal protection is still almost exclusively linked to sites and not to the individual trees. The guidelines for SSSI notification do not provide for individual trees, so Natural England cannot at present notify an ancient oak as an SSSI (Green, 2001). A case has to be made to amend the guidelines to make this at least legally possible. Tree Preservation Orders (TPO) served by local authorities are primarily concerned with amenity interests. These orders are served within the context of the Town and Country Planning Act 1990 and its subsequent amendments and regulations. They are generally not applied to trees which are not visible from the public domain and are rarely applied to trees in the open countryside away from public access. However, a good case can be made according to Andy Colebrook (see text box) for the more frequent use of this legal instrument to

How a Tree Preservation Order (TPO) could be applied to ancient oaks in the countryside.

1. The Tree Preservation Order (TPO) regulations do not define amenity, and as such, I believe that this term can encompass attributes other than solely public visibility.

2. One of the primary attributes that make ancient trees, and especially ancient oaks, important is the ecological diversity that they support; it follows that this diversity brings with it amenity in other forms. Two of the better-known texts on the subject of ancient trees refer to this (Read, 1999; Lonsdale, 2013).

3. The Government Planning Portal makes reference to the need to understand and record a tree's importance as a wildlife habitat.

4. Further protection of ecological attributes can stem from the Habitat Regulations 2010 and the Wildlife and Countryside Act 1981. For example, all wild birds and all bat species are protected and ancient oaks can often provide habitat for both — and thereby a form of amenity — which can be safeguarded further by use of a TPO.

5. Making and challenging TPOs is all about making a good case at the end of the day...

6. One well-used method for health-checking TPO decisions is TEMPO (Forbes-Laird, 2009) (www.flac.uk.com/wp-content/uploads/2014/12/TEMPO-GN.pdf). This provides a framework within which to make decisions, and provides a record of why these decisions were made in the event of a challenge to demonstrate the rationale, transparency and the consistency of decision-making. Using the following hypothetical tree as an example, we could apply a TPO to a tree with no public visibility as follows (please refer to the TEMPO pro-forma at the last page accessed via the above link).

Part 1: Amenity assessment

a) Condition score 3 (a tree in satisfactory condition, may display dieback and decline, possibly retrenchment to a secondary crown, plenty of dead wood and substantial decay but due to either self or artificial crown consolidation remains stable).

b) Retention span score 4 (ancient trees will frequently have many decades or even centuries of remaining life span ahead).

c) Relative public visibility score 1 (the lowest score possible for a tree with zero public visibility).

d) Other factors score 5 (per the method), having scored 7 or more, the tree qualifies for consideration under part d and merits the highest score awarded to trees with veteran status.

Part 2: Expediency assessment
Precautionary only score 1 (again we assume the lowest)

Part 3: Decision guide
(We scored 13), so we are in the middle of the 12–15 band that shows a TPO to be defensible.

One protected ancient oak with zero public visibility!

We could have dropped a point for retention span along the way and still made the grade; however, we could also have scored more for condition and retention span in many circumstances, so it would appear that public visibility is less of a determining factor in the potential protection of ancient oaks than is fully realized. I would be very happy to make a robust case for the protection of this hypothetical tree a) at first instance and b) at objection to use of TPO or c) at Appeal for refusal of consent to carry out inappropriate works or felling.

Andy Colebrook MICFor, Dip.Arb(RFS), M.Arbor.A
Chartered Arboriculturist
Arboricultural Consultant Greenspaces Team

---

protect ancient oaks. The concept of a natural monument as applied to ancient oaks, as used for example in Germany, Poland and Sweden, with concomitant protected status, does not exist in the United Kingdom. This ought to change, but perhaps the sheer numbers of ancient oaks in England are both an excuse for complacency and a reason for lack of legislation, considering the potential price tag of implementation and enforcement. The inherent biodiversity of the habitats provided by veteran and ancient oaks imposes a heavy responsibility upon the UK in the context of the Convention on Biodiversity (CBD), which directs the signatory states to draw up biodiversity conservation plans and to protect their own biodiversity. Perhaps there are options here under the Common Agricultural Policy (CAP) Countryside Stewardship scheme (or its replacement if the UK indeed leaves the EU) to assist farmers to protect ancient trees on their farmed land as 'provision of improved habitat management'. This requires recognition of an ancient tree (oak) as a habitat for biodiversity. In conclusion, stronger legal protection is a necessary first step to conserve the isolated ancient oaks of England. On a landscape or habitat scale, legal protection could be strengthened if pasture woodland (with ancient and veteran oaks) were to be added to the list of European natural habitat types in Annex I of the EU Habitats Directive (Plieninger *et al.*, 2015). Currently, the closest habitat type listed is 'Fennoscandian wood pasture', but this is obviously geographically defined and does not apply outside of Finland and Sweden.

With (better) legal protection, the unnatural decline and death of ancient oaks resulting from adverse land management, malpractice and vandalism could be more effectively prevented. Without it, no legal action can be taken against a farmer who allows his animals to damage an ancient oak or compact the soil around it, or who ploughs through the roots of a tree that stands on his land and not in a protected area. Legal action at any rate would be a last resort; with increasing awareness, we will see more farmers taking pride in such trees and providing protection voluntarily. It does not take much to reserve a limited but sufficient area around the tree that is safe from the plough and build a simple fence around it to keep animals away. It is this awareness that should be (discretely) increased, and various relevant NGOs as well as government have a role to play in this effort. Schools in the vicinity of ancient oaks (or even those with such oaks in their grounds) could be issued with relevant materials for a lesson on biodiversity and ancient oaks; many teachers would welcome this. 'Ownership' of an ancient oak by a class of children will turn vandalism into care. Adverse land management could be prevented by legislation if ancient oaks were more often considered in the planning permission process under the Town and Country Planning Act 1990, for example when a new golf course is proposed in a park. Golf courses are often seen as benign development in terms of landscape design, but can have detrimental effects on ancient trees. There are examples (such as Packington Park in Warwickshire) where the golf course appears to have been situated in the area with the greatest density of ancient and veteran oaks, when another part of the park might have been chosen instead.

The planting of conifers among ancient and veteran oaks is becoming a 'crime' of the past. The Forestry Commission certainly has stopped doing this, both on its own and on leased land. On private land, grant-aiding has also moved away from conifer plantations in lowland woods, especially in woods that are classified as ancient woodland. The economics of small-scale plantation forestry also favour moves towards cessation of replanting with conifers; it is often no longer seen as viable. Nevertheless, there is still work to do where conifers grow taller and are threatening ancient and veteran oaks. Haloing is being undertaken

in several plantations, such as those in Savernake Forest, Windsor Forest and Croft Castle Park, but other locations are being neglected or subject to interventions at a speed that may be too slow to save some oaks.

More widespread now, and more insidious, is spontaneous forest regeneration. At some sites, there is no active woodland management at all, as in the remaining park of Aldermaston Court, but in many others regeneration takes place at greater speed than the limited resources available for woodland management can cope with. Especially where beech is present, this can occur quickly and oaks are suppressed faster than they can be set free. Examples of this can be seen in many places and I mention just a few: Epping Forest, Swinley Park, Kingston Lacey (The Oaks), Scadbury Park, Duncombe Park and Birklands (Sherwood Forest). All such pasture woodlands were not only grazed in centuries past, but people would bring in their pigs at times of mast, cut firewood and poles from coppice and pollards, and collect bracken and gorse; in doing so, they kept the woodlands more open, with more diverse habitats (unless they were overused, resulting in the degradation of the vegetation to heathland).

While pasturing in open woodland still exists as a form of land use across Europe, it is becoming rare except in Portugal and Spain, where it mostly occurs with other species of oak (Plieninger *et al.*, 2015). Management for biodiversity executed by land owners, with or without subsidies from government or the EU, often cannot keep up with new natural forest regeneration, let alone work away the huge backlog. We need to involve the public on a much larger scale than before. The Woodland Trust has been showing the way forward with their Community Woods Initiative (see *Broadleaf 85*; Autumn 2015). Woodland 'ownership' can have various meanings for volunteer groups, from legal ownership through lease arrangements to having or sharing responsibility for management. The benefits to people, society and nature resulting from such ventures are great. (In helping with coppicing in ancient woodland only 2–3 weekends per year, I make a contribution, obtain unlimited firewood, and have some great hardworking days out in the woods with friends.) Where ancient oaks are to be set free from competing trees, the cautionary principle should apply and professional advice should be sought; a sudden change in microclimate can turn out to be detrimental (Alexander *et al.*, 2003).

There is now much more experience with 'restoration cutting' of native oaks than was reported at a meeting in Epping Forest on pollard and veteran tree management hosted by the City of London Corporation in 1993 (Read, ed., 1996; Read *et al.*, 1996). A similar meeting had taken

Figure 11-18. A natural pollard oak in Parham Park, West Sussex

crucial, as hot dry summers will put extra stress on the tree and increase the chance of oak mildew on the new foliage (Green, 1996). But will we ever be able to predict the weather of the next summer in Britain? 'Restoration pollarding' especially requires caution, sufficient limbs must always be left on the tree, and it may be better to begin with crown reduction and gradually working down towards the bolling. They know how to perform this process in Turkey (see Chapter 6, Figure 6-5) and we can do it, too, as demonstrated on an ancient oak in Bayfield Park, Norfolk (Figure 11-19) and on the 10.55 m girth pollard oak in the village of Ruckinge in Kent. Experience is also being gained by an increasing number of professional arborists with certifications from the Arboricultural Association, and this knowledge should be tapped into. If we can master and apply the art of restoration pollarding, we will be able to avoid propping up the ancient oaks with unsightly poles (now increasingly made of steel) more often.

The various pests and diseases mentioned in this chapter that have the potential to affect the survival of native oaks are apparently difficult to manage (Forestry Commission website: www.forestry.gov.uk/pestsanddiseases, accessed 12th October 2015). In particular on AOD,

place at Burnham Beeches in 1991 (Read, ed., 1991) and from these sessions, a cautionary approach was advocated due to regular failures. On the other hand, I have seen numerous oaks, from mature vigorous trees to veterans and ancients, where storms have done it for us and which have gone on to grow a new and healthy crown of foliage. It is possible that natural breakage forms a better stimulant for dormant epicormic shoot buds than the chainsaw. There is no reason to be shy, especially if there is a good stock of young oaks available. Local weather conditions can be

perhaps the most serious threat, there is still much research underway to obtain data about the possible vector(s) and to identify the bacteria involved. Monitoring of its spread is also ongoing and in part dependent on 'citizen science', as Forestry Commission staffing is inadequate for the task at hand. Recommendations for appropriate management have been given and involve the felling of infected trees if these occur among still healthy oaks, and information on how to treat the bark and sapwood in an attempt to stall the spread of the infection to neighbouring trees.

Figure 11-19. An ancient pollard oak re-cut in Bayfield Park, Norfolk

Oak mildew has been present in the UK for much longer and is a concern particularly in relation to oak regeneration. However, it is not listed among the top priority tree diseases and pests in the above cited website of the Forestry Commission. There is only anecdotal evidence that it causes a decline in oak regeneration, and how this could affect the ultimate abundance of veteran and ancient oaks is even less clear. Many of the young oaks that I encounter in the field, planted or spontaneous seedlings, are not affected, and once these trees become mature oaks, they often appear to be unaffected or only slightly affected. This disease is at most a growth inhibitor, not a cause of death, which makes assessment of its long-term effects on tree survival more complicated.

How many oaks have to survive to replace an ancient oak? The obvious answer is just one. To determine whether this is sustainable, we must be able to estimate the overall numbers of living oaks that might survive to the ancient stage and, given the time span involved, this will always be difficult. Monitoring and more research is what is needed most here.

References

Aberg, F. A. (ed.) (1978). Medieval moated sites. *Council for British Archaeology Research Report* No. 17. The Council for British Archaeology, London.

Ainsworth, A. M. (2001). Stipes and spores in two British *Ganoderma* species. *Field Mycology* 2: 64–65.

Ainsworth, A. M. (2004). Developing tools for assessing fungal interest in habitats 1: beech woodland saprotrophs. *English Nature Research Reports* No. 597. English Nature (Natural England), Peterborough.

Ainsworth, A. M. (2005). Identifying important sites for beech deadwood fungi. *Field Mycology* 6: 41–61.

Ainsworth, A. M. (2006a). *Survey for oak polypore and an initial assessment of dead beech habitat quality using fungal indicators at Epping Forest (2005).* Unpublished report for the City of London Corporation.

Ainsworth, A. M. (2006b). *Survey for oak polypore at Epping Forest (2006).* Unpublished report for the City of London Corporation.

Ainsworth, A. M. (2008a). *Phellinus (Fuscoporia) wahlbergii* new to Britain. *Field Mycology* 9: 131–135.

Ainsworth, A. M. (2008b). *Survey for oak polypore Piptoporus quercinus and other fungi at Ashtead Common NNR, Surrey (2007).* Unpublished report for the City of London Corporation.

Ainsworth, A. M. (2008c). *Survey for oak polypore at Epping Forest (2007).* Unpublished report for the City of London Corporation.

Ainsworth, A. M. (2008d). *Fungal survey at Langley Park, Buckinghamshire (2007).* Unpublished report for Buckinghamshire County Council.

Ainsworth, A. M. (2008e). *Survey for oak polypore at Epping Forest (2008) and general fungal update highlighting species of conservation concern.* Unpublished report for the City of London Corporation.

Ainsworth, A. M. (2008f). *Surveys, strongholds and revised red listing of the legally protected wood-inhabiting fungi oak polypore Piptoporus quercinus and bearded tooth Hericium erinaceus.* Unpublished report for Natural England.

Ainsworth, A. M. (2010). *Survey for oak polypore at Epping Forest (2009) and updated records for zoned rosette and other fungi of conservation importance.* Unpublished report for the City of London Corporation.

Alexander, K. N. A. (1992). *Trinodes hirtus* (F.) (Coleoptera: Dermestidae) new to Gloucestershire. *British Journal of Entomology and Natural History* 5: 89.

Alexander, K. N. A. (2002a). The invertebrates of living & decaying timber in Britain and Ireland – a provisional annotated checklist. *English Nature Research Report* No. 467. English Nature (Natural England), Peterborough.

Alexander, K. N. A. (2002b). Forthampton Oaks, Gloucestershire: a site of major importance for saproxylic invertebrates. *British Journal of Entomology and Natural History* 15: 63-64.

Alexander, K. N. A. (2003). *Globicornis rufitarsis* (Panzer) (Dermestidae) at Blenheim Park, Oxfordshire. *The Coleopterist* 12: 96.

Alexander, K. N. A. (2004a). *Gnorimus variabilis* (Linnaeus) (Scarabaeidae) in West Sussex. *The Coleopterist* 13: 92.

Alexander, K. N. A. (2004b). Revision of the Index of Ecological Continuity as used for saproxylic beetles. *English Nature Research Report* 574. English Nature (Natural England), Peterborough.

Alexander, K. N. A. (2005). Wood decay, insects, palaeoecology, and woodland conservation policy and practice - breaking the halter. *Antenna* 29: 171-178.

Alexander, K. N. A. (2007). Old Growth: ageing and decaying processes in trees. In: I. D. Rotherham (ed.). The history, ecology and archaeology of medieval parks and parklands. *Landscape Archaeology and Ecology* 6: 8-12.

Alexander, K. N. A. (2011). A review of the saproxylic invertebrate assemblages at Birklands and Bilhaugh Sites of Special Scientific Interest. Sherwood Forest, Nottinghamshire. *Natural England Commissioned Report* NECR072. Natural England, Peterborough.

Alexander, K. N. A. (2012). What do saproxylic (wood-decay) beetles really want? Conservation should be based on practical observation rather than unstable theory. Pp. 41-59 in: I. D. Rotherham, C. Handley, M, Agnoletti & T. Samojlik (eds.). *Trees Beyond the Wood Conference Proceedings*, Wildtrack Publishing, Sheffield.

Alexander, K. N. A. (2013). Ancient trees, grazing landscapes and the conservation of deadwood and wood decay invertebrates. Pp. 330-338 in: I. D. Rotherham (ed.). *Trees, forested landscapes and grazing animals. A European perspective on woodlands and grazed treescapes*. Routledge, Abingdon, Oxfordshire.

Alexander, K. N. A. (2014). A review of the scarce and threatened beetles of Great Britain. Buprestidae, Cantharidae, Cleridae, Dasytidae, Drilidae, Lampyridae, Lycidae, Lymexylidae, Malachiidae, Phloiophilidae and Trogossitidae. Species Status No. 16. *Natural England Commissioned Report* NECR134. Natural England, Peterborough.

Alexander, K. N. A. (2015). What do we really know about oak jewel beetle and acute oak decline? *Arboricultural Magazine* 169: 50-53.

Alexander, K. N. A., Dodd, S. & Denton, J. S. (2014). A review of the scarce and threatened beetles of Great Britain. The darkling beetles and their allies. Aderidae, Anthicidae, Colydiidae, Melandryidae, Meloidae, Mordellidae, Mycetophagidae, Mycteridae, Oedemeridae, Pyrochroidae, Pythidae, Ripiphoridae, Salpingidae, Scraptiidae, Tenebrionidae & Tetratomidae (Tenebrionoidea less Ciidae). Species Status No. 18. *Natural England Commissioned Report* NECR148. Natural England, Peterborough.

Alexander, K. N. A. & Green, E. E. (2013). The nature conservation work of the Crown Estate in Windsor Forest and Great Park. British Wildlife 24 (5): 305-315.

Alexander, K. N. A., Green, E. E. & Key, R. (1996). The management of overmature tree populations for nature conservation – The basic guidelines. Pp. 122-135 in: H. J. Read (ed.). *Pollard and veteran tree management II*. City of London Corporation.

Alexander, K. N. A., Stickler, D. & Green, E. E. (2003). Is the practice of haloing successful in promoting extended life? A preliminary investigation of the response of veteran oak and beech trees to increased light levels in Windsor Forest. *Quarterly Journal of Forestry* 104: 257.

Almond, R. (2009). *Daughters of Artemis. The huntress in the Middle Ages and Renaissance*. D. S. Brewer, Cambridge, UK.

Andrén, A. (1997). Paradise lost. Looking for deer parks in medieval Denmark and Sweden. In: H. Andersson, P. Carelli & L. Esgård (eds.). Visions of the past. Trends and traditions in Swedish medieval archaeology. *Lund Studies in Medieval Archaeology* 19: 469-490.

Baggs, A. P., Blair, W. J., Chance, E., Colvin, C., Cooper, J., Day, C. J., Selwyn, N. & Townley, S. C. (1990). Blenheim: Park from 1705. Pp. 460-470 in: A. Crossley & C. R. Elrington (eds.). *A history of the County of Oxford: Volume 12, Wootton Hundred (South) including Woodstock*. London [www.british-history.ac.uk/vch/oxon/vol12/pp460-470]

Bateman, J. (1883). *The great landowners of Great Britain and Ireland*. 4th edition, Harrison, London.

Banbury, J., Edwards, R., Poskitt, E. & Nutt, T. (eds.) (2010). *Woodstock and the Royal Park. Nine hundred years of history*. Woodstock and the Royal Park 900 years Association, with Chris Andrews Publications Ltd., Woodstock.

Baudouin, J. C., Spoelberch, P. de & Meulder, J. van (1992). *Bomen in België. Dendrologische inventaris 1987-1992*. Stichting Spoelberch-Artois.

Bengtsson, V. & Fay, L. (2009). *2009 veteran pollard survey, Ashtead and Epsom Commons, Surrey*. City of London Open Spaces Department, London (www.treeworks.co.uk/downloads/Ashtead-Common-Veteran-Tree-Survey-Report-09-05-15.pdf)

Bennett, K. D. (1988). Holocene pollen stratigraphy of central East Anglia, England, and comparison of pollen zones across the British Isles. *New Phytologist* 109 (2): 237-253.

Bevan-Jones, R. (2002). *The ancient yew; a history of Taxus baccata*. Windgather Press, Macclesfield, Cheshire.

Birks, H. J. B., Deacon, J. & Peglar, S. (1975). Pollen maps for the British Isles 5,000 years ago. *Proceedings of the Royal Society* B189: 87-105.

Blank, R. (1996). Beurteilung des Gesundheitszustandes der "1000jährigen Eichen" im Landschaftsschutzgebiet "Ivenacker Tiergarten" des Forstamtes Stavenhagen und Empfehlung von Ma nahmen zu ihrer Pflege und Erhaltung. Forstpathologisches Labor, Hardegsen. [Report November 1996]

Boddy, L. & Rayner, A. D. M. (1983a). Ecological roles of basidiomycetes forming decay communities in attached oak branches. *New Phytologist* 93: 77–88.

Boddy, L. & Rayner, A. D. M. (1983b). Origins of decay in living deciduous trees: the role of moisture content and a re-appraisal of the expanded concept of tree decay. *New Phytologist* 94: 623–641.

Boddy, L. & Thompson, W. (1983). Decomposition of suppressed oak trees in even-aged plantations. *New Phytologist* 93: 261–276.

Bond, C. J. (1981). Woodstock Park under the Plantagenet Kings: the exploitation and use of wood and timber in a medieval deer park. *Arboricultural Journal* 5: 201-213.

Bond, J. (1998). *Somerset parks and gardens, a landscape history.* Somerset Books, Tiverton.

Bond, J. (2004). *Monastic landscapes.* The History Press, Stroud.

Brewer, S., Cheddadi, R., Beaulieu, J. L. de & Reille, M. (2002). The spread of deciduous *Quercus* throughout Europe since the last Glacial period. *Forest Ecology & Management* 156: 27-48.

Brookes, P. C. & Wigston, D. L. (1979). Variation of morphological and chemical characteristics of acorns from populations of *Quercus petraea* (Matt.) Liebl., *Q. robur* L. and their hybrids. *Watsonia* 12: 515-324.

Brown, M. B. (1985). *Richmond Park: the history of a royal deer park.* Hale, London.

Buckland, P. I. & Buckland, P. C. (2006). *Bugs Coleoptera Ecology Package.* www.bugscep.com.

Buck-Sorlin, G. H. & Bell, A. D. (2000). Crown architecture in *Quercus petraea* and *Q. robur*: the fate of buds and shoots in relation to age, position and environmental perturbation. *Forestry* 73 (4): 332-349.

Buis, J. (1985). *Historia Forestis: Nederlandse bosgeschiedenis.* (2 volumes) HES Uitgevers, Utrecht.

Burgess, H. W. (1827). *Eidodendron, views of the general character and appearance of trees, foreign and indigenous, connected with picturesque scenery.* J. Dickins, London. [plate 11 = Chenies Oak]

Butler, J., Alderman, D. & Muelaner, B. (2015). Recording ancient trees. *British Wildlife* 26 (6): 398-405.

Cantor, L. M. (ed.) (1982). *The English medieval l*andscape. Croome Helm, London.

Cantor, L. M. (1983). *The medieval parks of England: a gazetteer.* Loughborough University of Technology, Leicester.

Cantor, L. M. & Hatherly, J. (1979). The medieval parks of England. *Geography* 64: 71-85.

Castle, G. & Mileto, R. (2005). A veteran-tree site assessment protocol. *English Nature Reports* No. 628. English Nature (Natural England), Peterborough.

CATE2 database (accessed Jan 2016). http://cate.abfg.org/

Chatters, C. & Sanderson, N. (1994). Grazing lowland pasture woods. *British Wildlife* 6 (1): 78-88.

Church, J. M., Coppins, B. J., Gilbert, O. L., James, P. W. & Stewart, N. F. (1996). *Red Data Books of Britain and Ireland: lichens. Volume1; Britain.* English Nature (Natural England), Peterborough.

Clifton, S. J. (2000). The veteran trees of Birklands and Bilhaugh, Sherwood Forest, Nottinghamshire. *English Nature Research Report* No. 361. English Nature (Natural England), Peterborough.

Coles, J. M. & Orme, B. J. (1976). The Sweet Track, railway site. *Somerset Levels Papers* 2: 34-65.

Cooke, R. C. & Rayner, A. D. M. (1984). *Ecology of saprotrophic fungi.* Longman, London.

Coppins, A. M. & Coppins, B. J. (2002). *Indices of ecological continuity for woodland epiphytic lichen habitats in the British Isles.* British Lichen Society, Wimbledon.

Crawley, M. J. (2005). *The Flora of Berkshire.* Brambleby Books, Harpenden.

Crockatt, M. E., Campbell, A., Allum, L., Ainsworth, A. M. & Boddy, L. (2010). The rare oak polypore *Piptoporus quercinus*: Population structure, spore germination and growth. *Fungal Ecology* 3: 94–106.

Denman, S. & Webber, J. (2009). Oak declines: new definitions and new episodes in Britain. *Quarterly Journal of Forestry* 103 (4): 285-290.

Denman, S. *et al.* (2015). *Report on Acute Oak Decline (AOD) research – April 2014 to April 2015.* Woodland Heritage / Forest Research, April 2015.

Dennison, E. (ed.) (2005). *Within the pale. The story of Sheriff Hutton Park.* William Sessions Ltd., York, England.

Dickson, G. & Leonard, A. (1996). *Fungi of the New Forest.* British Mycological Society, London.

Dowsett, J. M. (1942). *The romance of England's forests.* John Gifford Ltd., London.

Edlin, H. L. (1971). Woodland notebook: goodbye to the pollards. *Quarterly Journal of Forestry* 65: 157-165.

Eliasson, P. & Nilsson, S. G. (2002). 'You should hate young oaks and young noblemen' – The environmental history of oaks in eighteenth- and nineteenth-century Sweden. *Environmental History* 7 (4): 659-674.

Ellenberg, H. (1988). *Vegetation ecology of Central Europe.* Cambridge University Press, Cambridge, U.K.

Elwes, H. J. & Henry, A. (1906-13). *The trees of Great Britain and Ireland.* Privately published in seven volumes. [Vol. 2 contains accounts of "the English Oak" + photographs.]

Evans, S. E., Henrici, A. & Ing, B. (2006). *Preliminary assessment: the red data list of threatened British fungi.* www.fieldmycology.net/Download/RDL_of_Threatened_British_Fungi.pdf

Evelyn, J. (1664). *Sylva, or a discourse of forest trees.* London. [later editions 1670 to 1786]

Falinsky, J. B. (1986). Vegetation dynamics in temperate lowland primeval forests. Ecological studies in Białowie a Forest. *Geobotany* 8. Dr. W. Junk Publishers, Dordrecht.

Farjon, A. (2015). How old can a tree be? A rebuttal of claims of extraordinary ages of trees and other organisms in recent research and popular science literature. *The Linnean* 31 (2): 23-29.

Farjon, A. & Farjon, R. (1991). Naturnahe Laubwaldreste um Westerstede in der ostfriesisch-oldenburgischen Geest: Eine Vegetationsanalyse mit Berücksichtigung des Naturschutzes. *Tuexenia* 11: 359-379, Beilage.

Farmer, A. (1984, repr. 1996). *Hampstead Heath.* Historical Publications Ltd., London.

Fay, N. (2007). *Defining and surveying veteran and ancient trees.* Veteran Trees Initiative, Natural England, Peterborough.

Ferris, C., Oliver, R. P., Davy, A. J. & Hewitt, G. M. (1995). Using chloroplast DNA to trace postglacial migration routes of oaks into Britain. *Molecular Ecology* 4 (6): 731-738.

Firsov, G. A. & Fadeyeva, I. V. (2012). Trees and shrubs of Saint-Petersburg in the age of climate change. *International Dendrology Society Yearbook* 2011: 63-71.

Fleming, L. & Gore, A. (1982). *The English garden.* Mermaid Books, Michael Joseph Ltd., London.

Fletcher, J. (2011). *Gardens of earthly delight. (The history of deer parks).* Windgather Press, Oxford.

Forbes-Laird, J. (2009). *TEMPO, Tree Evaluation Method for Preservation Orders. Guidance note for users.* Forbes-Laird Arboricultural Consultancy, Dendron House, Blunham, Bedford MK44 3ND.

Foster, A. P. (1996). *Globicornis nigripes* (Fabricius) (Dermestidae) in Worcestershire. *British Journal of Entomology & Natural History* 9: 233.

Fowles, A. P., Alexander, K. N. A. & Key, R. S. (1999). The Saproxylic Quality Index: evaluating wooded habitats for the conservation of dead-wood Coleoptera. *The Coleopterist* 8: 121-141.

Fraiture, A. & Otto, P. (2015). *Distribution, ecology and status of 51 macromycetes in Europe.* Meise Botanic Garden, Meise.

Franklin, P. (1989). Thornbury woodlands and deer parks, part 1: the Earls of Gloucester's deer parks. *Transactions of the Bristol and Gloucester Archaeological Society* 107: 149-169.

FRDBI database (accessed Jan 2016). www.fieldmycology.net/ FRDBI/FRDBI.asp

Grant, R. (1991). *The Royal Forests of England.* Alan Sutton, Stroud.

Green, E. E. (1996). Thoughts on pollarding. Pp. 1-5 in: H. J. Read (ed.). *Pollard and veteran tree management II.* City of London Corporation.

Green, E. E. (2001). Should ancient trees be designated as Sites of Special Scientific Interest? *British Wildlife* 12 (2): 164-166.

Green, E. E. (2010a). Natural origin of the commons: people, animals and invisible biodiversity. *Landscape Archaeology and Ecology – End of Tradition* Vol. 8: 57-62.

Green, E. E. (2010b). The importance of open-grown trees; from acorn to ancient. *British Wildlife* 21 (3): 334-338.

Guy, A. (2000). Changes in a permanent transect in an oak-beech woodland (Dendles Wood, Devon). *English Nature Research Report* No. 347. English Nature (Natural England), Peterborough.

Hajji, M., Dreyer, E. & Marçais, B. (2009). Impact of *Erysiphe alphitoides* on transpiration and photosynthesis of *Quercus robur* leaves. *European Journal of Plant Pathology* 125 (1): 63-72. [Doi: 10.1007/s10658-009-9458-7]

Hallé, F. (2002). In praise of plants. *Timber Press*, Portland, Oregon.

Hammond, P. M. (2007a). *Pentaphyllus testaceus* (Hellwig) (Tenebrionidae): an established and perhaps native British species? *The Coleopterist* 16: 47-52.

Hammond, P. M. (2007b). *Dryophthorus corticalis* (Paykull), the 'Windsor Weevil' (Dryophthoridae): current status at Windsor and records for two further British localities. *The Coleopterist* 16: 73-79.

Hardin, G. (1968). The tragedy of the commons. *Science* 162: 1243-1248.

Harding, P. T. (1980). Shute Deer Park, Devon. A pasture-woodland of importance for the conservation of wildlife. *Natural Devon* 1: 71-77.

Harding, P. T. & Rose, F. (1986). *Pasture woodlands in lowland Britain; a review of their importance for wildlife conservation.* Institute of Terrestrial Ecology, Huntingdon.

Harding, P. T. & Wall, T. (2000). *Moccas: an English deer park.* English Nature (Natural England), Peterborough.

Harris, E., Harris, J. & James, N. D. G. (2003). *Oak, a British history.* Windgather Press, Macclesfield.

Hayman, R. (2003). *Trees, woodlands and Western civilasation.* Hambledon & London, London.

Hill, D. (1981). *An atlas of Anglo-Saxon England.* Oxford University Press, Oxford.

Hinde, T. (ed.) (1995). *The Domesday Book. England's heritage, then & now.* Coombe Books, Godalming, Surrey.

Hight, J. (2011). *Britain's tree story. The history and legends of Britain's ancient trees.* National Trust Books, London.

Hodder, K. H., Bullock, J. M., Buckland, P. C. & Kirby, K. J. (2005). Large herbivores in the wildwood and modern naturalistic grazing systems. *English Nature Research Report* No. 648. English Nature (Natural England), Peterborough.

Holden, E. (2003). *Recommended English names for fungi in the UK.* Plantlife International, Salisbury.

Hooke, D. (1998). Medieval forest and parks in southern and central England. Pp. 19-32 in: C. Watkins (ed.). *European woods and forests. Studies in Cultural History.* CAB International, Wallingford, UK.

Hooke, D. (2010). *Trees in Anglo-Saxon England. Literature, lore and landscape.* The Boydell Press, Woodbridge.

Hurst, C. (1911). *The book of the English Oak.* [reissued in paperback 2010] Lynwood & Co., Ltd., London.

Hyman, P. S. (1992). A review of the scarce and threatened Coleoptera of Great Britain. Part 1 [updated by M. S. Parsons]. *UK Nature Conservation* No. 3.

James, N. D. G. (1981). *A history of English forestry*. Basil Blackwell, Oxford.

James, P.W., Hawksworth, D. L. & Rose, F. (1977). Lichen communities in the British Isles: a preliminary conspectus. Pp. 295-413 in: M. R. D. Seaward (ed.). *Lichen Ecology*. Academic Press, London.

Johnson, C. (2008). Saproxylic Coleoptera from Calke, Chatsworth, Hardwick and Kedleston Parks, Derbyshire, 1881-1999. *The Coleopterist* 17: 35.

Kirby, K. J., Thomas, R. C., Key, R. S. & McLean, I. F. G. (1995). Pasture-woodland and its conservation in Britain. *Biological Journal of the Linnean Society* 56 (Supplement): 135-153.

Kirkham, T. (2009). The decompaction programme on trees at Kew. *International Dendrology Society Yearbook* 2008: 77-80.

Klose, F. (1985). *A brief history of the German forest – achievements and mistakes down the ages*. Eschborn, Germany.

Koop, H. (1981). *Vegetatiestructuur en dynamiek van twee natuurlijke bossen: het Neuenburger en Hasbrucher Urwald*. Centrum voor Landbouwpublikaties en Landbouwdocumentatie (Pudoc), Wageningen.

Krahl-Urban, J. (1959). *Die Eichen. Forstliche Monographie der Traubeneiche und der Stieleiche*. Verlag Paul Parey, Hamburg & Berlin.

Kremer, A. (2013). Evolutionary responses of European oaks to climate change. *International Oaks, the Journal of the International Oak Society* 24: 11-20.

Le Hard de Beaulieu, A. & Lamant, T. (2006). *Guide illustré des Chênes*. (2 volumes), Editions du 8ème, Paris.

Liddiard, R. (2003). The deer parks of Domesday Book. *Landscapes* 4 (1): 4-23.

Liddiard, R. (ed.) (2007). *The medieval park: new perspectives*. Windgather Press, Macclesfield.

Longstaffe-Cowan, T. (2005). *The gardens and parks at Hampton Court Palace*. (with photographs by V. Russell). Frances Lincoln Ltd., London.

Lonsdale, D. (2013). *Ancient and other veteran trees: further guidance on management*. The Tree Council, London.

Lonsdale, D. (2015). Review of oak mildew, with particular reference to mature and veteran trees in Britain. *Arboricultural Journal* 37 (2): 61-64.

Lott, D. A., Alexander, K. N. A., Drane, A. & Foster, A. P. (1999). The dead-wood beetles of Croome Park, Worcestershire. *The Coleopterist* 8: 79-87.

Loudon, J. C. (1838). *Arboretum et fruticetum brittannicum: or the trees and shrubs of Britain, native and foreign, hardy and half-hardy, pictorially and botanically delineated*. 8 volumes (second edition 1844). Longman *et al.*, London.

Mabey, R. & Jones, G. L. (1993). *The wildwood - in search of Britain's ancient forests*. Aurum Press, London.

Manwood, J. (1598). *A Treatise and Discourse of the Lawes of the Forrest*. Thomas White & Bonham Norton, London. [2nd ed. 1615; 3rd ed. 1665; 4th ed. 1717; 5th ed. 1741]

Marçais, B. & Desprez-Loustau, M.-L. (2014). European oak powdery mildew: impact on trees, effects of environmental factors, and potential effects of climate change. *Annals of Forest Science* 71: 633-642.

Marren, P. (1990). *Woodland Heritage, Britain's ancient woodland*. David & Charles Publishers, Newton Abbot, London.

Marren, P. (2014). The ancient oaks of Savernake Forest. *British Wildlife* 25 (5): 305-313.

Massingham, H. J. (1952). *The Southern Marches*. Hale, London.

Menzies, W. (1864). *The history of Windsor Great Park and Windsor Forest*. Longmans, London.

Menzies, W. (Jr.) (1904). *Windsor Park and Forest, being a short history of the park and forest and a description of the most interesting features to be found there*. Oxley & Son, Windsor.

Miles, A. (2007). *Hidden trees of Britain*. Ebury Press (Random House Group), London.

Miles, A. (2013). *The British oak*. Constable & Robinson Ltd., London.

Mileson, S. A. (2009). *Parks in medieval England*. Oxford University Press, Oxford.

Mitchell, A. (1966). Dating the ancient oaks. *Quarterly Journal of Forestry* 60 (4): 271-276.

Mitchell, A. (1974). *A field guide to the trees of Britain and northern Europe*. William Collins & Sons, London and Glasgow.

Mitchell, F. J. G. (2005). How open were European primeval forests? Hypothesis testing using palaeoecological data. *Journal of Ecology* 93: 168-177.

Morris, M. G. & Perring, F. H. (eds.) (1974). *The British Oak: its history and natural history*. Botanical Society of the British Isles / E. W. Classey Ltd., Faringdon, Berkshire.

Muir, R. (2000). Pollards in Nidderdale, a landscape history. *Rural History* 11 (1): 95-111.

Muir, R. (2005). *Ancient trees, living landscapes*. Tempus Publishing, Ltd., Stroud.

Muir, R. (2008). *Woods, hedges and leafy lanes*. Tempus Publishing Ltd., Stroud.

Naturvårdsverket (2006). *The oak, history, ecology, management and planning*. Proceedings from a conference in Linköping, Sweden, 9-11 May 2006.

Nicolson, N. & Hawkyard, A. (1988). *The counties of Britain, a Tudor atlas by John Speed*. Pavilion Books Ltd., in association with The British Library, London.

Nieto, A. & Alexander, K. N. A. (2010). *European Red List of Saproxylic Beetles*. Publications Office of the European Union, Luxembourg.

Odum, E. P. (1971). *Fundamentals of ecology*. Third edition, Saunders College Publishing, Philadelphia.

Oliver, J. & Davies, J. (2001). Savernake Forest oaks. *Wiltshire Archaeological and Natural History Magazine* 94: 24-46.

Ordnance Survey (1998). Explorer 160 map 1:25,000

Overall, A. (2008). *Richmond Park fungi survey report 2008*. http://londonfungusgroup.org.uk/surveyrichmond.pdf

Pakenham, T. (2003). *Meetings with remarkable trees*. Weidenfeld & Nicholson, London.

Palmer, J. (2003). The Major Oak, Sherwood Forest, England. *International Oaks, the Journal of the International Oak Society* No. 14: 25-28.

Parfitt, D., Hunt, J., Dockrell, D., Rogers, H. J. & Boddy, L. (2010). Do all trees carry the seeds of their own destruction? PCR reveals numerous wood decay fungi latently present in sapwood of a wide range of angiosperm trees. *Fungal Ecology* 3: 338–346.

Peterken, G. F. (1969). The development of vegetation in Staverton Park, Suffolk. *Field Studies* 3: 1-39.

Peterken, G. F. (1996). *Natural woodland: ecology and conservation in northern temperate regions*. Cambridge University Press, Cambridge.

Pilcher, J. (1998). Tell-tale bog oaks. *The Countryman* 103 (7): 38-39.

Piper, R. & Hodge, P. (2002). The rare species of UK *Cryptocephalus*: the current state of knowledge. *English Nature Research Report* No.469. English Nature (Natural England), Peterborough.

Pitt, J. (1984). Ancient pollards of Lullingstone Park, Kent. *Transactions of the Kent Field Club* 9 (3): 129-142.

Pittman, S. (1983). *Lullingstone Park. The evolution of a mediaeval deer park*. Meresborough Books, Rainham, Kent.

Plieninger, T., Hartel, T., Martín-López, B., Beaufoy, G., Bergmeier, E., Kirby, K., Montero, M. J., Moreno, G., Oteros-Rosas, E. & Uytvanck, J. van (2015). Wood pastures of Europe: Geographic coverage, social-ecological values, conservation management, and policy implications. *Biological Conservation* 190: 70-79.

Plimer, I. (2009). *Heaven and Earth: Global warming, the missing science*. Quartet Books Ltd., London.

Pollard, A. & Brawn, E. (2009). *The great trees of Dorset*. Dovecote Press, Wimborne.

Preston, E. & Wallis, S. (2006). *The Forest of Bere, Hampshire's forgotten forest*. Halsgrove, Tiverton, Devon.

Price, S. (2010). Report of the Derbyshire Limestone Field Meeting 9-12th October 2009. *British Lichen Society Bulletin* 106: 110-123.

Pryor, F. (2011). *The making of the British landscape*. Penguin Books, London.

Rabbitts, P. (2014). *Richmond Park – from medieval pasture to royal park*. Amberley Publishing, Stroud, Gloucestershire.

Rackham, O. (1976). *Trees and woodland in the British landscape*. J. M. Dent. London. [revised edition 1990; paperback edition 1996]

Rackham, O. (1980). *Ancient woodland; its history, vegetation and uses in England*. Edward Arnold, London.

Rackham, O. (1986). *The history of the countryside*. J. M. Dent, London.

Rackham, O. (1989). *The last Forest*. J. M. Dent, London.

Rackham, O. (1994). Trees and woodland in Anglo-Saxon England: the documentary evidence. Pp. 7-11 in: Rackham, J. (ed.). Environment and economy in Anglo-Saxon England. *Council for British Archaeology Research Report* 89.

Rackham, O. (2003a). *The illustrated history of the countryside*. Weidenfeld & Nicolson, London.

Rackham, O. (2003b). *Ancient woodland, its history, vegetation and uses in England*. New edition. Castlepoint Press, Colvend, Dalbeattie.

Rackham, O. (2004). Pre-existing trees and woods in country-house parks. *Landscapes* 5 (2): 1-16.

Rackham, O. (2006). *Woodlands*. The New Naturalist Library; a survey of British natural history. Collins, London. [2nd edition 2010]

Rackham, O. (2009). *Grimsthorpe Park, a preliminary report*. Internal report, unpublished. Grimsthorpe and Drummond Castle Trust Ltd.

Rayner, A. D. M. & Boddy, L. (1988). *Fungal decomposition of wood*. John Wiley, Chichester.

Read, H. J. (ed.) (1991). *Pollard and veteran tree management*. City of London Corporation.

Read, H. J. (ed.) (1996). *Pollard and veteran tree management II*. City of London Corporation.

Read, H. J. (1999). *Veteran trees: a guide to good management*. English Nature (Natural England), Peterborough.

Read, H. J., Frater, M. & Noble, D. (1996). A survey of the condition of the pollards at Burnham Beeches and results of some experiments in cutting them. Pp. 50-54 in: H. J. Read (ed.) *Pollard and veteran tree management II*. City of London Corporation.

Reddington, C. (1996). Lifting of the bog oaks. *The Countryman* 101 (1): 90-94.

Riches, C., Bailey, N. E. M. & Gaunt, C. (2010). *The Times Atlas of Britain*. Times Books, London.

Roberts, B. K. & Wrathmell, S. (2000). Peoples of wood and plain: an exploration of national and local

Roberts, J. (1997). *Royal Landscape. The gardens and parks of Windsor*. Yale University Press, New Haven & London.

Rooke, H. (1790). *Descriptions and sketches of some remarkable oaks in the park at Welbeck, in the county of Nottingham, a seat of His Grace, the Duke of Portland*. John Nichols (for the author), London.

Rose, F. (1976). Lichenological indicators of age and environmental continuity in woodlands. Pp. 279-307 in: D. H. Brown, D. L. Hawksworth & R. H. Bailey (eds.): *Lichenology: Progress and Problems*. Academic Press, London.

Rose, F. (1993). Ancient British woodlands and their epiphytes. *British Wildlife* 5 (1): 83-93.

Rose, F. & Wolseley, P. (1984). Nettlecombe Park, its history and its epiphytic lichens: an attempt at correlation. *Field Studies* 6: 117-148.

Rotherham, I. D. (2007a). The ecology and economics of medieval deer parks. *Landscape Archaeology and Ecology* 6: 86-102.

Rotherham, I. D. (2007b). *The history, ecology and archaeology of medieval parks and parklands*. Wildtrack Publishing, Sheffield.

Sanderson, N. A. (2010). Lichens. Pp. 84-111 in: A. C. Newton (ed.). *Biodiversity in the New Forest*. Pisces Publications, Newbury, Berkshire.

Schumer, B. (1999). *Wychwood. The evolution of a wooded landscape*. The Wychwood Press, Charlbury, Oxfordshire.

Shirley, E. P. (1867). *Some account of English deer parks, with notes on the management of deer*. John Murray, London.

Sibbett, N. (1999). Ancient tree survey of Staverton Park and The Thicks SSSI, Suffolk. *English Nature Research Reports* No. 334. English Nature (Natural England), Peterborough.

Singh, G., Dal Grande, F., Cornejo, C., Schmitt, I. & Scheidegger, C. (2012). Genetic basis of self-incompatibility in the lichen-forming fungus *Lobaria pulmonaria* and skewed frequency distribution of mating-type idiomorphs: Implications for conservation. *PLoS ONE* 7 (12): e51402.

Small, D. (ed.) (1975). *Explore the New Forest. An official guide by the Forestry Commission*. Her Majesty's Stationery Office, London.

Smith, A. (1776). *An enquiry into the nature and causes of the wealth of nations*. W. Strahan & T. Cadell, London.

Smith, C. (2004). *The Great Park and Windsor Forest*. Bank House Books, New Romney, Kent.

Smith, J. H., Suz, L. M. & Ainsworth, A. M. (2015). Red List of Fungi for Great Britain: *Bankeraceae*, *Cantharellaceae*, *Geastraceae*, *Hericiaceae* and selected genera of *Agaricaceae* (*Battarrea*, *Bovista*, *Lycoperdon* & *Tulostoma*) and *Fomitopsidaceae* (*Piptoporus*). Submitted to JNCC.

Spake, R., Linde, S. van der, Newton, A. C., Suz, L. M., Bidartondo, M. I. & Doncaster, C. P. (2016). Similar biodiversity of ectomycorrhizal fungi in set-aside plantations and ancient old-growth broadleaved forests. *Biological Conservation* 194: 71–79.

Spooner, B. & Roberts, P. (2005). *Fungi*. Collins New Naturalist Library. Collins, London.

Stebbing, E. P. (1916). *British forestry. Its present position and outlook after the war*. John Murray, London.

Stebbing, E. P. (1928). *The forestry question in Great Britain*. John Lane / The Bodley Head Ltd., London.

Steel, D. (1984). *(Shotover) The natural history of a Royal Forest*. Pisces Publications, Oxford.

Strutt, J. G. (1826). *Sylva Britannica, or portraits of forest trees, distinguished for their antiquity, magnitude or beauty*. Folio edition, privately published by subscription.

Strutt, J.G. (1830). *Sylva Britannica, or portraits of forest trees, distinguished for their antiquity, magnitude or beauty*. Second (quarto) edition, Longman, Rees, Orme, Brown and Green, London.

Stubbs, A. E. & Drake, M. (2001). *British Soldierflies and their allies*. British Entomological & Natural History Society, Reading.

Suz, L. M., Barsoum, N., Benham, S., Dietrich, H. P., Fetzer, K. D., Fischer, R., García, P., Gehrman, J., Kristöfel, F., Manninger, M., Neagu, S., Nicolas, M., Oldenburger, J., Raspe, S., Sánchez, G., Schröck, H. W., Schubert, A., Verheyen, K., Verstraeten, A. & Bidartondo, M. I. (2014). Environmental drivers of ectomycorrhizal communities in Europe's temperate oak forests. *Molecular Ecology* 23: 5628–5644.

Taylor, W. L. (1946). *Forests and forestry in Great Britain*. C. Lockwood & Son, London.

Tittensor, R. M. (1978). A history of The Mens: a Sussex woodland common. *Sussex Archaeological Collections* 116: 347-374.

Tubbs, C. R. (1986). *The New Forest*. The New Naturalist Library; a survey of British natural history. Collins, London.

Tubbs, C. R. (2001). *The New Forest. History, ecology and conservation*. [2nd revised edition] New Forest Ninth Centenary Trust, Lyndhurst.

Tyler, M. (2008). *British oaks; a concise guide*. The Crowood Press, Marlborough.

Vera, F. W. M. (2000). *Grazing ecology and forest history*. CABI Publishing, Wallingford.

Voß, E. & Rüchel, F. (2014). *Ivenacker Eichen. Tausendjährige Eichen im Wandel der Zeit*. Forstamt Stavenhagen; Förderverein Ivenacker Eichen.

Ueckermann, E. & Hansen, P. (1968). *Das Damwild. (Naturgeschichte, Hege und Jagd)*. Verlag Paul Parey, Hamburg, Berlin.

Ullrich, B., Kühn, S. & Kühn, U. (2009). Unsere 500 ältesten Bäume: Exklusiv aus dem Deutschen Baumarchiv. BLV Buchverlag GmbH & Co. KG, München.

Walwin, P. C. (1976). *Savernake Forest*. Privately Published, Cheltenham.

Watkins, C., Lavers, C. & Howard, R. (2003). Veteran tree management and dendrochronology [in] Birklands & Bilhaugh cSAC, Nottinghamshire. *English Nature Research Report* No. 489. English Nature (Natural England), Peterborough.

Welch, R. C. (1987). *Gnorimus variabilis*. Pp. 179-180 in: D. B. Shirt (ed.). *British Red Data Books: 2. Insects.* Nature Conservancy Council, Peterborough.

Werf, S. van der (1991). *Bosgemeenschappen. Natuurbeheer in Nederland*, Deel 5. Pudoc, Wageningen.

Whitaker, J. (1892). *A descriptive list of the deer parks and paddocks of England.* Ballantyne, Hanson & Co., London.

White, J. (1995). Dating the veterans. *Tree News* Spring/Summer 1995. The Tree Council, London.

White, J. (1997). What is a veteran tree and where are they all? *Quarterly Journal of Forestry* 91 (3): 222-226.

White, J. (1998). Estimating the age of large and veteran trees in Britain. *Forestry Commission Information Note* 12. Forestry Commission, Edinburgh.

Whitehead, D. (1996). Brampton Bryan Park, Herefordshire: a documentary history. *A Hereford Miscellany*: 163-177. The Woolhope Club.

Whitehead, D. (2001). *A survey of historic parks & gardens in Herefordshire.* Hereford and Worcester Gardens Trust, Hereford.

Whitlock, R. (1979). *Historic forests of England.* Moonraker Press, Bradford-on-Avon, Wiltshire.

Whittock, M. (2009). *A brief history of life in the Middle Ages.* Constable & Robinson, London.

Wilks, J. H. (1970). The Meavy Oak. *Quarterly Journal of Forestry* 64 (3): 280-281.

Wilks, J. H. (1972). *Trees of the British Isles in history and legend.* Frederick Muller, London.

Williamson, T. (2013). *Environment, society and landscape in early medieval England – time and topography.* The Boydell Press, Woodbridge.

Wiltshire, M., Woore, S., Crips, B. & Rich, B. (2005). *Duffield Frith: history and evolution of the landscape of a medieval Derbyshire Forest.* Landmark Publishing Ltd., Ashbourne.

Woodland Trust (2008). *Ancient Tree Guide no. 4: What are ancient, veteran and other trees of special interest?* Woodland Trust, Grantham, Lincolnshire.

Yahr, R., Coppins, B. J. & Ellis, C. J. (2014). Quantifying the loss of lichen epiphyte diversity from the pre-industrial Exmoor landscape (south-west England). *Lichenologist* 46 (5): 711-721.

Young, C. R. (1979). *The Royal Forests of medieval England.* University of Pennsylvania Press, Philadelphia.

Glossary

acid grassland: grassland on acidic soil (pH 4–5.5) characterised by grasses and herbs that thrive on nutrient-poor and well-drained substrates such as sand and gravel; such grasslands are *unimproved* and usually lightly grazed

acid rain: rain contaminated by aerial pollutants such as sulphur dioxide from industrial burning of fossil fuels (especially coal) which caused 'lichen deserts' and, in extreme cases, the death of trees, especially conifers

Acute Oak Decline (AOD): a bacterial disease or infestation affecting native oaks, destroying cambium and potentially lethal, especially in mature oaks of modest size (see Chapter 11)

adventitious buds: buds formed 'accidentally', i.e. as a consequence of or in reaction to an injury, such as a cut stem or branch

afforestation: a) in the Middle Ages, a legal act declaring a tract of land to be administered under *Forest Law*, not Common Law, in order to protect the *venison* and *vert* for the king's hunting pleasure; b) in modern times, to plant a tract of land with trees in order to grow them for timber, with management for wildlife at best a subordinate purpose

ancient oak: an oak with characteristics that show it to be in the final stage of its life (see Chapter 1) but here more specifically all oaks with girths ≥6.00 m which could be >400 years old

Ancient Tree Inventory (ATI): an online database (www.ancient-tree-hunt.org.uk/) of recorded ancient and veteran trees in the UK, administered by the Woodland Trust, the Tree Register of the British Isles (TROBI) and the Ancient Tree Forum. The data are collected and checked by hundreds of volunteer recorders and verifiers and include details about the trees, often including photographs. The trees are georeferenced, allowing mapping on OS maps and Google Earth aerial imagery of the landscape. It is the primary data source used for this book

annual ring: the total amount of wood, in the seasonal climate of England distinguished into early and late wood, that a tree lays down in a year in a concentric ring around a thereby expanding trunk, as well as on branches and roots (Figs 2-1, 2-11)

assarts: clearings, often illegal, made in *Royal Forests* by would-be settlers (or squatters) for the purpose of agriculture; the activity is called assarting

biodiversity: the sum total of diverse species of animals, fungi and plants in a *habitat*; usually expressed as a list of species (or the number of species) living in a particular location

bolling: the permanent trunk of a *pollard* as contrasted with the trunk of a *maiden* which is often referred to as the bole

brown rot: wood rot involving the decay of *cellulose* leaving the *lignin* intact and caused by specific fungi (see Chapter 10) (Fig. 2-2)

burr: irregular growth of bark and wood on the trunk and/or larger branches, often forming a bulging outgrowth; they are caused by the activation of numerous *dormant* and hidden buds, most of which do not succeed in breaking through the bark, distorting the regular cambium layers that produce wood and bark on a tree

cellulose: a polysaccharide (C_6H10O5) of glucose units constituting the chief part of plant cell walls

chases: areas of open (not enclosed) mostly uncultivated land, usually partly wooded and made exempt from *Forest Law*, where magnates favoured by the king were allowed to hunt the 'beasts of the forest' without royal supervision and could impose sanctions on trespass (see Chapter 5)

commons: areas of mostly open (not enclosed) land used for pasture, wood cutting (often in enclosed *coppices*) and collecting of firewood and other materials, and where appropriate *pannage* was communal rather than private (see Chapter 5)

coppice: a) (*underwood*) trees which are cut to near ground level every few years and then grow again from the *stool*; b) *woodland* that is managed by coppicing, i.e. periodic cutting near ground level

coppice stool: the living stump of a tree that remains after coppicing, sprouting new shoots that grow into multiple stems which can be cut again

covert: a (planted) thicket hiding game

current annual increment (CAI): the amount of wood added each year after the mature stage in the life of a tree (i.e. optimum crown size) has been reached; this amount remains constant as long as the optimum crown size is maintained

deer parks:

  medieval deer parks: deer parks created during the medieval period; in England, these date from a mention in Domesday Book (1086) to 1485 (see Chapter 4)

  Tudor deer parks: deer parks created during the reign of the Tudors (1485–1603) (see Chapter 4)

dendrochronology: the science of constructing a chronology of identifiable years by matching *annual ring* patterns (mainly variations in width) in wood of known age with older wood with partially overlapping years, thereby extending the chronology back in time and, subject to certain limitations, providing a means by which an ancient piece of wood with sufficient *annual rings* can be dated

Domesday Book: a survey of land tenure conducted by order of William I and issued in 1086; it covers most of England and is of great historical importance as it indicates and quantifies land use parish by parish, with mention of ownership before and after the Norman Conquest (1066)

dormant buds: buds that lie inactive under the bark until some undetermined cause triggers their growth; if successful, these buds produce shoots that grow into branches

endophyte: (adjective: endophytic) growing inside, here pertaining to fungi that have entered by spores in an earlier life stage of a tree before the *mycelium* began to develop and spread

epicormic growth: growth of shoots from buds located under the bark of the trunk or larger branches; see also *dormant buds*

forest: a) with capital 'F' — land put under *Forest Law* by the Crown to serve as a royal hunting reserve (Latin *forestis*), it could be wooded (in part) or virtually lack trees and may have included settlements (see Chapter 5); b) with lower case 'f' — 1) land dominated by trees with a mostly closed canopy (barring disturbances) and in Europe now almost always planted or of planted origin, 2) in a medieval or earlier context, a forest with lower case 'f' could be, for example, an Anglo-Saxon forest not yet under *Forest Law* in which the trees were never planted

forestis (from Latin *foris* meaning 'outside of', see also the English word foreigner) see *forest*

Forest Law: rules and regulations pertaining to *Royal Forests* imposed by the Crown that excluded such areas from Common Law, restricting the use of the land in order to promote the 'beasts of the forest' for the royal hunt; this law had its own jurisdiction and courts and was imposed by the invading Normans after the Conquest of 1066

forestry: the activity of growing and manageing trees for timber. In the Middle Ages, the management of *coppice woods* (with or without oak *standards*) would have been almost the only form of *forestry*; planting and growing trees for timber did not play a role before *c.* 1550. Only very recently has a multi-use concept in *forestry*, including amenity, ecology and recreation, become prevalent in *forests* managed by public bodies

FRDBI database: Fungal Records Database of Britain and Ireland (administered by the British Mycological Society)

girth: the measurement (in metres and centimetres in this book) around the trunk of a tree, conventionally at 1.50 m above the ground (see Chapter 2)

habitat: the natural home or environment of a species of animal, fungus or plant

haloing: cutting and removing younger trees around an ancient or veteran tree to create a 'circle of light' (halo) and thereby releasing the ancient tree from competition

heartwood: the dead core wood of a tree; in large oaks, this is all of the wood that occurs inside a limited number of most recently formed annual rings through which liquid transport occurs (see *sapwood*)

lammas: growth of new leafy shoots from *dormant buds* under the bark of the trunk or large branches; it can occur in summer in reaction to earlier defoliation by insects (from Lammas Day, 1st of August, a Roman Catholic Church feast)

landscaping: the creation of a visual landscape by design, usually in a park; at the height of this fashion, this could include planting of (double) avenues, planting of groups of trees, opening up of vistas, creation of broad lawns and damming of streams to create lakes; the mansion usually formed the centrepiece in these schemes

lignin: an amorphous polymeric substance related to *cellulose*; together these substances form the woody cell walls of plants

maiden: when applied to trees, it refers to a tree that has never been cut for regrowth, either as *coppice* or as *pollard*. A natural pollard is technically a maiden and is not treated as a *pollard* in the statistics in this book

manors: in the medieval period in England, a manor was a farming community under the feudal system of land tenure, with the lord of the manor living in a manor house and the dependent peasants usually in a village; it had lands with various uses supporting the community and providing income for the lord (see Chapter 5 and Fig. 5-28)

mast years: years in which oak and beech (in this country) produce larger than average crops of acorns and nuts; such abundant crops occur irregularly every few years

moats: a broad ditch surrounding a building or buildings, usually a dwelling of some kind; there were dry *moats*, usually in hilly terrain, and wet *moats* filled with water and both had a defensive purpose

multi-stemmed: with more than a single stem, where the stems separate below the level where a single stem (bole or *bolling*) would be evident and could be measured; such trees are normally *coppice* trees whereas a *pollard* is not classified as *multi-stemmed* because it has a distinct *bolling* that can be measured at a minimal height of 1.30 m above the ground

mycelium (plural mycelia): thread-like fungal bodies forming masses of woven microscopic tubular filaments (hyphae) which extend at their tips, branch and fuse together into a living network

mycorrhizas: the symbiotic connective parts of fungal hyphae and root tips of plants where nutrients are traded between fungus and plant (here an oak tree)

occlusion: blocking a cavity; in oaks (and other trees), new wood and bark are formed from the edges of the gap, for example growing inwards from the circumference of a broken branch, closing the gap

pannage: the collection of acorns and beech nuts to feed pigs, or the driving of pigs into woodland that produces these nuts

park pale: earth bank with inside ditch and wooden fencing enclosing a *deer park* to prevent the deer from escaping (see Fig. 4-1 for remains)

pasture woodland: (also known in the literature, for example in Oliver Rackham's books, as 'wood pasture') *woodland* open to grazing animals such as deer and domestic cattle, or if no longer grazed, retaining a typical semi-open structure with small areas of closed canopy, solitary trees in parkland and open grassland or heathland in a mosaic pattern; including, besides oaks and other major native trees, characteristic species like field maple and hawthorn or blackthorn

podzol: a soil profile in acid sand caused by leaching of minerals by slow downward percolation of slightly *acid rain* water and characterised by leached grey sand near the surface down to dark brown sand stained by precipitating iron oxides. In NW Europe, such a leached profile is evidence of ancient deforestation and replacement of woodland by heathland

pollen diagram: a diagram giving a visual record of the abundance of pollen, expressed as a percentage of total pollen counts, through a depositional column and divided by tree species or genera, grasses and other plants. The column is peat or sediment in which pollen has been preserved and in which there is a time sequence from bottom to top, and of which samples of peat or sediment with pollen counts can be dated by 14C methods (see Fig. 8-2)

pollard:

managed: a tree that has been (repeatedly) cut at a height above ground beyond the reach of browsing animals (e.g. deer, cattle, horses) and that has grown new branches from about this height

natural: a tree that has been broken well above the ground by natural causes (storms or lighting strikes) and has lost (most of) its crown, and that has grown new branches from about this height

radiocarbon dating: a dating method using the known rate of decay of the radioactive isotope of carbon (14C) into the neutral isotopes (12C, 13C) in dead but preserved organic matter, such as wood, which ceased taking up new carbon when it died; the amount of remaining 14C in the sample set against its 'half-life' gives a date of the material's death in 14C years BP (before present = 1950)

reiteration: (adjective: reiterative) growth of secondary branches below the earlier primary branching; these secondary branches originate from axillary buds, from buds that were *dormant* under the bark (in *ancient oaks* often from *burrs* on the trunk or on the thickest limbs) or from *adventitious buds* (e.g. after cutting (see *coppice* and *pollard*))

retrenchment: a reduction of crown size by gradual die-back of major branches; in this manner, the tree 'economizes' and reaches a new balance, probably between the remaining functional root system and the total leaf surface and annual increment of wood. A smaller crown is then often formed by *reiteration*

ridge and furrow: a pattern of low, rounded and often slightly curved ridges interspersed with shallow furrows, visible from the air and/or on the ground, now mostly in grassland that has not been ploughed in later times. This pattern results from a medieval manner of ploughing and is evidence of an arable open field system with no trees at the time of ploughing

Royal Forest: see *forest*

saprotrophic fungi: fungi feeding on dead organic tissues

saproxylic: wood consuming; organisms, in our context mainly fungi and invertebrates, that consume dead wood are *saproxylic*

sapwood: the outer wood under the bark with living wood cells through which liquid transport takes place; on the inside of this zone, wood cells die forming *heartwood*, on the outside, new cells are formed; the thickness of the *sapwood* zone varies with the size of the branch, trunk or root

senescence: the final stage of life when dieback and decay proceed faster than new growth, the tree is in terminal decline, eventually leading to death

shredding: cutting the branches off a tree but leaving the bole and branches high in the crown intact; this method of wood cutting was apparently rare, at least with the oaks in England

stag-horn oaks: oak trees with dead or dying branches in the upper crown signifying *retrenchment* and to be distinguished from pathological die-back, in which case no new foliage branches are formed below the *stag-horn* branches

standards: trees within *coppice* woods, usually oaks, that were left to grow to a useful size for timber. In the Middle Ages, these trees were not planted but selected from spontaneous seedlings and protected against browsing animals; they stood more or less evenly spaced and free from other such trees for a few decades before being cut

sustainable forestry: developed in eighteenth century Germany and adopted elsewhere in continental Europe, this was a 'scientific' method of growing and harvesting trees which ensured continuation of the timber resource, as opposed to the destructive practices that had led to deforestation previously

underwood: trees and shrubs growing to a height below the main canopy of a *wood*; in managed *woods*, these are the *coppiced* trees and shrubs

unimproved grassland: grassland that has not been mechanically levelled or dragged and/or re-sown and has not been treated with fertilizers, herbicides or frequent mowing; the vegetation is composed of native species including those of old and stable grassland communities and grazing is light and often patchy

venison: meat of deer; but in terms of *Forest Law* it would refer to both this and the living deer

vert: (French for green) the trees in a *Royal Forest* protected by *Forest Law*

veteran oak: an oak developing the characteristics of an *ancient oak* but still deemed to be in the second (mature) stage of its life. Here. the term veteran is ignored when applied in the database to an oak with a *girth* of ≥6.00 m, and the tree is considered ancient

white rot: wood rot involving the decay of *lignin*, leaving the *cellulose* intact and caused by specific fungi (see Chapter 10)

wildwood: a hypothetical form of *woodland* (hence put between inverted commas in the main text) postulated for a period in prehistory when human influence was minimal and hence had little or no impact on the composition and structure of the tree-dominated vegetation; it is thought to have predated the arrival of agriculture and husbandry, in Britain around 6,000 years ago

woodland: land dominated by trees; here it applies to all land with native trees at least in part unplanted, ranging from open parkland to closed canopy (but see *forest* and *woods*). Of special interest in this book is *pasture woodland* (see Chapter 8)

woods: here distinguished from *woodland* in a more general sense as managed *woods*, i.e. *coppice* with or without standard trees (mostly oaks); in *woods*, trees, apart from *coppice stools*, never grow large and ancient. In modern times, coppicing may have been replaced by planting and harvesting trees, often conifers (see *forest*)

List of illustrations and tables

Acknowledgements

The making of this book can be said to have gone through a prehistoric phase long before the historic phase (writing it) was even contemplated. When I came to live and work in England in the summer of 1993, I had no knowledge whatsoever of its ancient oak trees. I arrived from The Netherlands to take up a fixed-term research job at the Department of Plant Sciences of the University of Oxford. The project lasted two and a half years. It had to do with pine trees in Mexico, not with oak trees in England, but I knew something about the English countryside and decided to take up temporary residence in the village of Eynsham, about six miles from the Department, which distance I biked on most working days. I also took my bicycle out into the countryside on Sundays with fine weather, and on one of these wanderings I passed through a back gate of Blenheim Park. This was Combe Gate which connects, via a public footpath past High Lodge and through Springlock Gate, with the landscaped part of Blenheim Park. Along this tarmacked road, one can spot some of the park's ancient oaks and my curiosity was alerted. I parked my bike and wandered in a bit further along a track and what I saw then amazed me. I returned to my bike and went on through the park to exit at a gate near Bladon. Along the way were some more ancient oaks, but nothing as impressive as in the wilder part of the park I had just left. During my time in Oxford I did not pursue this any further, perhaps I was too focussed on conifers.

On 1 February 1996 I was appointed by the Royal Botanic Gardens, Kew as a research scientist studying the taxonomy of the world's conifers. I moved and bought a house in Brentford, near my new permanent job. Even though there is less countryside around this location, I still took my bike out on Sundays. Richmond Park was one of the better places close enough to visit regularly and there I found the ancient oaks again! I began to realize

that this was quite special, there is nothing like it in The Netherlands, even though that country has its share of the same native oaks, including extensive woodlands on sandy soil. Then I found Windsor Great Park, an even more impressive site for ancient oaks than Richmond Park, with some oaks much bigger in girth. Now my interest had been activated enough to make it a 'hobby', a word I despise as a real interest ought to be more serious. When, in 1997, the Duchess of Devonshire wrote to RBG Kew to ask if someone could help her determine how old the oaks of Chatsworth are, her letter landed on my desk. I was invited to visit, prepared some kind of an answer by looking at growth rates in sawn-off oak limbs in Richmond Park, and travelled to Derbyshire. The Duchess herself took me around Chatsworth and we went into the Old Park. Another great site with many ancient oaks revealed itself and I was now hooked.

During the ensuing years I made many visits to the ancient oaks of England, building up a library of high-quality photographs and field notes of the oaks I had seen. However, my research at RBG Kew was on conifers, for which I travelled around the world and about which I wrote many papers and 11 books. It occurred to me that perhaps after retirement I might do something similar about the ancient oaks, but for a long time I had no clear ideas about how to approach the subject. I knew I would not be the first to write a book about them and it seemed pointless just to duplicate the effort of others. Yet it would be good to learn more in the meantime, so I read these books and added the more informative ones to my personal library. After retirement from RBG Kew and completion of my conifer projects, I was free to switch to the ancient oaks. As an Honorary Research Associate at RBG Kew, I retain a base there and can use the facilities as if I were still a member of staff. Perhaps the most significant benefit for this project was that my association with RBG Kew opened the doors to virtually every site in the country that I wished to visit to see the oaks. I am grateful for the trust RBG Kew has invested in me to use this passport with discretion.

At some point in time I came across the Woodland Trust's website *Ancient Tree Hunt* (www.ancient-tree-hunt.org.uk/), which allowed me to locate many more ancient oaks that I could visit. Then I realised that if I could obtain the data on ancient and veteran oaks from the Woodland Trust as an Excel spreadsheet, I could analyse this and formulate a research project using the data. I am grateful to David Alderman of The Tree Register/Woodland Trust, who manages the *Ancient Tree Hunt* (ATH) database (now the Ancient Tree Inventory, ATI) for agreeing to my request. Slowly it dawned on me that, with all the records geo-referenced and with various information, often including photographs, attached to each tree record, I could try to add still more information. Each tree could be plotted on the Ordnance Survey map, both the nineteenth century first edition and the current map, as well as on Google Earth imagery. I could find out things about the landscape and its history in which these trees stood. At a meeting with Jill Butler of the Woodland Trust I outlined my plans and she agreed that I could use the data for this purpose; a research project was born. I thank the Woodland Trust and the Tree Register of the British Isles (TROBI), whose data are incorporated in the ATI, for this and for subsequent updates.

Meanwhile, I decided to become a tree recorder myself and in due course added hundreds of ancient and veteran oaks to the ATI, as well as hundreds more photographs. The annotated database of all living oaks with girths of 6.00 m and greater in England which I built from these records formed the backbone of my research; without it, this book could not have been written. In addition to the data in my annotated database of oaks derived from the ATI records, I have made use of survey data with significance for ancient oaks from several estates, which were compiled by professional consultants. The National Trust, the Crown Estate Windsor, Blenheim Palace, Chatsworth Estate and Staverton Park all made their tree survey data available, and I have made use of these datasets to analyse in more detail the populations of ancient and veteran oaks on the relevant sites. I thank Brian Muelaner (National Trust until 2014, now Ancient Tree Forum Chair), Ted Green (Windsor), John Smith of Vantage Point Cartographics Ltd. (Blenheim), Steve Porter (Chatsworth) and Gary Battell (Staverton) for sharing these data with me. Andrew Oakley shared his data on named historic oaks, which formed a useful addition and update to lists compiled

in the published literature. I thank him for this and subsequent email discussions. There seemed to be a subject for yet another book on oaks in the information shared by Andrew, so I have made a restrained use of this type of data in this volume.

My research on the ancient oaks has used two approaches. It often began with a desktop investigation using the database and the ATI website facility, which led to further mostly historic research on the internet, supported by library literature and similar sources. There are very good web-based sources on relevant history, such as British History Online (www.british-history.ac.uk/), websites on the history of English counties and parishes, and historical maps going back to the sixteenth century (which are both online and published in hard copy), and I have made extensive use of these. Wikipedia, The Free Encyclopedia (www.wikipedia.org/) is improving with time and when consulted with care has yielded valuable information. As a Fellow of the Royal Geographical Society I had access to the resources of the Society's old maps collection; the assistance given with my searches by staff is appreciated. I thank Felix Lancashire of Richmond Local Studies at the Richmond Library for permission to use an image of a seventeenth century map of Richmond Park. Other aspects of the ancient oaks needed different kinds of information.

Beyond what is available in the published literature I found several people who have concerned themselves with estimating the age of oaks and who have used interesting approaches. I am thankful to one of them, David Lovelace of Hereford, for information about his interesting and detailed experiments on growth rates in veteran oaks, as well as for sharing other information on ancient parks and oaks in Herefordshire. Torsten Moller of Saffron Walden, Essex sent me his data and calculations on the age of the Panshanger Oak. These approaches ultimately rely on (disappeared) annual rings in the wood and I needed a good image of this, which was readily provided by RBG Kew's wood anatomist, Peter Gasson. With the wealth of data from the ATI, I knew I could analyse the distribution of ancient oaks in England and show the results on maps. I was very fortunate that Justin Moat, who runs the Geographical Information System (GIS) unit at RBG Kew, was prepared to help. He experimented with mapping methods and GIS tools until he could present the answers to my questions, and the distribution maps in this book were produced by him with my data. I am very grateful for Justin's work on this aspect of the book.

The second approach that I used to study ancient oaks was field visits. I substantially intensified my visits to ancient oaks all over England, traveling to see numerous oaks. I wanted to visit as many sites of significance as possible in addition to the largest oaks that stand solitarily. These oaks occur on land owned by public bodies and non-governmental organisations (NGOs), on private estates, in parks that are open to the public to various degrees, on farms, at restricted MOD sites and on yet other types of property. (The tree symbols generated on the map from the ATI data are of two colours: green on land with public access and red on private land often requiring permission to visit. This information is helpful in planning a visit but needs to be checked as it is not always accurate or up to date.) Across privately owned land, there are often public footpaths or right of ways and if oaks were near these, or on land generally open to the public, permission to view was not required. Despite this, I made my intentions for field work known to bodies such as the City of London Corporation, the Crown Estate Windsor, the Forestry Commission, the National Trust and the Royal Parks London. I am particularly grateful for advice and, where needed, special permission to Adam Curtis of the Royal Parks, Jeremy Dagley of the City of London Corporation, Brian Muelaner of the National Trust and Daniel West of Crown Estate Windsor. Access to some sites with SSSI status needed help or mediation from Natural England, and this was provided by Saul Herbert for Moccas Park and by Nicola Orchard for Yardley Chase; Nicola joined me in the field and I thank both for their efforts. A large number of private land owners opened their parks or estates to me and I am grateful to all; many are listed below and I am hoping not to have omitted anyone who should be included. On some occasions, hospitality was extended beyond the opportunity to visit oaks in the field, perhaps the most memorable being put up for the night in a four-poster bed at Kimberley Park. I am grateful to all land owners, including the anonymous people who just said "yes, go ahead" after I had pulled the doorbell or sounded the knocker.

Thanks to James H. Bathurst of Eastnor Castle, for permission to visit the deer park on two occasions; Michael Baxter, Estate Manager, Albury Park, for permission to visit and for showing me the oaks; Ray Biggs, Access Manager, Grimsthorpe Park, for unlimited access to the park on two days; Robert ('Robbie') and Iona Buxton of Kimberley Park, for hospitality and showing me the oaks in the park; Nichola Cariss of Eaton Manor in Shropshire, for hospitality and showing me a 10.00 m ancient oak; Alistair and Sarah Carr of The Grove, Frostenden in Suffolk, for hospitality and showing me the ancient and veteran oak pollards; David Chivers, Estates Bursar, Hazlegrove Preparatory School, for showing me the two big oaks in the school grounds; The Cowdray Estate, for permission to visit the golf course and adjacent parklands to see the oaks; Roy Cox, Rural Enterprises Manager, Blenheim Palace, for permitting repeated access to the amazing High Park and doing field research on the ancient oaks there; Lord Cranbrook and his son Jason (Lord Medway) for showing me ancient oaks near Glemham House; Jules Curtis, Head Gardener of Longleat, for showing me the ancient oaks in the park; Lord Devonport, for hospitality in the far north and discussions about gentry land ownership; Mrs Enthoven of Spye Park in Wiltshire, for showing me the two great oaks in the park; Mrs J. H. Garnier of Lydham Manor in Shropshire, for showing me the Lydham Manor Oak; Kay Gleeson, Estate Agent, Packington Park, for permission to access and giving me a tour of the park; John Grimshaw, Castle Howard and Yorkshire Arboretum, for organising my visit to see the ancient oaks on the estate and for alerting me to the significance of Brampton Bryan Park; Mr T. Gwyn-Jones and Hannah English of Hamstead Marshall Park, for permission to access and arranging visits to the park, with help in the field from Alun Jones, Head Gardener; Edward and Victoria Harley of Brampton Bryan Park, for hospitality and permitting access to the park; Lord Hertford of Ragley Park, for permission to visit the oaks in the park; Major P. W. Hope-Cobbold of Glemham Park, for permission to visit the oaks in the park; Chris Hoyes, Park Ranger, Grimsthorpe Park, for taking me around the former deer parks and discussing management of these sites; Sir Thomas Ingilby of Ripley Castle, for permission to visit the ancient oaks in the deer park; John King of Abbotswood Farm, Gloucestershire, for permission to visit the oaks in former Brockworth Park; Andrew Kinnear, Manager of Parham Park, for permission to visit the deer park and the oaks; Mrs Plumtre of Fredville Park in Kent, for showing me the Fredville Oak in October 2004; Steve Porter, Head of Gardens and Landscape, Chatsworth, for permission to visit Old Park on two occasions after my initial visit with the Duchess in April 1997; Sarah Pratt of Ryston Park, for showing me 'Kett's Oak'; Lord Rotherwick of Cornbury Park, for permission to visit the oaks in the park on two occasions; Hektor Rous for allowing repeated access to Henham Park in Suffolk; Lord Salisbury of Hatfield House, for permission to visit the oaks on the Hatfield Estate; John Shields of Donington Park in Leicestershire, for permission to visit the deer park and for help with finding some of the oaks; Stoneleigh Deer Park Golf Club, for permission to visit the oaks in Stoneleigh Park; Lord Tollemache of Helmingham Hall, for *ad hoc* permission to visit the park off season and showing me some of the oaks; The Hon. Charlotte Townshend of Melbury Park, for permission to see the Billy Wilkins Oak and other oaks in the park with the deer keeper; and Robert Williams of Haughley Park in Suffolk, for showing me the ancient oak in the private garden.

In the field, I was sometimes joined by other specialists and/or tree recorders. Brian Jones, ATI verifier for Herefordshire and Gloucestershire, deserves special mention here for taking a full day out of his busy schedule to accompany me to sites in Gloucestershire and Herefordshire, and also for alerting me to new finds of large oaks in Herefordshire. Archie Miles of Stoke Lacy, Herefordshire, himself an author of books on trees, took me to see the hidden 'Gospel Oak' and we had many discussions about ancient oaks on that day. I thank Susan Pittman of the Kent Archaeological Society for coming with me on a second visit to Lullingstone Park, for giving me a copy of her book on the park and for helpful discussions. I am also grateful to Benedict Pollard, of Eynsham, Oxfordshire for his help with some of my ancient oak surveys in High Park, Blenheim and for lively discussions on why these trees are so special. Martyn Rix deserves thanks for kindly taking me to see some interesting sites in Devon and Somerset.

The success of a book depends on publicity, and this involves the internet as well as other media. Denis Filer of Oxford University helped me with a website on ancient oaks as part of his BRAHMS-based library of botanical research websites: http://herbaria.plants.ox.ac.uk/bol/ancientoaksofengland. I have been asked to give a lecture on the topic of ancient oaks to many associations and societies, and I thank all for inviting me. My presentation at the Linnean Society of London on 18th February 2016 was hailed as "the most populated event in the history of the Linnean Society" and I thank the Society for organizing it. Other websites have paid attention to my research, and I particularly thank Rob McBride for sharing information on ancient oaks on his website, www.treehunter.co.uk. I thank Robert Shaw of Sunbury, Surrey for organising an exhibition of some of my photographs of ancient oaks at the Sunbury Embroidery Gallery and for giving me the printed photos afterwards; these were put on display in the foyer of the Herbarium Building at RBG Kew where many visitors have seen them.

Before I began to research and write this book, I explored ideas with several people, one of whom was Thomas Pakenham. We met for a meal in a pub on Kew Green and had an amicable and interesting exchange of approaches. This discussion helped me to determine the direction I should take, which given the kind of data I had collected, was closer to that adopted by Oliver Rackham than to that taken by Archie Miles or Thomas Pakenham. And so I worked on the book, chapter by chapter, simultaneously gathering and analysing data and writing. This work was interspersed with field trips and sometimes with attending meetings, such as a discussion led by Robert Liddiard in London on the rise and fall of medieval deer parks. Chapters emerged from the data that I analysed and the questions they posed, and I sought answers in the analysis of more or other data.

As the book began to take shape, I realized from reading yet more literature how important the ancient oaks are for biodiversity. Here was a subject requiring specialist attention if justice were to be done to it. I am fortunate to have found three such specialists who were interested in my book and willing to contribute to it. They wrote the chapter on biodiversity related to ancient oaks, using invertebrates (Keith Alexander), lichens (Pat Wolseley) and non-lichenised fungi (Martyn Ainsworth) as example organisms. All three also read drafts of the other chapters, making themselves familiar with the book to which they were to contribute and, especially Keith and Martyn, correcting errors and commenting on various aspects and issues raised while going through it. Where opinions were expressed, I took advice and criticisms seriously, but I remain responsible for the final text and the result does not imply that they agreed with all of it. I am grateful not only for the chapter they contributed, but for the professionalism with which they treated my text.

Torsten Moller read Chapter 2 and commented, with examples, on the calculations of ages of oaks using John White's method; he also produced the girth-related graphs shown in this book. Jill Butler read Chapter 11 on the conservation of ancient oaks and put me right on some mistakes. Where we might have understood an issue differently, I took a second look at the relevant parts but the final text is my responsibility only. Here again, I am grateful for the time Jill took to listen and discuss. For the same chapter, and after discussions over a glass of wine at the Linnean Society, I asked Andy Colebrook, a chartered arboriculturist, to provide a text box on his views about the suitability of Tree Preservation Orders (TPO) to protect ancient oaks. He responded positively and I am grateful for his contribution. David Lonsdale made helpful comments on my text about oak mildew in Chapter 11. Other chapters, or sections of these, were read by Owen Johnson of TROBI; by Helen Read of the City of London Corporation, who focussed on text on Ashtead Common, Burnham Beeches and Epping Forest;

by Andrew Scott of the City of London Corporation, who looked especially at text on West Wickham Common; by John Lock of Richmond, who focussed on text on Richmond Park; and by the Dutch ancient tree specialists Jeroen Pater and Jeroen Philippona, who looked in particular at text on oaks in continental Europe, about which they know far more than I do. To all I express my gratitude for the removal of errors, for pointing out factoids, for suggestions of how to write about complicated issues clearly, and for offering insights of which I was only vaguely aware. Clarity of expression was also enhanced

by my copy-editor, Sharon Whitehead, whose efforts are much appreciated. Nevertheless, any remaining errors are my responsibility entirely. I also thank the staff at Kew Publishing for their professional and sympathetic collaboration in seeing this book through to publication. Last but not least, I found my good friend Lawrence Banks of Kington in Herefordshire prepared to write a foreword. Lawrence knows a great deal about trees, owns a country estate (with a few ancient oaks) and a famous arboretum, and has promoted and supported research on trees for many years. I am most grateful for his encouragement.

Index

Words given a definition in the glossary are printed in **bold**. Page numbers in *italic* are references to illustrations. Main entries for names of organisms are in English or Latin according to whichever is used preferentially in the text. *Quercus robur* is thus cross-referenced to 'pedunculate oak' but organisms without a common English name are referenced to their Latin name only.

Acer campestre see field maple

Acer pseudoplatanus see sycamore

acid grassland 88, 103, 105, 221, 226, 228, 229, 232, 234–35, 241, 248, 252, 257, 265, 272, 330

acid rain 287, 330, 332

Aconbury Forest 138

acorns 6, 8, 13, 22–23, 104, 149, 205, 211–12, 217, 231, 236, 246, 331–32

Acton Round Oak 136, 165

Acute Oak Decline (AOD) (see Chapter 11) 247, 313, 317, 320, 330

adventitious buds 16, 330, 332

Aesculus hippocastanum see horse chestnut

afforestation 55, 122, 191–92, 194–96, 330

age of oaks (see Chapter 2)
  anecdotal ages 25
  annual ring (counts) *25*, 28–30, 36–37, *38–39*, 41, 45, 48, 130, 202, 315, 330–31
  proxy data (or 'proxy') 4, 26, 29–30, 36, 315

agriculture 55, 57, 69, 101, 114, 124, 131, 134, 138, 148–49, *157–58*, 160, 189, 212, 247, 272, 287, 299, 306, 308, 330, 333

Agrilus biguttatus see two-spotted oak bupestrid beetle

Albury Park 92, *95*, 341

Alces alces see elk

alder 203, 205, 207, 231–32, 234, 254

Aldermaston Park 70, 301, 319

Alfoxton Park *56–57*, 165

Alice Holt 123, 138, 140, 299–300

Alnus glutinosa see alder

Althorp Park 110–11,

Ampedus cardinalis see cardinal click beetle

Ampedus nigerrimus 293, 295

ancient countryside 57–58, 161, 211

ancient oaks (see Chapter 1) 3, 6, 11, 13, 15, 19, 25–27, 31, 33, 36,
42, 45–47, 48–49, *52–53*, 57, *59*, *62–63*, 65, 70, 102, 138–39, 141, 164–66, 168, 174–78, 199, 218–19, *236*, 269, 272, 274, 298, 315, 318, *passim*

Ancient Tree Forum 4, 27, 50, 266, 330

Ancient Tree Hunt (ATH) *see* Ancient Tree Inventory

Ancient Tree Inventory (ATI) 4, 11, 26, 50, 70, 138, 167, 169, 218, 227, 232, 236, 298, 330, 340

Anemone nemorosa see wood anemone

Anitys rubens 292

annual rings 25, 28–29, 30, 36–37, *38–39*, 41, 45, 48, 130, 315, 330–31

ash 16, 69, 94, 101, 104–05, 119, 130–31, 135, 139, 147–49, 161, 202, 204–05, 211, 226, 231–32, 238, 241, 243, 246, 252, 254, 268, 288,

Ashdown Forest 124, 138

Ashridge Park 82, 150, 157

Ashtead Common (see Chapter 9) *15–16*, *148*, 161, 150, 209, 214, 218, 222–23, 269, 273, 280–81, 292, 294, 311

Ashton Court Park (see Chapter 9) 102, 165, 218, 224–25, 226, 269, 280

Askerton North Moor Chase 143

Aspal Close (see Chapter 9) 88, 102, 209, 214, 218, 220–21, 269–70, 273, 280–81

assarts (assarting) 55, 122, 133–34, 136, 141, 144, 260, 330

aurochs 207, 209

Austria 174

Badminton Park 80

Baginton Park 98

Barnsley Park 80

Bayfield Park 320–21

beaver 207, 209

beech 11, 16, 28, 94, 103, 106–07, 119, 124, 126, 128–29, 130–31, 133–34, 137–41, 145, 147, 149, 150–51, *152*, 156–57, 192, 194, 196, 202–06, 209, 211–15, 220, 226, 231–32, 236, 242–43, 246, 252–54, 260–61, 268, 276, 279, 286, 290, 294, 300–01, 314, 319,

beech woodland 261, 279

beefsteak fungus *11*, 12, 224, *278*, 281

beetles (see Chapter 10) 12, 208, 226, 231, 237, 255, 292–95, 296–97

Belgium 174, 178

Bere Oak 35, *122–23*, 164,

Berkhamstead Common 82, 150

Bernwood Forest 125, 138

Betulo-Quercetum roboris see oak-birch woodland

Betula pubescens see (downy) birch

Białowieża National Park 178, 182, 205–06

Biatora veteranorum 288

Big Belly Oak 134, 166, 260

bilberry 222, 228, 262–63

Bilhaugh (Sherwood Forest) 45, 135–36, 226–28

Billy Wilkins Oak *32–33*, 164, 342

Binswood 151, 157

biodiversity 4–5, 24, 103, 116, 119, 141, 191, 198, 205, 215, 235, 238, 246, 252, 255, 266, 274–76, 302, 317–19, 342

birch 8, 28, 52, 125, 128, 134–35, 138, 144, 150, 160, 194, 202–05, 209, 211, 222, 224, 226–28, 231–32, 234, 243, 254, 261–62, 265, 302,

Birklands (Sherwood Forest) (see Chapter 9) 60, 135–36, *140*, 208, 213–14, 218, 226–28, 236, 241, 269, 280, 314, 319,

Bison bonasus see wisent

Black Death 68, 109, 136, 188

black poplar 207

blackthorn 206, 222, 243, 332

Blenheim Park (see also High Park) *12–13*, 29, 59–60, 68, 101, 109, 139, 169, *170*, *204*, *206–07*, 213–14, 218, 220, 242–43, *244–45*, 246–47, 269, 280, 289, 292–93, 295, 302, 315, 317, 339, 341–42

bluebells 137, 237, 246, 268

Boconnoc Park 103

bolling 13, 15–16, 19–20, 23, 32, 35, 40–42, 44, 48, 183, 213, 216, 251, 309, 311–12, 315, 320, 330, 332

Bolton Abbey Park 111

Bos taurus see aurochs

boundary oaks 47–48, 162, 166, 186

Bourton Oak 164

Bowthorpe Oak 15, *46–47*, 49, 165

Brachygonus ruficeps 293

Brachypodium sylvaticum see false brome

bracken 75, 94, 109, 116, 150, 194, 205, 209, 215, 220, 224, 228–29, 231–32, 234, 237–38, 241, 243, 246, 252, 254, 257, 262, 268–69, 305, 311, 319

Bradgate Park (see Chapter 9) 11, 68, 101, 103, 218, 228–29, 266, 269,
280, 293

brambles 206, 224, 237, 262, 313

Brampton Bryan Park (see Chapter 9) 103–04, 110, 164, 213, 217–18, *230–31*, 269, 280, 341

Braxted Park 77

Braydon Forest 125, 138

Breame Oak 165

Bringewood Chase 143, 164

British History Online 129, 160, 300, 340

Brocket Park 82–83, 102, 165, 272, 302

Brockhurst Park 252

Brockton Coppice *144*

Brockworth Park 80–*81*, 164, 341

Brookhill Plantation 151

brown rot (see Chapter 10) 12, *26*, 38, 279, 281, 292–95, 297, 311, 330

Buckholt Forest 124–25, 138

Bucknell Hill 151

Buellia hyperbolica 288, *290*

Buglossoporus (Piptoporus) quercinus see oak polypore

Bulstrode Park 73, 308, 313

Burnham Beeches *150–51*, *210*, 214–15, 311, 320, 342

burrs 14, 16, 30–32, *33*, 35–36, 42, 47, 82, *240–41*, 330, 332

Burry Maiden Oak 164

Burton Park 94

Bushy Park 69, 166, 294

Byrkley Park 143–44,

Calicium adspersum 289–*90*,

Calicium quercinum 289

Calke Abbey Park (see Chapter 9) 75, 231, *232–33*

Calluna vulgaris see heather

Cannock Chase 144, 165

Capreolus capreolus see roe deer

Carbrooke Oak 165

cardinal click beetle 257, 292–93

Carpinus betulus see hornbeam

Cassiobury Park 83

Castanea sativa see sweet chestnut

Castle Howard (estate) 105, 129, 341

Castor fiber see beaver

Cathedral Oak 166, 260, *263*

cedar 225, 231, 255, 302

Cedrus see cedar

Cedrus libani see Lebanon cedar

cellulose 12, 45, 275, 278–79, 292, 330–31, 333

Ceriporia metamorphosa 279–81, *284*